OUR ENERGY FUTURE

INTRODUCTION TO RENEWABLE ENERGY AND BIOFUELS

CARLA S. JONES

STEPHEN P. MAYFIELD

UNIVERSITY OF CALIFORNIA PRESS

University of California Press, one of the most distinguished university presses in the United States, enriches lives around the world by advancing scholarship in the humanities, social sciences, and natural sciences. Its activities are supported by the UC Press Foundation and by philanthropic contributions from individuals and institutions. For more information, visit www.ucpress.edu.

University of California Press
Oakland, California

Library of Congress Cataloging-in-Publication Data

Jones, Carla S., 1981- author.
    Our energy future : introduction to renewable energy and biofuels / Carla S. Jones and Stephen P. Mayfield.
        pages   cm
    Introduction to renewable energy and biofuels
    Includes bibliographical references and index.
        ISBN 978-0-520-27877-6 (pbk. : alk. paper)
        ISBN 0-520-27877-1 (pbk. : alk. paper)
        ISBN 978-0-520-96428-0 (ebook)
        ISBN 0-520-96428-4 (ebook)
    1. Biomass energy—Textbooks.   2. Biomass energy—Environmental aspects—Textbooks.   3. Renewable energy sources—Textbooks.   4. Fossil fuels—History—Textbooks. 5. Fossil fuels—Environmental aspects.   I. Mayfield, Stephen P., 1955– author.   II. Title.   III. Title: Introduction to renewable energy and biofuels.
    TP339.J65 2016
    662′.88—dc23

                                                              2015022634

Manufactured in China
25 24 23 22 21 20 19 18 17 16
10 9 8 7 6 5 4 3 2 1

The paper used in this publication meets the minimum requirements of ANSI/NISO z39.48-1992 (R 2002) (*Permanence of Paper*).

OUR ENERGY FUTURE

To all those striving for a sustainable future

CONTENTS

ACKNOWLEDGMENTS

This book was a vision we had while developing and teaching an introductory to bioenergy course at the University of California, San Diego, as part of the Educating and Developing Workers for the Green Economy (EDGE) program. We have many people to thank for their contributions, insights, and assistance. We would like to thank Robert Pomeroy, Jason Pyle, James Rhodes, Byron Washom, and Josh Graff Zivin for contributing their expertise and support as lecturers in the initial bioenergy course that served as a source of information and an outline for this book. We would also like to thank the EDGE Biofuels Science Technician students for providing valuable feedback, particularly Sophia Tsai for feedback and editing, Elena Ceballos for assisting with graphics, updates, and referencing, and Paula Morris for her expertise in graphic design. Special thanks to William Leung and Benjamin Miller in Mark Jacobsen's lab for their perspectives and comments relative to the economics of energy, and to Adam Jones, Stephen Bolotin, and Travis Johnson for providing overall editorial support and suggestions. We greatly appreciate Chuck Crumly and Merrik Bush-Pirkle at UC Press for taking on this new and exciting project. This book would not be possible without the support of the University of California, San Diego, the University of California, San Diego, Extension Program, and the California Department of Labor. Finally, behind every successful project there is a great coordinator and we must thank Wendy Groves for her endless patience, incredible organization and coordination skills, and her many other valuable contributions.

PREFACE

*Energy will be the immediate test of our ability to unite this nation, and
it can also be the standard around which we rally. On the battlefield of
energy we can win for our nation a new confidence, and we can seize
control again of our common destiny. . . . We are the generation that will
win the war on the energy problem and in that process rebuild the unity
and confidence of America.*

PRESIDENT JIMMY CARTER, 1979

In July of 1979, President Jimmy Carter gave a televised speech in response to the energy crisis plaguing the United States during this era. This crisis spawned a realization of the country's complete dependence on the availability of foreign energy supplies and the need to develop new energy sources within the United States. President Carter refers to these developments as a "war on the energy problem" and a test of the country's ability to unite behind a common cause. Unfortunately, since that speech, we have increased our need for energy and made only modest diversification to our energy sources. This was brought about mainly because the price of oil stabilized after the 1970s oil shock, and remained relatively low for the next 35 years. However, our energy problems were hardly solved; they were just lying dormant, and starting in 2006 the price of oil again began to spiral upward. At the same time, it became clear that climate change caused by the emission of carbon dioxide from the burning of fossil fuel was not only quite real but also coming much quicker than previously understood. With the high cost of oil and the realization of the consequences of climate change, we again started to make investments to diversify our energy sources. However, this new war on energy, like the one started in the 1970s, appears to again be short lived. Today, although we still face diminishing supplies of inexpensive fossil-fuel-derived energy sources, and continue to rely on the importation of petroleum from regions of economic and social insecurity, these are not the headlines in the news reports. Instead, we hear about US energy independence and a new energy revolution brought about by "fracking" that will make the United States an energy-exporting country in just a few years. Even the looming environmental consequences of global climate change and their adverse impacts on every aspect of our economy, environment, and lives are largely downplayed. Headlines are one thing, reality is another, and the facts can be ignored for only so long. We need new sources of sustainable renewable energy that do not contribute to climate change, and we need to develop them now.

Previous generations used inexpensive fossil fuel energy to allow for the rapid expansion of the world economy, but this expansion of both the economy and world population has placed the planet on an unsustainable path of environmental imbalance. Current times present an opportunity to change this path if we are able to develop technologies for the generation and utilization of environmentally clean, sustainable, and renewable energy resources. It will not be easy. With the era of cheap energy gone and energy food shortages on the horizon for many countries, coupled with atmospheric carbon dioxide levels continuing to rise, the energy issues that the world will face over the next 50 years are certain to be severe and to have enormous consequences on lives present and in the future. Despite this challenge, we have the opportunity to change the way we produce and utilize energy and in so doing put us on a path to a sustainable energy future.

There are many options to expand the use of renewable energy resources to help limit the use of fossil fuels, reduce reliance on wasteful energy practices, and prevent further environmental damage. These options include solar photovoltaics, wind power, geothermal power, hydropower, tidal energy, and others; however, all of these options are primarily replacements for electrical power generation, which accounts for only about half of our fossil fuel use. We also need to replace the energy-dense and easily transportable liquid petroleum based fuels that make up almost 30% of energy consumption, mostly used for transportation, and the options to replace this energy sector are limited. However, a solution to our reliance on these environmentally unfriendly and dwindling resources may lie on the horizon in the form of biofuels, liquid fuels created from living biological organisms.

Following an introduction to fossil-fuel-based energy sources and their role in the environment and society, this book will introduce the many sources of biomass that can be used to create bioenergy, the means by which they are being developed, and the impact they could have on the environment and the future of the planet. Whether one was previously aware of the looming energy crisis, the environmental damage being incurred, and the development of alternative renewable energy resources to combat these issues, or this is one's first introduction to these topics, this book will provide a resource to understanding the current benefits and limitations of bioenergy production and the impact that developing these new energy sources could have on society now and in the future.

The aim of this book is to awaken in the reader the knowledge that developing high-yielding, economically viable biofuel-producing organisms is critical to the energy stability of the future. Reading this book is the first step to becoming aware of these issues and working to change the future and thereby becoming part of "the generation that will win the war on the energy problem."

CHAPTER ONE

Overview of Energy Usage in the United States and the World

To begin, take a moment to stop and listen. Try to observe all sounds: the tick of a clock, the hum of a computer, the whirl of a passing car, the whisper of a heater, or perhaps the growl of an airplane overhead. These sounds, so common that they often go unnoticed, would likely sound thunderous to societies living 200 years ago. What makes society today so different from those historical societies? The answer is simple: **fossil fuel**.

The availability of easily accessible and cheap energy has breathed technological life into society. It has influenced the way we transport ourselves, the houses we live in, the offices we work in, the classrooms we study in, and even the food we eat. Energy has single-handedly changed the face of society from one of short walking commutes, locally farmed foods, and face-to-face personal communication to a society where travel distances are unlimited, fresh foods are always available regardless of season, and communication, over even vast distances, is instantaneous and done using battery-operated handheld devices. Although these technological changes have led to a society of convenience, they have not come without cost to our energy supply, our food security, national security, and the environment.

Today, fossil-fuel-based energy is critical to sustaining the lifestyle that many countries have become accustomed to. And the numbers can be overwhelming; today in 2015 we burn 93 million barrels of oil per day, or 1.4 trillion gallons of oil per year, and oil is only about one-third of the energy we consume every year. To maintain this level, let alone allow for increased energy use, we will need to develop new alternative sources of energy if we are going to preserve this lifestyle and prevent further reduction to energy supply, security, and the environment. As a basis for this entire book, this introductory chapter will explore the basic concepts of energy production and utilization and its critical role in modern society.

Understanding Energy

Energy, by definition, is the ability to do work on a physical system. Yet, what does this really mean? Since energy is the capacity to do work, one must first understand the concept of work. Work is defined as force multiplied by distance. A force alone is not enough to constitute work; it must be combined with a movement, a term equivalent to the distance component of the equation. Consider the example of someone pushing a shopping cart and running into a wall. No matter how hard this person pushes the cart or how much force he or she exerts, if the cart is not changing position then no work is being done. Therefore, a source of energy needs to allow for both force and motion. In addition, this energy source must be replaced to maintain the work that is occurring. The energy is replaced in a car by filling the tank with gasoline, or in a body by eating food, such as an apple. The apple is a source of energy because it provides nutrients powering the body with food calories and allowing for continued work.

Energy is measured in many different ways. In the previous example of food, energy was measured as the number of food calories in the apple. Food calories are really kilocalories (and take the capitalized form, Calories); one Calorie is equal to 1,000 calories. A calorie is defined as the amount of energy that will raise the temperature of one gram of water by one degree Celsius. In addition to the food calorie and calorie, there are many other metrics of energy important to our understanding of the concept of energy as shown in Table 1. Energy is often measured in terms of joules. A joule is equivalent to a Newton times a meter (remember force times distance). In this book, energy will also be referred to in British thermal units (BTU). A BTU is equivalent to 1,055 joules and is often used when discussing energy sources used to heat water or other substances. In many cases, energy is produced or consumed at such a

TABLE 1

Energy conversion units

	J	kWh	BTU	kcal
1 joule (J)	1	2.8×10^{-7}	9.5×10^{-4}	2.4×10^{-4}
1 kilowatt-hour (kWh)	3.6×10^6	1	3,412	860
1 British thermal unit (BTU)	1,054	2.9×10^{-4}	1	0.252
1 kilocalorie (kcal)	4,184	1.2×10^{-3}	3.97	1
1 tonne oil equivalent (toe)	4.5×10^{10}	11,630	4.2×10^7	1.1×10^7

NOTE: Each publication may discuss energy in terms of different units such as J, BTU, kcal, kWh, and toe. This table is designed to help understand the relationship between these units and allow for the conversion between different energy-related units.

DATA FROM: APS Panel on Public Affairs (2014).

TABLE 2

Comparison of energy content for varying energy sources including fossil fuels and common food sources using differing units of comparison

	J	kWh	BTU	kcal
1 barrel oil (42 gallons)	6.1×10^9	1,713	5.8×10^6	1,468,800
1 ton coal	2.7×10^{10}	7,560	2.5×10^7	6,480,000
1 therm gas	1.1×10^8	29	1.0×10^5	25,200
1 gallon gasoline	1.3×10^8	36.0	1.2×10^5	31,248
1 bushel corn (56 pounds)	8.6×10^7	24.5	8.1×10^4	20,453
1 fast-food hamburger	2.8×10^6	0.8	2.7×10^3	670

DATA FROM: APS Panel on Public Affairs (2014); US Department of Energy (2014).

large scale that a very large standard of measurement is needed. An example of such a large consumption of energy is the consumption or production of fossil fuels around the world. Energy on this very large scale is often measured by comparing it to the burning of 1 ton of petroleum also known as a tonne of oil equivalent (toe). One toe is equivalent to about 42 gigajoules of energy (APS Panel on Public Affairs, 2014). In order to get a better grasp on how these various energy measurements relate to one another, Table 2 shows energy equivalents of some common energy sources.

The various ways of measuring energy are due in large part to the many different forms of energy. There are fundamentally five forms of energy: chemical, electromagnetic, mechanical, nuclear, and electrical. The first form of energy, **chemical energy**, is energy that occurs due to a chemical transformation or chemical reaction. The notion of a chemical reaction might bring to mind a picture of flasks containing colored liquid bubbling over a Bunsen burner. Nevertheless, chemical energy is actually quite varied and common, and includes things like burning wood in a campfire for warmth or eating a slice of pizza to satisfy one's appetite. In the case of burning wood, coal, or any other substance, the solid wood or coal is broken down and heat and ash are created. As the chemical bonds are broken within the wood or coal, energy is released in the form of heat. This heat can be used simply for direct warmth or it can be used to heat another substance like water to produce steam, a key reaction in the production of electrical power. Although a slice of pizza is not burned in the same way that wood or coal is burned, it is also broken down through digestion into smaller components like amino acids, carbohydrates, and lipids, changing the pizza's chemical state. This change in chemical state results in the biological components needed for metabolic respiration and the creation of adenosine triphosphate (ATP), a molecule that provides fuel for the reactions occurring within the body.

Electromagnetic energy is another form of energy. Electromagnetic energy (or radiation) is the energy of light. Most of the natural light received on Earth comes from the sun. Electromagnetic radiation results from oscillating energy particles called photons moving in wavelike patterns. These photons are a result of thermonuclear reactions that occur within the core of the sun and then radiate outward toward the Earth. Depending on the amount of energy these pho-

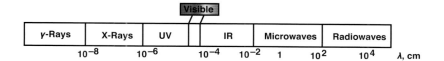

| γ-Rays | X-Rays | UV | Visible | IR | Microwaves | Radiowaves |

$$10^{-8} \qquad 10^{-6} \qquad 10^{-4} \quad 10^{-2} \quad 1 \quad 10^{2} \qquad 10^{4} \qquad \lambda, cm$$

Gamma (γ-Rays) – Nuclear Fission and Fusion
X-Rays – Diagnostic Radiology
Ultraviolet (UV) – Formation of Vitamin D and Sunburn
Visible – Photosynthesis and Light
Infrared Radiation (IR) – Heating, Night Vision, and Short Distance Communication (Remote Controls)
Microwaves – Communication, Radar and Power
Radiowaves – Communication

FIGURE 1.1 Chart showing the wavelength ranges for the various particles of electromagnetic radiation. The amount of energy contained within each of these particles varies based on this wavelength. The longer wavelengths have less energy, while the shorter wavelengths have more energy. Each range of wavelengths has a unique name that can also be associated with different and unique functions. For instance, ultraviolet wavelengths cause sunburns, while radio wavelengths are used for communication.

tons contain, the wavelengths change and they are classified as different types of radiation as shown in figure 1.1. For instance, wavelengths classified as visible radiation (visible light) allow human beings to see colors, while wavelengths classified as ultraviolet radiation cause their skin to sunburn at the beach. Electromagnetic radiation is very important to the balance of the Earth not just because this energy source provides light but also because it provides heat. Chapter 3 will cover how electromagnetic energy enters the Earth's atmosphere and how a significant portion of it becomes trapped within the atmosphere. This trapped energy is what gives the Earth its mild temperatures and one of the factors that allows life to be sustained on this planet.

Another important form of energy is mechanical energy. **Mechanical energy** is often classified into two categories: potential and kinetic. **Potential energy** is the energy of an object due largely to its position in relation to the force of gravity, while kinetic energy is the energy of motion. Take, for example, a rollercoaster. Many know the gut-clenching feeling brought on by the click, click, click sound as the rollercoaster car is brought up to the summit of the first big hill. When this car is sitting at the top of the hill, it has a lot of potential energy. The car has the "potential" to roll down the hill due to gravitational forces and thereby gain great speed. As the rollercoaster car accelerates down the hill, it transfers the potential energy into **kinetic energy**. Since kinetic energy is the energy of motion, the faster the car moves, the more kinetic energy it contains. You can also think of kinetic energy in terms of an automobile accident. A car that crashes while moving slowly does not suffer nearly as much damage as a car going extremely fast. The fast car has more kinetic energy meaning that more energy will be used to damage the car when it comes to an abrupt stop.

Nuclear energy is another form of energy that is important to society. Nuclear energy results from reactions that change the structure of an atom's nucleus. Nuclear reactions occur in the interior of the sun and in the interior of the Earth as well as in nuclear reactors used for energy

production. The first type of nuclear reaction is the fusion reaction. This reaction takes place in the sun and occurs when two atomic nuclei fuse to form a single atomic nucleus such as two hydrogen nuclei combining to form a helium atom. The second reaction is the fission reaction. Fission reactions take place in nuclear power plants and, in this case, the nucleus of a single atom breaks into two atoms such as when uranium splits apart. Both of these reactions are capable of releasing a lot of energy; thus, they are an important component in the consideration of sustainable energy development. Nuclear energy will be discussed in more detail in Chapter 4.

The final form of energy is **electrical energy**. Atoms are made of subatomic particles that include positively charged protons and neutrally charged neutrons within the nucleus, and negatively charged electrons floating around the edges of the nucleus. Electrical energy occurs when electrons are passed from one atom to another, and this occurs when an electric field is applied to metals. Because all electrons are negatively charged, they repel one another. By maintaining this repulsive relationship down a wire, these electrons create an electric current. Once the electrons make it to a resistor, they react with the atoms within the resistor releasing either heat or a magnetic field. This final release of heat or creation of a magnetic field is what results in power generation. There are several ways to generate electricity, including renewable sources like wind, solar, and water, and these will be discussed in a later chapter. However, one essential concept of electrical energy is that this energy source must be used immediately. Electric energy must be consumed as it is produced or it simply dissipates as heat. Storage of electrical energy in batteries is possible, but as many of you know, batteries are able to store only limited amounts of energy and must be recharged often.

All of these energy sources play an important role in modern society. Yet, to really understand the value of energy, one cannot consider these forms of energy individually but must consider them as a whole. The reason for this is due to the **First Law of Thermodynamics**. This physics law states that *"energy can neither be created nor destroyed; it can only be transformed from one state to another."* This means that all of these various forms of energy are not individually being created but rather they are simply energy in one form that is being converted to another form. Let us consider a couple of examples. Thermonuclear reactions within the interior of the sun release electromagnetic energy from the creations of new atoms by nuclear fusion. This energy travels to the Earth via photons (electromagnetic radiation energy), and these photons can be absorbed by chlorophyll in plants, such as in the leaves of trees. The trees use the energy from these photons to fix carbon dioxide and produce sugar that eventually becomes the wood of the tree. This wood can then be burned to create heat for a home or to produce steam to drive a turbine. In this energy transformation example, the nuclear energy from the sun becomes electromagnetic energy in the form of photons and then becomes chemical energy in the form of wood. Another example that can be considered is hydroelectric power generation. In this case, the electromagnetic energy in the form of photons heats the Earth's atmosphere, causing the evaporation of water, which in turn forms rain or snow. This snow melts or rain fills up a lake located behind a dam where the water represents potential energy. As the water travels from the top of the dam to the bottom of the dam through a tunnel, the potential energy is converted to kinetic energy that can be used to turn a turbine. The turning of the turbine generates an electric field that creates electricity, which travels to homes to power electrical devices. These examples demonstrate how energy is never created nor destroyed; it is just transferred from one form to another. The First Law of Thermodynamics is not to be confused with what we call "energy production." Energy production is not the same as energy creation. In energy production, people go out and find an energy source that already exists and then extract this source. Once extracted the source can be converted into other types of energy resources. Most of the energy produced in the world is derived from fossil fuels including petroleum, coal, and natural gas, all resources already available, and all we really do is extract these existing energy sources from the Earth.

The First Law of Thermodynamics is very critical in understanding energy and the development of renewable energy sources because there are a limited number of primary sources that transfer energy on to the surface of the Earth. These include nuclear, solar, geothermal, and tidal energy, which will be discussed further in Chapter 4. All other energy sources including fossil fuels are ultimately derived from one of these four sources. Three of these primary energy sources (solar, geothermal, and tidal) represent another key concept, renewable energy. **Renewable energy** is defined as energy that comes from one of the primary sources. The ideal renewable energy source is also sustainable meaning that the resource will be replenished with-

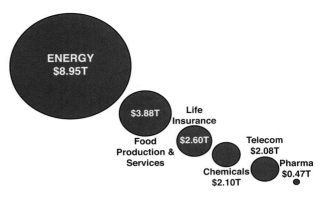

FIGURE 1.2 Comparison of market values for large industries around the world. The world's energy market valued at an estimated $8.95 trillion is over twice as large as the food production and services market valued at an estimated $3.88 trillion. The energy market also dwarfs other large markets including life insurance, chemicals, telecom, and pharma. As the energy market continues to expand, we will likely see an even greater separation of value (data from Bloomberg, 2014).

out a detriment to the environment. While solar, geothermal, and tidal energy are considered both renewable and sustainable, nuclear energy still has significant environmental issues. Although elements such as uranium that are used in nuclear reactions are not replenished on the Earth, the resources of these elements on Earth are very large and the amount of energy generated is so high that it is believed that nuclear energy will be available for thousands of years, leading many people to mistakenly consider nuclear energy a traditional sustainable source. Like hydroelectric, photovoltaic, and wind energy, bioenergy is directly produced from solar energy and is therefore a renewable energy source.

Energy, Population, and Standard of Living

Energy is one of the most influential and important aspects of modern society. To understand the magnitude of its importance, consider its value economically. By comparing the energy market to other major markets such as food production, telecom, and insurance as shown in figure 1.2, one can begin to understand the incredible value of energy. Energy is easily the largest global market valued at nearly $9 trillion worldwide with other large industries ranking well below this enormous number (Bloomberg, 2014). The value of energy in the United States is also very large at $1.2 trillion, roughly one-tenth of the gross domestic product (GDP; US Department of Commerce, 2012). In addition to its huge raw value, the energy market also literally fuels many other markets, resulting in its value to the United States and world economies being much higher. As the world's largest market, energy is a main driver of the world's economy.

Despite its enormous value, the world's energy market is not stagnant, nor has it reached a plateau. As an example, the US energy market is expected to increase in value from $1.2 trillion in 2010 to $1.7 trillion in 2030, an increase of about 30% in 20 years (US Department of Commerce, 2012). The world energy market will increase at an even faster rate, largely due to an expanding population and an increase in the standard of living around the world. The world's population is currently over 7 billion and is expected to reach 8 billion by 2030 (US Census Bureau, 2012). As more people populate the Earth, more energy is consumed to sustain the population with basic resources such as food, transportation, and housing. All of these basic resources have become critically dependent on the availability of energy resources. Even if the world's population were to plateau, the energy market would continue to increase due to the link between energy and standard of living. An indicator of a country's standard of living is its **GDP** per capita. The GDP of a country represents the value of all goods and services produced in that country, thus indicating a country's wealth. When this overall wealth is divided by its population, a rough estimate of the standard of living for individuals within that country results. For example, a large developed nation like the United States has an annual GDP of about $15 trillion and a population of about 307 million, giving the United States a GDP per capita of $48,859. We can then compare this to a large developing nation such as China. China has an annual GDP of almost $6 trillion and a population of about 1.3 billion people, giving China a GDP per capita of only $4,615 (CIA, 2012). Since the GDP per capita roughly estimates the annual income available to individuals, it is clear that the standard of living in the United States is much higher than that of China. Again, there is a link between standard of living and energy use per capita where a higher standard of living equals a higher use of energy per capita. Figure 1.3 compares primary energy consumption per capita to the GDP per capita of many countries around the world. This graph shows that countries with high standards of living (measured as high GDP per capita) including Qatar, Singapore, Norway, and the United States are also some of the highest primary energy consumers, while nations with lower GDP per capita such as China, India, and Egypt are also the nations that consume the least amount of energy per capita.

While this comparison of GDP per capita to primary energy consumption gives insight into how some countries are able to maintain much higher standards of living than others, it also offers a preview of what the future of energy consumption may entail. Countries such as the United States have set the bar high for standard of living and many other countries want to reach that same standard. Envisioning a world where all people are living in conditions similar to the United States in terms of food availability, housing, and water availability may seem like an ideal situation, but this lifestyle will have a significant impact on energy use. As an example, China's economy is growing, which will ultimately lead to a higher standard of living for the people of China; however, as China's economy grows, so does its use of energy. The BP Statistical Review of World Energy states that developing nations like China and India accounted for 90% of the net increase in energy consump-

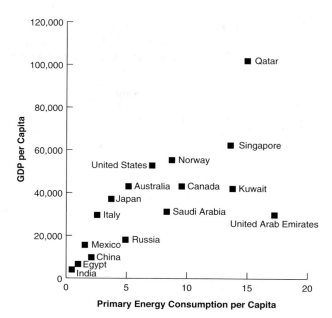

FIGURE 1.3 Graphical comparison of worldwide gross domestic product (GDP) per capita to primary energy consumption per capita, 2013–2014. This comparison provides evidence for the link between the consumption of energy and a higher standard of living. Many countries that consume higher levels of energy such as Qatar and Singapore also have a much higher GDP per capita. Countries such as Egypt and India have low primary energy consumption per capita and a low GDP per capita. A higher GDP per capita is associated with a higher level of income within each household (data from CIA, 2014; British Petroleum, 2014).

tion in 2012 and China's use of energy increased by 11.2% between 2009 and 2010 (British Petroleum, 2011, 2013). Many other countries are also seeing a rise in energy consumption as the people in these countries gain better living conditions. The huge use of energy in countries with high GDP per capita and the growing use of energy in those countries with increasing GDP per capita represent a significant challenge for the future of energy production and utilization, and enormous challenges for the environment. Worldwide fossil energy sources are finite and getting more expensive to extract every day. In order to balance a growing population with the energy needs of higher standards of living around the world, it is critical to develop new technologies enabling the more efficient use of fossil fuels, while we simultaneously develop renewable resources to supplement and replace the finite fossil energy resources.

Energy Resources Today

Today, a majority of the world's energy **consumption** comes from fossil fuel sources including coal, natural gas, and petroleum as shown in figure 1.4. Together, these sources make up about 87% of all energy used in the entire world. The remaining 13% of energy consumption is derived from alternative energy sources including nuclear, hydropower, and other renewable sources (British

(A)

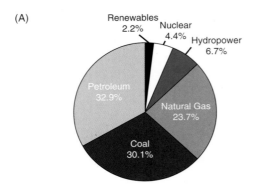

(B)

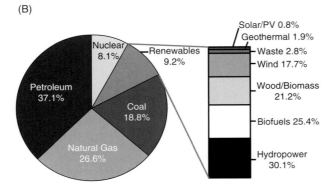

FIGURE 1.4 Comparison of world (A) and US (B) energy consumption by energy source. The fossil fuels including oil, coal, and natural gas dominate both global and US energy consumption at 86.7% total consumption and 82.5% total consumption, respectively. In the United States, renewable resources like hydropower, biomass, biofuels, and wind make up 9.2% of energy consumption (data from British Petroleum, 2014; EIA, 2014a).

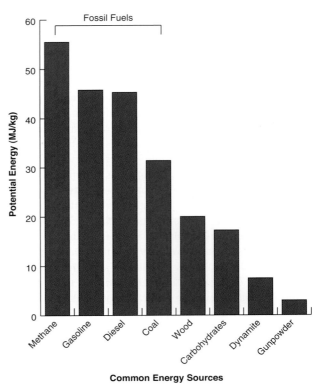

FIGURE 1.5 Graphical comparison of chemical potential energy levels among common energy sources. Fossil-fuel-based sources and products are the most energy dense of these substances (data from Elert, 2012).

Petroleum, 2013). The energy consumption patterns in the United States are similar to that of the world where 82.5% of energy consumption is from fossil fuels. It is apparent that the use of non-hydropower renewable energy sources in both the United States and the world is minimal at only 6.4% and 2.2%, respectively. The US energy consumption graph shows the breakdown of renewable energy consumption as spread between these different resources including solar, photovoltaic, geothermal, energy from waste, wind power, biofuels, and wood (EIA, 2014a).

Fossil fuels dominate the energy market largely because they are energy dense and historically relatively cheap. Figure 1.5 shows a comparison of energy density for a few common energy sources including fossil fuel resources, carbohydrates, and explosives (Elert, 2012). As shown by this comparison of energy per kilogram of material, fossil fuels are considerably more energy dense than even substances thought of as very powerful such as dynamite or gunpowder. Another advantage of fossil fuels is that they are relatively cheap. Compare the petroleum product gasoline to that of a common college food, pizza. One gallon of gasoline contains 116,275 BTU of energy and costs about $3.50. To get the same amount of energy from pizza, one would have to eat over 108 slices and it would cost about $220.

While this much pizza is undoubtedly out of the price range of an average college student, buying a gallon of gasoline is often not a problem. This simple comparison clearly illustrates how on a price per unit energy basis, fossil fuels are much cheaper than most products including the food people eat.

Competitive pricing and high-energy densities have resulted in a steady increase in the consumption of all three fossil fuels both around the world and in the United States. In 2012, the United States consumed about 17.4 quadrillion BTUs of coal, 25.9 quadrillion BTUs of natural gas, and 32.5 quadrillion BTUs of petroleum. These figures are drastically higher than the consumption levels in 1950 of 12.3 quadrillion BTUs of coal, 6.0 quadrillion BTUs of natural gas, and 13.3 quadrillion BTUs of petroleum. This increase of 41.5%, 331.7%, and 144.4% for coal, natural gas, and petroleum, respectively, is largely due to the development of technologies heavily dependent on the use of these fossil fuels such as personal cars, individual home climate control, and the many electrical devices people have come to rely on every day (British Petroleum, 2013). Obviously, these increases in consumption have also required a steady increase in **production** of each of these fossil fuels. Figure 1.6 shows a comparison of production and consumption levels of petroleum in the United States since 1950.

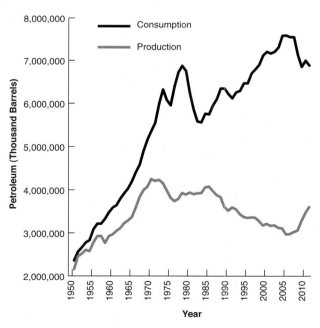

FIGURE 1.6 Comparison of US production and consumption of petroleum from 1950 to 2011. Beginning in the 1950s, the consumption of petroleum in the United States quickly began outpacing the production. This resulted in an increase in petroleum being imported from foreign sources (data from EIA, 2012).

Beginning around 1970, the consumption level is much higher than the production level. To meet the rising demand that the increased consumption level represents, the United States imports petroleum from foreign nations to fulfill the gap between production and consumption levels.

Over the last century, global production of fossil fuels has kept pace with the rising levels of consumption every year. However, keeping up with the projected growth in the future is likely to come at a significantly increased price. Fossil fuels are a finite resource and the extraction of these resources can vary leading to fluctuations in their costs. For instance, as sources of petroleum that are easier to extract are depleted, the new sources that come on line like tar sands and shale oil are likely to require more infrastructure or increased use of energy resources and result in generally higher costs for a barrel of oil. Eventually, this cost may become higher than most people are willing or able to handle. Chapter 2 will discuss how even though fossil fuels were created hundreds of millions of years ago, we will likely burn through these reserves in just a few hundred years. Once these resources are depleted, there is no way to speed up the natural process of creating fossil fuels.

To understand fossil fuels as a finite resource, it is important to consider an estimation of the remaining resources available on Earth using a term called **proved reserves**. A

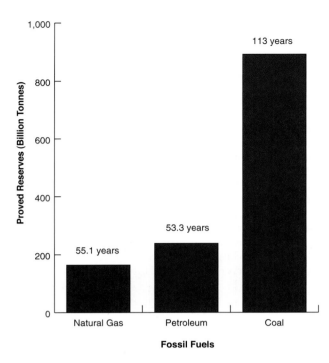

FIGURE 1.7 Global proved reserves for oil, natural gas, and coal, 2013. Reserve-to-production (R/P) values representing the number of years remaining for each fuel at current production levels are listed above each column. Estimates predict that globally coal will last the longest, but these values are likely to change as energy resources are consumed and new reserves are identified and accessed (data from British Petroleum, 2014).

proved reserve is the estimated quantity of a natural resource like a fossil fuel available for extraction on Earth using the technologies available today. These reserves estimate the total amount of a fossil fuel resource available for production and ultimately consumption. By dividing the proved reserve by our current production levels for that fossil fuel, the number of years remaining for a given fossil fuel at current production levels can be estimated, known as the **reserve-to-production ratio** (R/P). Figure 1.7 shows the global proved reserves and R/P values for coal, natural gas, and petroleum as of 2013. The R/P value for both natural gas and petroleum indicate that these resources will be significantly depleted within the next 50 years, a period likely within the lifetimes of many people today. Coal is estimated to outlast both natural gas and petroleum; however, with current energy infrastructure, there may be an increased use of coal as natural gas and petroleum are depleted. This option will ultimately decrease the number of years of availability remaining for coal and could have vast environmental consequences.

As these fossil fuels become depleted over the next few decades, the world will likely see a drastic increase in cost associated with a decline in production levels, a fluctuation we see even today for petroleum. The limited availability and less competitive pricing for some of the world's most important commodities will undoubtedly have a huge impact on society in terms of both security and economics. The best way to manage the impact of these changes is to diversify worldwide energy production and consumption by developing alternative and renewable forms of energy.

The Future of Electricity and Transportation

The diversification of energy utilization from fossil fuels to renewable energy technologies requires understanding how we use each of these fossil fuels within modern society today. Figure 1.8 graphs the consumption of various forms of energy based on different sectors within society. Through this graph, it is evident that coal, natural gas, nuclear, and renewable sources are more commonly used for residential, commercial, industrial, and electrical applications, while petroleum dominates the transportation sector. One of the main reasons why these industries require different fossil fuel resources is explained through two related terms: energy and power. Earlier in the chapter, the five types of energy and the definition of energy as being the capacity to do work were discussed. It was explained that energy cannot be created nor destroyed but only transferred from one form to another. This transformation is what ultimately differentiates energy and **power** because this transformation allows energy to take on a form that can be stored and used when needed. Power is slightly different. Power uses energy but has an associated time component. Power is energy divided by time and is usually expressed in terms of watts where a watt is equivalent to the transfer of 1 joule of energy per second. While oil is an energy source, electricity is a power source. A key difference between these two is that an energy source such as oil can typically be stored, while a power source such as electricity is used as it is generated. This is an important concept when considering that in the United States and most nations around the world, daily infrastructure relies on the availability of both a stored energy source and a power source.

Today, burning coal and natural gas are the two most common methods used to produce power. The heat from burning these fossil fuels is used to create steam. The pressure of the steam will turn a turbine providing energy to a generator that then generates electricity. **Electricity** is the flow of negatively charged electrons from one atom to another usually induced by a magnetic field. In an electrical generator, the turbines move magnets and a good electron conductor like copper together to create an electrical current. This current of electrons runs in a circuit, and as long as nothing blocks this circuit, the electrical energy continues to flow. However, if the circuit is broken, then the flow of electrons stops and so does the electrical energy. One can envision this by thinking about a wall socket for an electrical device like a light with a plug. When the plug is put into the wall socket, the two prongs create a circuit and the light will turn on. But, as soon as the light is unplugged from the

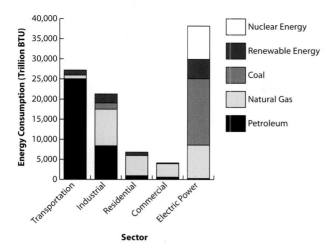

FIGURE 1.8 US primary energy consumption by source and sector, 2013. The transportation sector relies heavily on the availability of petroleum. The industrial sector also uses petroleum and consumes significant quantities of natural gas. Natural gas is an important component of both commercial and residential sectors as well. Electric power is the largest consumer of energy derived almost entirely from coal, natural gas, renewable resources, and nuclear energy (data from EIA, 2014b).

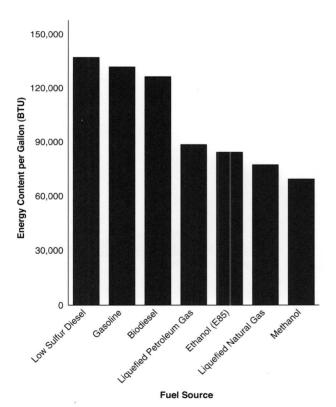

FIGURE 1.9 Comparison of the energy content contained within various types of fuels. Liquid hydrocarbon-based fuels have the highest energy contents per gallon when compared to alternative liquid fuels like methanol and ethanol (data from US Department of Energy, 2011).

wall, the light goes out because the circuit has been broken and the electrons are no longer flowing.

Electricity is one of the most highly used sources of energy in the world and is produced primarily from coal and natural gas. According to the International Energy Agency, in 2012 40.4% of worldwide electricity was produced from coal, 22.5% from natural gas, 16.2% from hydropower, 10.9% from nuclear power, and only 10% from all other sources including petroleum (IEA, 2014). Coal and natural gas are ideal sources for steam generation because they contain a lot of energy that can be burned at high temperatures to create steam. However, both coal and natural gas are finite resources as mentioned earlier, and they are responsible for environmental damage in the form of releasing harmful atmospheric gases. The value of electricity to society and the desire to lower the environmental impact of electricity usage has led to the development of other ways to generate electricity including many renewable sources. Both nuclear energy and hydropower have been used for the production of electricity for decades, while newer electricity sources such as wind and solar power technologies are continuing to advance and grow in popularity around the world. These renewable energy sources will be discussed in Chapter 4.

While the residential, commercial, and industrial sectors rely heavily on coal and natural gas for power, the transportation sector relies almost completely on the availability of the third fossil fuel, petroleum. Petroleum is naturally occurring crude oil derived from decayed oceanic algae and cyanobacteria, and is the precursor to many items common in everyday use including gasoline, plastics, and motor oil. The chemical and biochemical properties of petroleum will be further discussed in Chapter 2. However, there are two main reasons why petroleum has become the fossil fuel used for the transportation sector. First, petroleum and petroleum products have very high-energy contents. In figure 1.9, one can see that, when compared to other products being developed as fuel alternatives including ethanol and liquefied natural gas, the petroleum products of gasoline and diesel have much higher energy densities. Secondly, the transportation sector requires an energy source that is transportable. Transportation relies on the ability of vehicles to move from one location to another, making the use of power generation from coal or natural gas more difficult to take advantage of in a vehicle. However, as many petroleum products such as gasoline are liquids, they can be pumped into a vehicle and stored in the gas tank until needed by the engine. These characteristics of petroleum make it absolutely ideal for use in the transportation sector and extremely difficult to replace with other alternative energy sources.

The Future of Petroleum

Petroleum is an extremely valuable commodity in modern society, a commodity that is often taken for granted. This is largely due to the value it has in the transportation sector and its function in allowing for increases in the number of personal vehicles on the road, distances traveled by airplanes, and the constant availability of basic goods like food and household products. There is no better place to see the impacts of petroleum than in the United States where people drive personal vehicles long distances regularly, rarely use public transportation, and stock fresh produce in grocery stores even when out of season, thanks to the ability to transport food all over the world by trucks and airplanes. Over the past century, the reliance on petroleum in the United States has resulted in massive increases in its consumption. In 1949, the United States consumed about 5.7 million barrels of petroleum a day, but in 2013, the country consumed about 19 million barrels of petroleum every day, an increase of 235% (EIA, 2014a). Unfortunately, these increases in consumption have not completely matched with increases in domestic production levels; however, recent years have shown an increase in domestic production largely due to the development of infrastructure for the extraction of shale oil. But shale oil is generally more expensive than the conventional sources of foreign oil that have been imported for many years.

Figure 1.10 shows the production of petroleum in the United States since 1860. Notice that in the 1970s the level of petroleum production in the United States was at its highest. This time period is called **peak oil**. Peak oil is the point at which a country or the world is producing the maximum number of barrels of oil per day possible. While this appears to have happened in the 1970s for the United States, the world's peak oil point was likely reached sometime around 2007 (Nashawi et al., 2010; EIA, 2014a). The importance of peak oil is that it is the maximum; there will no longer be increases in production levels and this usually indicates that production levels will begin to decrease after this point. Unfortunately, looking back at figure 1.6 one will also see that, while the United States has surpassed peak oil, consumption levels have certainly not peaked. Recently the United States has seen an increase in oil production using hydraulic fracturing or fracking. This increase in oil production was brought about not because new oil fields were discovered, but rather because the price of oil rose to a level sufficient to support oil production using fracking technologies. Although this increase has been dramatic over the last five years with almost 3 million barrels of oil per day added to US production levels, even this new expensive oil production cannot get us back to a new peak oil level or even keep pace with the world's energy consumption increase (British Petroleum, 2014).

Consumption of petroleum in the United States declined significantly during the economic downturn of 2008, but this decline has reversed as the economy has improved. Today even with the increase in fracking-based oil production, the United States still has a large gap between the amount of production within the country and our consumption levels. In order to fill in this gap, the United States continues to import significant amounts of oil, about 7.4 million barrels per day or 40% of the consumption levels in 2012, the lowest average since 1991 (EIA, 2014a).

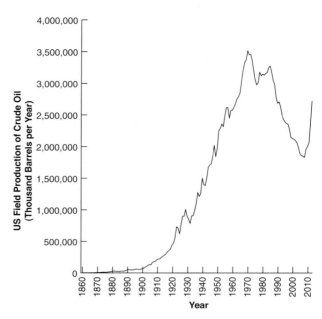

FIGURE 1.10 US annual production of crude oil between 1860 and 2013. Graph shows that peak oil production occurred in the 1970s. Recent years have shown an increase in crude oil production likely due to the development of shale oil extraction techniques (data from EIA, 2014c).

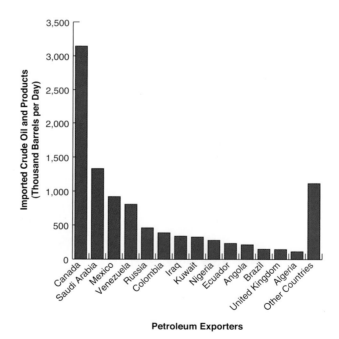

FIGURE 1.11 Annual US petroleum imports by country of origin for 2013. Canada is the most important source of foreign oil imported into the United States followed by Saudi Arabia (source: EIA, 2013).

Figure 1.11 shows that Canada is the major source of imported oil in the United States followed by Saudi Arabia, Mexico, and Venezuela. The importation of such a significant amount of foreign oil opens the door to two major problems: supply and energy security. Since the United

States is clearly dependent on a steady flow of oil from foreign locations, there is a significant chance that those supply lines could be disrupted in times of foreign instability. An example of this situation is seen when oil prices rise due to unrest in the Middle East, Africa, or other oil-exporting countries. The second problem is energy security. The economy of the United States and the world depends on the availability of fossil fuel resources including petroleum. If the supply of petroleum is disrupted, or the price of petroleum rises dramatically, then national economies suffer and this impacts national security.

While petroleum plays a critical role in the economy and societies of the world, it is also a fossil fuel that similar to coal and natural gas has a negative impact on the environment. When petroleum is used either in the engine of a car or in industrial processes, it releases harmful environmental pollutants including greenhouse gases. Chapter 3 will discuss how these gases are leading to global warming and climate change.

Unlike coal and natural gas used for electricity production, the opportunities to develop renewable energy sources to replace petroleum and lower its negative impacts on society are few. However, scientists are working on technologies that can produce fuels that have high energy densities, a low environmental impact, and the capacity to be easily transported like petroleum. One of these technologies is biofuel. Biofuels are a source of chemical energy derived from the sun and sequestered in plants or algae. Plants and algae can be used to produce a number of different biofuels including bioethanol and biodiesel that may represent alternatives to liquid petroleum products.

Plants and algae are capable of absorbing massive amounts of carbon dioxide and utilize sunlight to efficiently convert this carbon dioxide into stored chemical energy in a process called photosynthesis. Since all fossil fuels are originally derived from decaying plants and algae, it seems natural to look at these organisms as a replacement for crucial fossil fuel resources.

The next several chapters will cover more about the history and importance of fossil fuels, particularly petroleum to society, and the renewable energy technologies that are being developed to help replace these fossil fuels. These chapters will provide the background needed to understand the latter parts of this book focused on the various bioenergy technologies, their development, and their potential impact on the energy market and the environment. By the end of this book, one will understand the importance of developing alternative energy technologies and the value that bioenergy could have in modern society.

STUDY QUESTIONS

1. Explain the definition of energy. What types of energy exist? Do we create energy?
2. Is petroleum considered a primary energy source? Why or why not? What are the four primary energy sources on Earth?
3. Why will the energy market continue to expand over the next several decades?
4. If GDP per capita and primary energy consumption per capita are linked, then what are the implications of a rising standard of living in many countries around the world?
5. What makes fossil fuels such a valuable commodity on the global market for energy?
6. How have US energy consumption and production changed in the last 60 years? What happens when the US consumption falls below its production level? Who do we rely on to fill these gaps?
7. How do we calculate how much of a given fossil fuel is remaining? How many years do we have remaining for coal, natural gas, and petroleum?
8. Why are different energy resources used in different sectors?
9. What is peak oil and what does this tell us about the future of petroleum?

CHAPTER TWO

Why Fossil Fuels Energize Our Society

Energy is the world's largest and one of its most important markets, influencing society in a myriad of ways. Fossil fuels supply the electricity to our homes and gasoline to our cars, enables modern agriculture, dominates national security and the health of the economy, and impacts the Earth's fragile environment. Most of this energy market is controlled by a single finite resource, **fossil fuels**. As explained in Chapter 1, the world's insatiable need for energy and society's dependence upon fossil fuels to fulfill this need will likely lead to economic and environmental consequences of unprecedented proportion in the coming century. Yet, the battle has just begun as we are only now starting to appreciate what a lack of these resources could mean for our everyday lives, and how climate change will impact both the economy and the environment.

The realization that fossil fuel supplies are limited has resulted in the exploration of renewable resource technologies to replace these supplies before they are exhausted; however, to truly understand the potential of renewable energy and its applications within society, one must take a step back to better understand fossil fuels themselves. First, we will consider the biochemical characteristics of these fuels and then build an understanding for how they were created and transformed into such an important component of modern society. With this knowledge, we will have a solid foundation upon which to continue discussing renewable energy resources for the future.

Fossil Fuel Energy

To begin, let's consider two questions: What do the molecular structures of fossil fuels look like? And what makes them such a great source of energy? The three major types of fossil fuels—**coal, petroleum,** and **natural gas**—all fulfill a role as critical energy resources even though they are physically very different. Coal is a solid, petroleum is generally a liquid (sometimes a very thick liquid), and natural gas is obviously a gas. Despite these physical differences, each of these fossil fuels still contains a huge amount of chemical energy that results from similarities in their chemical structures, mainly the use of the same atomic components: carbon and hydrogen. It is basically the number of carbon and hydrogen atoms that separates the three fossil fuels from one another (Williams, 2011).

Figure 2.1 illustrates the basic chemical structure of common fuel molecules and the energy density of these fuels. As one can see, these molecules share some striking similarities largely due to the dominance of hydrogen and carbon atoms. Their hydrogen and carbon molecular backbone places these fuels into the chemical family known as hydrocarbons. **Hydrocarbons**, as their name suggests, are molecules primarily composed of hydrogen and carbon. Hydrocarbons are quite easily combusted because the carbon–hydrogen (C-H) bonds trap a lot of energy and when broken release that energy usually as heat. From the figure you can see that chains of carbon and hydrogen as shown for gasoline and diesel molecules have higher energy densities than the smaller liquefied natural gas (LNG) as well as the renewable liquid transportation fuel, ethanol, that contains an oxygen atom (Demirel, 2012).

In order to fully understand energy, we need to consider some very basic laws like the **Second Law of Thermodynamics**. This law states that systems tend toward disorder and randomness, and that idea can help us understand the relationship between hydrocarbon structure and energy density. While a completely disordered system will contain very little energy, an ordered system will contain a much higher quantity of energy and chemical bonds make for an ordered system. Consider two atoms such as a hydrogen atom and a carbon atom that come together to form a chemical bond. While they are separated, these two atoms

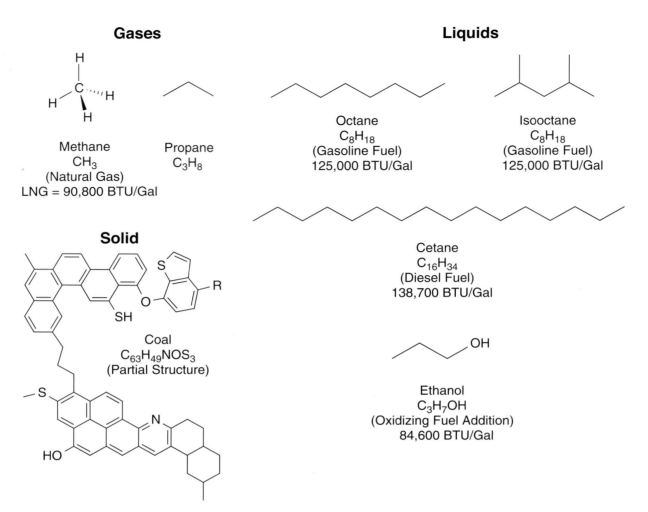

Gases

Methane
CH_3
(Natural Gas)
LNG = 90,800 BTU/Gal

Propane
C_3H_8

Liquids

Octane
C_8H_{18}
(Gasoline Fuel)
125,000 BTU/Gal

Isooctane
C_8H_{18}
(Gasoline Fuel)
125,000 BTU/Gal

Cetane
$C_{16}H_{34}$
(Diesel Fuel)
138,700 BTU/Gal

Ethanol
C_3H_7OH
(Oxidizing Fuel Addition)
84,600 BTU/Gal

Solid

Coal
$C_{63}H_{49}NOS_3$
(Partial Structure)

FIGURE 2.1 Basic chemical structures of common fuel molecules and median energy density for gaseous and liquid fuels. Fossil-fuel-based gases and liquids such as natural gas, propane, gasoline, and diesel are made entirely of carbon and hydrogen atoms. Coal, a solid type of fossil fuel, also contains oxygen, nitrogen, and sulfur atoms in addition to the carbon and hydrogen atoms. Ethanol, a renewable form of liquid fuel, also contains oxygen. Energy content is a value representing the energy in a fuel based on volume (BTU/gallon), a comparison particularly important when considering transportation of fuels. Diesel has a higher energy content than gasoline, while ethanol has a much lower energy content than either of these traditional fossil-fuel-based liquid fuels. Liquefied natural gas has a higher energy content than ethanol but still much lower than either gasoline or diesel. *Within the diagram each line without a designated atom represents a carbon–carbon bond where each carbon contains an appropriate number of hydrogen atoms to fulfill octet vacancy. Generally, this means that the carbon at the end contains three hydrogen atoms and a carbon in the middle contains two hydrogen atoms* (source: Demirel, 2012).

have a much larger degree of disorder and thus contain very little energy individually; however, once these two atoms form a bond, the degree of disorder is lowered (because now they are required to stay right next to each other) and energy is trapped within the connecting bond. Because this energy is dependent on the configuration and individual atoms within the molecule, bond energies vary based on the two individual atoms being disassociated from one another. For instance, a carbon–carbon (C-C) bond has a bond energy of about 376 kilojoules per mole, which is lower than the bond energy for a C-H bond at about 438 kilojoules per mole. This means that there is more energy trapped in a C-H bond than a C-C bond (McMurry, 2000).

Calculating the potential energy of a hydrocarbon is not as simple as summing all of the individual bond energies largely because, in any chemical reaction including combustion, there are bonds being made and bonds being broken at the same time. If the overall breaking of the atomic bonds within the molecules requires more energy than is being released by the formation of new bonds, then the reaction is termed an **endothermic reaction**; however, if the opposite is true and more energy is released during this reaction then it is termed an **exothermic reaction**. To calculate the overall energy change for a reaction, the endothermic energy must be subtracted from the exothermic energy to find the **enthalpy** of the reaction (McMurry, 2000). If this enthalpy is negative, the overall reaction is exothermic and energy is being released during the reaction—the overall goal of combustion. Consider the combustion of octane (C_8H_{18}), a molecule in gasoline outlined below.

Energy of Combustion

$$C_8H_{18} + 12.5O_2 \rightarrow 8CO_2 + 9H_2O$$

C_8H_{18} C–H = 18 bonds × 418 kJ/mol/bond = 7,380 kJ/mol
C–C = 7 bonds × 347 kJ/mol/bond = 2,429 kJ/mol

$12.5O_2$ O=O = 12.5 bonds × 494 kJ/mol/bond = 6.175 kJ/mol

$8CO_2$ C=O = 16 bonds × 799 kJ/mol/bond = 12,784 kJ/mol

$9H_2O$ O–H = 18 bonds × 460 kJ/mol/bond = 8,280 kJ/mol

Enthalpy = (7,380 kJ/mol + 2,429 kJ/mol + 6,175 kJ/mol)
Endothermic Reaction

– (12,784 kJ/mol + 8,280 kJ/mol) = −5,080 kJ/mol
Exothermic Reactions

The **combustion reaction** is an oxidation reaction often associated with burning. When octane is ignited, hydrogen (H) and carbon (C) atoms react with oxygen (O_2), recombining to form the products water (H_2O) and carbon dioxide (CO_2). By adding up the individual bond energies within the octane molecule and oxygen molecules and comparing it to the bond energies in the formation of the carbon dioxide molecules and water molecules, the carbon dioxide and water products have more energy. Thus, an additional 5,080 kilojoules of energy is released through exothermic reactions during combustion than is used for endothermic reactions to break atomic bonds in octane and oxygen. Therefore, the combustion of octane is an exothermic reaction that releases energy and this energy can be used to power a car.

Below is another example of a combustion reaction using the natural gas methane.

$$CH_4 + 2O_2 \rightarrow CO_2 + 2H_2O$$

One molecule of methane (CH_4) reacts with two molecules of oxygen ($2O_2$). The reaction between methane and oxygen is endothermic, meaning energy must be added to the system to initiate the reaction and generate the products. In the combustion of fossil fuels, this energy is usually a spark or flame. Once the spark adds energy, the atoms within methane and oxygen recombine into a new chemical pattern forming one molecule of carbon dioxide (CO_2) and two molecules of water ($2H_2O$). Once again, the excess exothermic energy released in this reaction, compared to the endothermic energy input, results in a net energy release in the form of heat. This heat can be used to warm a pan of water or to turn water into steam that powers a generator for electricity.

Chemical equations like this can be deceiving as it is easy to think larger molecules release more energy in combustion; however, this is not the case because as a molecule gets larger the amount of endothermic energy needed to initiate the reaction also increases. In fact, the amount of

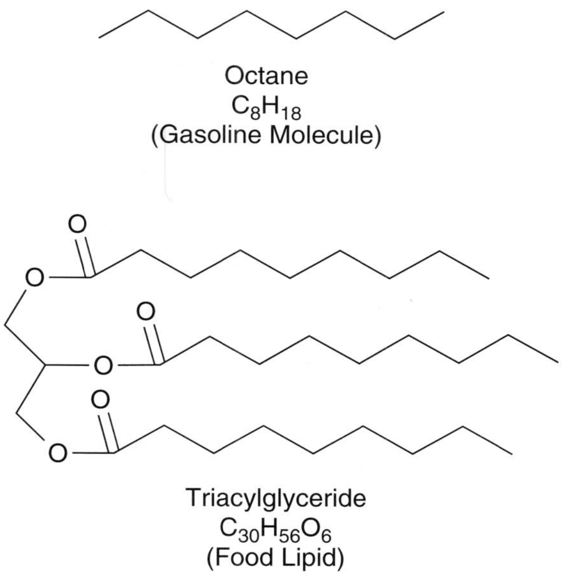

FIGURE 2.2 Comparison of the basic chemical structure of a molecule of gasoline with the structure of a triacylglyceride, a common lipid storage form. Here it is evident that the long hydrocarbon chain of gasoline is very similar to each of the hydrocarbon chains found in a triacylglyceride.

energy released by a given fossil fuel depends largely on two factors: (1) the hydrogen/carbon ratio and (2) the state of oxidation. In other words, molecules with more hydrogen, fewer carbons, and even fewer oxygen atoms will release the most energy. This is why more heat is released from methane than diesel fuel on a per mass unit basis.

Lipids: Another Fuel Source

The value of hydrocarbon molecules is not confined only to their role in the formation of fossil fuels. While the chemical structures of coal, petroleum, and natural gas may be most familiar, another type of hydrocarbon is even more ubiquitous: lipids. Lipids are one of three major macronutrients (along with carbohydrates and proteins) that are important in human health. Lipids are an important component of the human body acting in all cell membranes, and lipids also serve as key energy storage components of the body, often in the form of fat! These fat reserves are obtained primarily through the overconsumption of foods combined with a lack of exercise. You might be asking what fats and fossil fuels have in common. Well, despite the seeming differences between lipids that run bodies and fossil fuels that run cars, these molecules have strikingly similar molecular structures. Figure 2.2 juxtaposes the chemical structure of gasoline with that of **triacylglycerides**, the main storage form of lipids in the human body. The fatty acid tail of the triacyglyceride is almost exactly the same as a petroleum product on the molecular level, varying only in the level of oxidation and carbon chain length. Even though lipids in foods like cheese and fossil fuels like gasoline are considered in very different contexts, they actually have nearly the same function. Just as fossil fuels are packed with energy to power homes and drive cars, lipids are packed with energy to power our bodies and drive our metabolism. Lipids are in fact the most energy dense storage unit for the body, providing long-lasting fuel reserves.

This comparison between fossil-fuel-derived hydrocarbons like gasoline and the metabolic lipids like fats is the key to understanding renewable energy. The structural similarity and energy content of plant-derived lipids and fossil fuels have resulted in far-reaching interest in using plant-derived lipids to replace fossil fuels. Unlike fossil fuels, plant-derived lipids come from growing plants and do not require long, geological processes to form. Because they come from growing plants, these energy sources are renewable year after year. Energy sources from or derived from living matter like plants are collectively termed **bioenergy**, and this topic will be the focus of the latter parts of this book. For now, it is important to remember that these sources of bioenergy are chemically and energetically very similar to some of the fossil fuel resources that are voraciously being consumed by society.

Formation of Fossil Fuels

If coal, petroleum, and natural gas are all hydrocarbons just like plant-derived lipids, then why do we call them fossil fuels? Are they really fossils? Did fossil fuels, as is popular belief, come from dinosaurs? These are all common questions when considering the genesis of the name "fossil fuel." Despite the different physical forms and chemical shapes of the three major fossil fuels, they were actually all created in a very similar manner hundreds of millions of years ago from plants and algae—not dinosaurs. It is their prehistoric creation story lending them the name "fossil" fuel.

The formation of all fossil fuels began 300–400 million years ago when the Earth was covered in dense swampy vegetation and the oceans teemed with tiny microscopic photosynthetic organisms. Just like today, these phototrophs would live, grow, and die. Upon their death, they would either be consumed to complete the proverbial "circle of life" or be buried by dirt, silt, and sand. The decomposition of these plants upon burial during these prehistoric times represents the beginning of all the fossil fuel energy sources consumed today (Nersesian, 2010; Williams, 2011). Every gasoline molecule that powers a car and every ounce of coal burned to power an air conditioner are nearly 300 million years old. It is hard to imagine something so common being so old.

The long time period required for the creation of these fuel sources stems from the slow method of decomposition and **diagenesis** required to transform the dying green plants into hard black coal or thick oily petroleum. While decomposition is largely a biological action occurring at or just under the surface of the Earth, diagenesis is the set of abiotic (nonliving) reactions brought about by changes in heat and pressure that occur to organic material as it sinks deeper into the Earth. These diagenetic reactions are ultimately what results in the various physical and chemical forms of the three fossil fuels (Barnes et al., 1984). However, the process of differentiating between these fuel types really began with the specific vegetation that resulted in their creation. Petroleum and natural gas are derived primarily from microscopic oceanic photosynthetic organisms like algae, while coal is derived primarily from the higher plants found in prehistoric swamps such as trees and ferns. The higher complexity of these swampy plants, especially their high cellulose content, provided the framework to create solid coal, while the simplicity and high lipid content of the microscopic organisms provided the necessary organic material to make liquid petroleum (Nersesian, 2010).

Let us consider the formation of coal more closely. In prehistoric times, land-based vegetation died and fell into the oxygen-depleted environment of a swamp. This anoxic environment led to a much slower rate of biological decay and resulted in the formation of peat. Due to the lack of complete decomposition, peat built up over time to form peat bogs. **Peat** represents the least evolved form of coal and is still an important source of energy in some areas of the world including the United Kingdom, Russia, and the United States (Nersesian, 2010). Peat can be either directly burned for fuel or burned to create charcoal, which is then used for fuel. In these ancient bogs, the peat was eventually buried by silt and dirt. As the peat was buried deeper and deeper over time, temperature and pressure squeezed out the water, dehydrating the original peat material to higher and higher carbon content levels. Since the original material was largely from higher plants such as trees, this process resulted in long dense units that typify the first form of coal, **lignite**. Upon increasing temperature and pressure that comes with geologic depth, the formation of lignite was followed by **subbituminous**, **bituminous**, and finally **anthracite** forms of coal. Anthracite contains the highest carbon content and thus the highest energy content of the four types. This process of geologic compaction required an estimated seven feet of biological material to form one foot of coal (Nersesian, 2010; Waskey, 2011). Today, the thickest coal seam in the United States is Lake DeSmet in Wyoming with a thickness of 250 feet. This seam likely required over 1,750 feet of original organic material to form (DOI, 2013).

Despite being derived from simple microscopic organisms, the formation of petroleum and natural gas may be perceived as slightly more complex than that of coal. The formation of these fuels began when microscopic photosynthetic oceanic organisms like algae grew and then died in the ancient oceans. These microscopic phototrophs then sank to the depths of the ocean, where they were either consumed on the journey down or buried under silt and sand at the bottom. Once the biological material reached the seafloor, biochemical reactions from microbial decomposition continued to decay the material. The rate of biological degradation is one of the defining characteristics determining whether or not fossil fuels were formed. The faster living matter is degraded biologically, the less likely it would be for it to become a fossil fuel. This may explain the patchiness of fossil fuel resources seen on Earth today. For instance, consider the formation of petroleum from marine algae. Like today, ancient algae usually lived and thrived at the surface of the ocean where sunlight is found in abundance. Upon death, these algal cells sank to the bottom of the ocean. When a tiny fraction of these cells (1–10%) reached the ocean floor and were covered by silt and sand to create an anoxic environment, the biological breakdown of these cells slowed, leading to diagenesis to form petroleum and natural gas (Nersesian, 2010). Diagenesis begins as pressure and temperature increases basically cooking the biological material and forming it into **kerogen**. Kerogen is an irregular polymer formed when complex biological molecules like carbohydrates, proteins, and lipids are broken down into their monomeric units. These monomeric units were then randomly recombined to form kerogenic geopolymers. Once kerogen reached a depth between 3,500 and 10,500 feet below the Earth's surface where the temperature ranged between 50 and 150 degrees Celsius, the process of **catagenesis** began. Catagenesis is the process of cracking or breaking down complex geopolymers into simpler units. In this case, the kerogen was broken down into hydrocarbon fragments containing 5–50 carbons. These hydrocarbon fragments are classified as petroleum today. The formation of petroleum hydrocarbon fragments required a very specific temperature range known as the oil generation window. As depth increased, catagenesis continued by breaking the hydrocarbon fragments into even smaller fragments. These fragments resulted in the third and final fossil fuel: natural gas. Extra deep reactions occurring below the catagenesis stage are known as **metagenesis**, and they result in the formation of the simplest hydrocarbon, methane, containing only one carbon atom. Beyond this point, hydrocarbon material is broken down into its individual elements of carbon and hydrogen (Barnes et al., 1984).

Overall, the process of fossil fuel formation began with biological material and took hundreds of millions of years. Increasing pressure and temperature ultimately broke this material down into shorter and shorter hydrocarbons known today as fossil fuels. Despite the 300 million years required to create these energy dense resources, current trends indicate they will be completely consumed in just over 400 years of major human use. The tremendous increase in fossil fuel consumption over the past 200 years has resulted in the declining availability of these resources as discussed in Chapter 1. Unfortunately, the slow timescale for the creation of these resources leads to the unfortunate conclusion that fossil fuels are a finite resource and cannot be created in the short timescale needed to solve the looming energy crisis. The ability to solve this crisis will undoubtedly fall on the overall lowering of energy consumption and the development of renewable energy technologies in the future. But to understand these technologies, we must understand how fossil fuels became so engrained in the fabric of society.

Transitioning to a Fossil Fuel-based Society

With fossil fuels present throughout our history, it would seem possible that humans could have been consuming large amounts of fossil fuel energy for thousands of years. Indeed, humans have long dabbled in understanding the natural phenomena like the ignitable springs that seeped from the ground in places like Greece and Persia, yet it was not until the mid-1700s and the Industrial Revolution that humans really began to utilize large quantities of fossil fuel resources (Williams, 2011).

The first humans focused primarily on foraging with the sole aim of sustaining their lives on a daily basis. There were no tools other than hands, no farming, and certainly no grocery stores; every person was responsible for gathering food and sustaining his/her own life. The first real understanding of energy came not as a way to provide more energy to the community, but as a way to conserve more energy in their own bodies. The use of simple tools helped humans extract more energy from their food sources and later the use of fire improved upon this extraction. Controlled fire allowed people to cook food, enabling the extraction of more energy from the food and allowed people to warm their environment, thus likely lowering the caloric expenditure required to regulate their warm-blooded bodies in cold environments (Carmody et al., 2009).

During these prehistoric times, humans were steadily focused on finding ways to lower their energy expenditures, thereby reducing the amount of food needed and also finding methods for easier food collection. Another significant change in energy consumption by prehistoric humans occurred with the development of traditional agriculture about 12,000 years ago. There is no clear understanding for why this transition occurred in different parts of the world, possibly a change in climate promoting annual plants or a decreasing supply of natural food sources, but regardless of the reason, humans began to understand that growing food near their homes eliminated the need to forage and opened the door for the development of societies. Agricultural productivity continued to increase as humans became more adept at domesticating plants and animals. For instance, the use of livestock to replace human labor in the fields increased the productivity of farming while decreasing the energy expenditure of humans. Fewer people were required to produce more food, allowing for population growth and increased sizes in communities (Diamond, 2002; National Geographic, 2015). Suddenly humans had the time to develop significant lifestyle changes including infrastructure, cooking, and clothing. This radical change in how societies were organized set the stage for the Industrial Revolution and thus the explosion in demand for fossil fuels.

Industrialization was largely driven by people's desires to build communities. Communities required the construction of many large buildings, which in turn created a market for wood and metal. The development of fossil fuels is partly attributed to this need for construction resources.

The preparation and shaping of metal requires a tremendous amount of thermal energy for smelting the ore. Initially, this thermal energy was derived from biomass such as wood or peat. However, as societies grew the availability of wood began to decrease due to extreme deforestation and massive quantities of smoke polluted the air in the growing cities. Eventually, coal, a product used in smaller quantities for thousands of years replaced the use of wood in industrial processes. The large-scale production of coal was further supported with the development of the steam engine as it helped increase the productivity of mining. An early limitation in the use of coal was its transportation. The invention of the steam-driven train revolutionized energy by allowing coal to be transported to regions far from the coal mine. Eventually the steam engine gave rise to the steam turbine used today for electricity generation and powered primarily by coal and natural gas (Rhodes, 2007; Nersesian, 2010).

Unlike coal, natural gas has a long history rooted in its ability to catch fire easily. Thousands of years ago, people had already documented natural springs that burned when ignited. These fires are thought to be the first discoveries of natural gas. The first human use of natural gas as an energy resource was likely by the Chinese in 500 BC where they used the piped gas to evaporate seawater to acquire salt. In the United States, natural gas was first discovered by Native Americans who were curious about the exploding bubbles seeping from many rivers in the eastern regions. In 1816, it was used as a method to illuminate streets in Baltimore, and in 1821, William Hart drilled the first natural gas well in New York near the edge of a river. The natural gas from the well was piped to a grocery store and another small building in the surrounding town to provide light. Natural gas attempted to compete with manufactured gas and kerosene in the lighting market; however, the invention of the lightbulb by Thomas Edison at the end of the nineteenth century powered by coal-driven electricity severely reduced the market for natural gas lighting. Natural gas also played an important role in heating and cooking, a role that it continues today. Today, the role of natural gas in heating water to form steam and generate electricity is becoming increasingly valuable. Natural gas burns cleaner than the other fossil fuels, thus reducing atmospheric pollution during electricity generation and improving its reputation as a mainstay fuel source. Natural gas has also been developed for use in cars; however, the natural gas auto market is very small when compared to the petroleum-powered automobile market (Nersesian, 2010; Williams, 2011).

The last fossil fuel to really come into fruition was petroleum. However, the development of petroleum as an energy resource had a somewhat surprising start. The commercial value of oil really began with whales, the right whale to be exact. Hunters along the coast would harpoon these whales due to their slow speeds and proximity to shore. The meat could be eaten and the whales' blubber or fat could be burned for light and heating. Fortunately for the right whale, their oils let off a foul odor, making it unpleasant to use, so people began to hunt sperm whales. The sperm whale contained a significant amount of oil that could be harvested particularly from the head region and when burned for lighting did not stink. Although still only a minimal energy resource compared to coal, in its glory days in the mid-1800s, people harvested 550,000 barrels of whale oil every year equaling about 19 million gallons of oil (Pees, 2004). Compared to the nearly 18.9 million barrels used in the United States every day now, this was a very small harvest, but in terms of the animal cost, this amount of oil required the death of 14,250 whales every year (British Petroleum, 2014). This level of whaling was quickly found to be unsustainable for the large-scale production of oil, setting the stage for the development of "rock oil" or crude oil in 1855.

The initial discovery of rock oil in the United States stemmed from salt mining. When salt miners dug into their salt mines they would find their salt contaminated with an oily, green liquid that would at times explode. One salt miner, Samuel Kier, realized that he could take this rock oil from the salt mines and sell it as a medicinal agent for the treatment of tuberculosis and as a skin balm—not a particularly appealing medicinal agent. Eventually, George Bissel and colleagues realized rock oil could be commercialized not only as a medicinal agent but also as a substance for illumination. They assumed if rock oil was welling up in the salt mines, then the oil itself was held deeper in the Earth, and to access large quantities, drilling would be required. He formed the Pennsylvania Rock Oil Company in 1854 where he hired Col. Edwin Drake to begin explorations of the Pennsylvania area for crude oil. In 1859, Col. Drake struck oil in Pennsylvania, producing 25 barrels of oil a day and altering the path of the future (Tarbell, 1904).

In its early stages, the market for crude oil as an illuminant was not large due to the availability of both the lightbulb and natural gas; however, in one of the greatest market rebounds in history, the invention of the assembly line by Henry Ford and affordable personal automobiles resulted in crude oil becoming one of the most valuable commodities in existence.

Coal: Powering an Electric Society

While petroleum continues to drive the automobile industry, coal and natural gas dominate the electricity market. Coal is a sedimentary rock located at an average depth of about 300 feet under the Earth's surface. A coal seam is usually 2–8 feet thick but in some cases can be over 200 feet thick. As discussed earlier, there are four types of coal that include lignite, subbituminous, bituminous, and anthracite. The oldest and most valuable of these coal sources is anthracite, which has a carbon content of 86–98% and a particularly high calorific value or energy density. However, anthracite is not as prevalent as the other coal sources. In the United States, bituminous with a carbon content of 45–86% makes up the largest percentage of coal followed closely by subbituminous with a carbon content of 35–45%.

HOW AMERICANS GOT HOOKED ON GASOLINE

Travis L. Johnson

Before electricity, Americans used kerosene to light their lamps, a fuel derived from petroleum. In the late nineteenth century, oil companies were distilling petroleum to create kerosene for this purpose, and in the process they created what they thought was a waste product: gasoline. Refineries would dump thousands of barrels of gasoline into rivers and streams because they did not know what to do with it. Standard Oil decided to use this waste product as a fuel source in their kerosene distillation pro-cess, and when Henry Ford produced the Model T automobile at the start of the twentieth century, gasoline was a cheap fuel and Standard Oil was poised to produce enough of it to meet the demand of the growing American automobile industry. As Standard Oil became the dominant oil refiner and distributor in the United States, gasoline became the standard automobile fuel. By 1920, there were 9 million vehicles on the road powered by gasoline, and gas stations were opening all across the country.

SOURCES: Mayfield (2014); Roque (2013); EIA (2015).

Lignite has the lowest calorific value of the types of coal with a carbon content of only 25–35% and makes up only a small portion of the coal mined in the United States (Coal, 2002; Nersesian, 2010). Globally, coal is found in a wide number of locations, but the largest proved reserves are located in North America, Europe and Eurasia, and the Asia-Pacific region as shown in figure 2.3. The reserve-to-production (R/P) ratio for coal in most regions of the world ranges between 126 and 254 years; however, it should be noted that the Asia-Pacific region only has an estimated 54 years remaining despite having large reserves of coal (British Petroleum, 2014).

As a solid sedimentary rock, coal must be mined rather than piped from the ground. There are two primary methods used to mine coal: surface mining and underground mining. In surface mining, miners uncover the coal from the surface. This is also referred to as strip mining or open-pit mining because the surface soil is stripped away. In some cases, this type of mining is referred to as mountaintop removal and can have obvious and profound effects on the local ecosystems. Surface mining allows for the recovery of about 90% of an available coal seam and is commonly used in the United States, Australia, and Russia. The second mining method is underground mining. This method is often the most publicized due to its risk in trapping miners below the surface of the Earth. In underground mining, miners dig tunnels into the Earth to reach the coal seams. They then excavate the coal from the surrounding rock and haul it back to the surface. While valuable in some locations, this method of mining usually results in a total recovery of only 50–60% of the available coal (Pierce, 2012; UK, 2012).

Once excavated, either through surface mining or underground mining, the coal is transported by train to a power plant and pulverized into a fine powder. This powder is blown into a boiler and burned at high temperatures. Inside the boiler, water within pipes heats to the point of becoming steam. The generated steam has a high level of energy that will build up pressure and force the steam down the pipes and through a turbine. As the turbine spins, it creates an electric current within a generator to produce electricity. This electricity is then easily channeled through electric power lines to residential, commercial, and industrial consumers.

Today, coal remains the primary resource for power generation in the world, although in the last few years natural gas has been increasingly used. Coal is responsible for about 40% of the power generated in the United States (EIA, 2014), although in some states, like California, coal is no longer used due to its very high release of greenhouse gases compared to natural gas. In 2013, the world produced 3,881.4 million tonnes oil equivalent of coal with the United States producing about 13% of this total coal at 500.5 million tonnes oil equivalent (British Petroleum, 2014). The United States is estimated to have enough coal to sustain current national production levels for about the next 266 years, yet the total resources of coal in the United States are predicted to be much higher, leading to the speculation that with continued advancements in technology more coal will be available in coming years. However, as

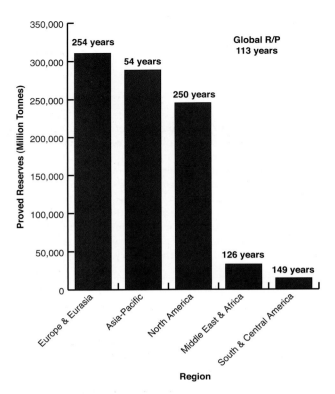

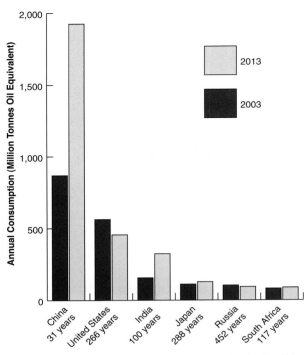

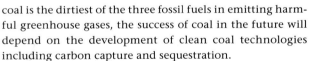

FIGURE 2.3 Regional proved reserves and reserve-to-production ratios for coal, 2013. Europe and Eurasia, Asia-Pacific, and North America have the largest supply of coal reserves. The reserve-to-production (R/P) ratio, equivalent to the number of years of coal remaining based on current production levels for each region, is listed above each column. Asia-Pacific has a large reserve of coal, but their production is also very high, leading to the lowest number of years remaining without the need to import coal from other regions (data from British Petroleum, 2014).

FIGURE 2.4 Comparison of coal consumption between 2003 and 2013 for the top six coal-consuming countries. Coal consumption in Japan, Russia, and South Africa has remained steady, while consumption in the United States has decreased. The biggest changes in coal consumption during this time period occurred through increases seen in India (+107%) and China (+122%). The numbers of years remaining for coal in each country based on production levels in 2013 are shown under the X-axis (data from British Petroleum, 2014).

coal is the dirtiest of the three fossil fuels in emitting harmful greenhouse gases, the success of coal in the future will depend on the development of clean coal technologies including carbon capture and sequestration.

When comparing the consumption of coal for the top six coal consuming countries between 2003 and 2013 as shown in figure 2.4, it is evident that China and India have increased their consumption, while the other countries including the United States, Japan, Russia, and South Africa have maintained or even lowered their consumption over this decade. While R/P ratios indicate the United States will have coal for over 200 years, this number fails to account for countries like China who, getting about 70% of their energy from coal, will run out in about 60 years, thereby importing more coal from countries like the United States and hastening the depletion of North American reserves (British Petroleum, 2014).

Natural Gas: A Cleaner Fossil Fuel

As the fossil fuel with the lowest harmful emissions, natural gas is a common and growing resource for electricity generation. Globally about 23% of primary energy con-

sumption came from natural gas in 2013, and in the United States over one quarter of all primary energy consumed in industrial, residential, and commercial sectors is derived from this resource (British Petroleum, 2014).

Natural gas is typically associated with sources of petroleum and coal due to its formation from these sources. In some cases, simply drilling into the Earth will allow natural gas to flow freely to the surface, but in other cases there is not enough pressure to drive the gas to the surface. In these instances, the gas must be pumped from the underground reservoirs. Natural gas molecules are small and can creep into crevices in the rocks, becoming trapped and difficult to extract. One method currently being used to help dislodge natural gas from these crevices is termed **hydraulic fracturing**. Fracking, as it is commonly called, is a process where a well is drilled and then flooded with high-pressure water containing chemicals and sand. The high-pressure water–chemical mixture causes fracturing of the rock and produces cracks and crevices that open up slightly allowing enough room for the particles of sand to be inserted. Once the water pressure is stopped, the sand keeps the cracks open enough to allow the gas or liquid to escape. The gas and liquids, called

DEEP INJECTION OIL WELLS AND WATER

Travis L. Johnson

There is more than just oil that comes out of a deep injection oil well. There is also a lot of water (called produced water), used during the hydraulic fracturing process as diagrammed in the figure. In 2013, Colorado and Wyoming produced roughly 128 million barrels of oil and with that more than 2.4 billion barrels of produced water. That means that for every barrel of oil produced, there were nearly 19 barrels of produced water. This water has chemical additives, including gels, acids, soluble hydrocarbons, and other naturally occurring substances, and must be treated and disposed. Much of this wastewater is filtered and injected back into the earth, although many oil companies are recycling the water back into wells. Throughout this process, there are concerns of potential contamination of ground and surface water, which could impact drinking water resources. The figure shows how water is inputted

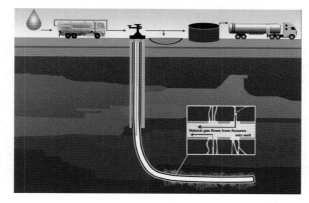

into a deep well and then both water and oil are pumped from the well and separated before the oil is sent by truck to refineries.

SOURCES: Paterson (2015); EPA (2015). Image provided by US Environmental Protection Agency.

condensates, are then pumped to the surface. The process of natural gas fracking increases the efficiency of extracting natural gas from the ground, but its potential risk to the environment remains a subject of debate (Gleick, 2014). Despite the potential negative impacts associated with the technology, fracking combined with horizontal drilling has opened up vast reserves of natural gas and oil, especially those associated with previously unobtainable shale oil deposits. Fracking has greatly increased the production of natural gas over the last several years, and with this increased production came lower prices, which has altered the consumption of natural gas in the United States and has shifted the electrical generation energy markets towards the use of this cleaner fossil fuel.

Once natural gas reaches the surface, a significant obstacle in its large-scale use is transportation. Transporting a gas is not as simple as a liquid or solid due to its larger volume at normal pressures. On land, natural gas can be transferred directly by pipeline or pressurized into a liquid that can be shipped by rail or truck. The natural gas associated with oil reservoirs found offshore are often more difficult to transport due to the need to load the gas onto a ship for transport

from the oil rigs; therefore, this gas is often flared, or burned, rather than retained.

Natural gas is used to generate power in a similar way to coal. Natural gas is burned to heat water that forms steam, which drives a turbine and turns a generator to create electricity. Natural gas is also piped directly into many residential, commercial, and industrial buildings to be used for heating and cooking. Most of the global natural gas reserves are located in the Middle East and Eurasia. As of 2013, Russia had the largest proved reserves with 1,162 trillion cubic feet followed by Iran at 1,187 trillion cubic feet. The proved reserve in the United States is 350 trillion cubic feet, but that number has increased significantly with the additional reserves from fracking. Despite having a lower quantity of proved reserves, the growth in production over the last five years has resulted in the United States being the largest producer of natural gas. Just between the years 2011 and 2012, the United States increased its production of natural gas by 4.1% (British Petroleum, 2014).

Today, electricity generation is attributable mainly to the production of coal and natural gas resources and has allowed for tremendous advances in technology over the past 100 years. However, with a decreasing supply of these resources and their adverse environmental impacts especially from coal, renewable sources of electricity generation are needed. Fortunately, significant advances have also been made in developing technologies that could play a role in replacing coal and natural gas. While electricity generation can be replaced with sources such as solar, wind, and waterpower, as will be discussed in Chapter 4, the heating and cooking applications of natural gas require the development of renewable biogas from either waste energy or bioenergy. We will discuss biogas in Chapter 8.

Black Gold to Gasoline: Transportation's Dependence on Petroleum

At the beginning of the twentieth century, the fate of petroleum as a valuable transportation fuel was sealed and resulted in it becoming the most consumed fossil fuel, earning the common nickname "Black Gold." Today, the uses of this profitable resource extend beyond those of just the transportation industry into almost every facet of our lives including the formation of plastics, rubbers, and other important chemicals like fertilizers.

Petroleum is found interspersed throughout the world. Figure 2.5 shows the proved global reserves for petroleum. Not surprisingly, the Middle East has the highest quantity of proved reserves and their supply is what pushes the global R/P ratio to about 52.9 years. If we were to consider only the proved reserves of petroleum in the United States, the availability of petroleum would be extremely short-lived at about 12.1 years at the production levels in 2013. In the United States, we have about 44 billion barrels of proved reserves coming from a variety of states including Alaska, California, Louisiana, North Dakota, and Texas. While the global production and consumption of petroleum are about equivalent, the United States produces only about 10 million barrels daily compared to its consumption of almost 19 million barrels daily (Murkowski, 2013; British Petroleum, 2014). This difference, also discussed in Chapter 1, has led the United States to importing a large percentage of its petroleum resources from foreign countries for many years, leading in many cases to the instability seen in the price of gasoline.

Petroleum deposits, like coal and natural gas, are found deep within the Earth. In the past 50 years, the ability to match production with rising consumption has required the location of new sources of oil. Geologists locate petroleum deposits using advanced imaging systems that can visualize rock formations many miles deep. These systems use sonic vibrations that bounce back in different patterns depending on the composition of the rock below the Earth's surface. For instance, sonic vibrations that bounce off solid rock are reflected differently than vibrations bouncing off thick liquid petroleum. Using this system, geologists can

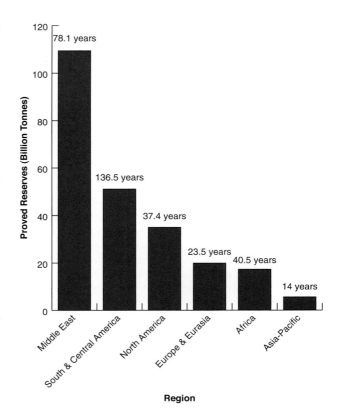

FIGURE 2.5 Regional proved reserves and reserve-to-production ratios for petroleum, 2013. The Middle East has the largest supply of petroleum reserves followed by South and Central America. The reserve-to-production ratio of petroleum based on current production levels for each region is listed above the column. Despite having the third highest level of proved reserves, North America is predicted to deplete its oil supply within the next 40 years based on current production levels and available technology (data from British Petroleum, 2014).

find oil deposits both on land and under the seafloor. Once geologists find an oil deposit, the oil must be accessed through drilling and then lifted (pumped) to the surface. All of this is accomplished using an oil rig that allows a pipe to be drilled to the level of the oil deposit, sometimes miles below the surface. The differential between the pressure in the pipes and that of the oil and natural gas deposit within the surrounding rock drives the oil into the pipe. Once the pipe reaches the deposit, about 33% of the oil will be pushed to the surface of the Earth by natural geologic pressure known as the recovery factor. This factor can be increased by creating an injection well, a nearby separate well either specifically drilled or no longer in use for oil production. By injecting water, a process also known as pumping, natural gas, or even carbon dioxide into the injection well, the pressure within the oil reserve increases and forces more oil and natural gas to the surface through the original well. This injection process can increase yields to about 40–50% of the total natural deposit (Nersesian, 2010). Pumping oil requires the use of energy to force the air or

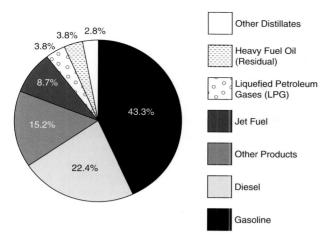

▢	Other Distillates
▦	Heavy Fuel Oil (Residual)
◎	Liquefied Petroleum Gases (LPG)
■	Jet Fuel
▨	Other Products
▢	Diesel
■	Gasoline

FIGURE 2.6 Average composition of products made from a barrel of crude oil. Gasoline and diesel are high-value products that result in over 65% of the total products obtained (data from Bryan, 2011).

fluid down into the injection well. The more viscous the crude oil, the more pressure is needed, and thus the more power required. The need for more power offsets some of the energy production and will lower the overall value of the oil. For this reason, viscosity is one of the characteristics used to classify crude oil around the world. Imagine trying to drink water versus molasses through a straw; clearly, the molasses will require a larger amount of force to reach the top of the straw. Likewise, when comparing different types of oil deposits around the world, a light crude oil requires less pumping power, whereas thick Canadian tar sands (bitumen) require a much greater pumping power (Veil and Quinn, 2008).

In any case, most crude oils are thick, flammable, yellow-to-black mixture of gaseous, liquid, and solid hydrocarbons. This sludge-like material contains a mixture of organic hydrocarbons and inorganic molecules like metal ions, sulfur, and nitrogen. Some inorganic molecules like sulfur are considered contaminants and must be removed from the crude oil during refining. These contaminants lower the overall value of the unrefined product. The amount of sulfur in crude oil is another characteristic used to classify crude oil around the world. If crude oil contains less than 0.5% sulfur contaminants, it is referred to as "sweet," while those resources that contain more than 0.5% sulfur are considered sour (Leffler, 2008).

In total, crude oil contains 100,000–1,000,000 different molecules. Figure 2.6 shows the typical composition of products from crude oil. There is a small amount of gas followed by about 43% gasoline molecules containing between 4 and 10 carbon atoms, about 22% diesel molecules containing 12–20 carbon atoms, about 8% jet fuel type molecules containing 10–12 carbons, and about 4% heavy fuel oil molecules containing 20–40 carbons. The remaining molecules in a barrel of oil are used for other products (Bryan, 2011). While these figures represent an average

across crude oils, oil deposits around the world are quite varied in the ratios of these molecules. Depending upon the composition of the crude oil, the density changes in comparison to water. The American Petrochemical Institute (API) uses this difference in density to classify petroleum based on API gravity. The less dense the crude oil, the greater the API value. The lightest crude oil has an API of greater than 38, the medium crude oils have an API of 22–38, and the heavy crude oils have an API of 10–22, while the extra heavy crude oil like the tar sands has an API of 0–10. The API gravity of water for comparison is 10 (Veil and Quinn, 2008).

With crude oil, as with any complex starting material, the goal is to extract as much value-added product as possible. Even though crude oil is used in a wide array of industrial applications—such as for the ethylene used in plastic grocery bags, butadiene used in synthetic rubbers in shoes and golf balls, and toluene used as a varnishing agent and adhesive—transportation fuels such as gasoline and diesel are still the highest valued products. However, since crude oil typically contains only about 40% molecules classified as gasoline, the crude oil usually undergoes a series of processing steps known as refinement to enhance the quantities of gasoline and diesel produced from each barrel of oil.

The first step in petroleum refining is **distillation**. In the distillation process, crude oil is placed in a boiler attached to a distillation column. As the boiler is gradually heated to higher and higher temperatures, different molecules hit their boiling temperatures and turn to vapor. The vapor then travels up the distillation column and condenses at different levels depending upon the temperature at which they return to liquid, draining out into separate tanks. Heavy oils will condense first followed by lubricating oil, diesel, kerosene, and finally gasoline (Nersesian, 2010).

To further increase the quantity of gasoline or diesel components from crude oil following distillation, there is a second step of refinement called conversion or **cracking**. During this second step, larger hydrocarbons are broken (or cracked) into smaller molecules by breaking C-C bonds. For example, a hydrocarbon containing 30 carbons may be cracked to produce a hydrocarbon with 20 carbons and one with 10 carbons to create a diesel molecule and a gasoline molecule. This process can occur by two different methods, **thermal cracking** and **catalytic cracking**. During thermal cracking, the larger molecules are heated to between 650 and 1,500 degrees Fahrenheit to obtain temperatures hot enough to actually break the bonds. In catalytic cracking, metals such as aluminum oxide and silica act as a catalyst to speed up the reaction of cracking from heat (Nersesian, 2010).

The final step in refinement is the removal of inorganic contaminants. Contaminants such as sulfur and nitrogen oxide as well as other metals and organic salts are dissolved

CRUDE OIL AROUND THE WORLD

Travis L. Johnson

Crude oil can be found in many parts of the world, and the composition of this oil can differ based on the origin. As mentioned in this chapter, petroleum is categorized by its viscosity and its sulfur content. Higher American Petroleum Institute (API) gravity measurements equate to lower viscosity (light oil) and lower API measurements mean higher viscosity (heavy oil). Oil with a sulfur percentage lower than 0.5% is known as sweet, while a sulfur percentage greater than 0.5% is known as sour. Low-viscosity and sweet crude oil is preferred because it requires less refining, thus decreasing the cost of production. A few examples of crude oil from different parts of the world include Russian Export Blend which is sour and has medium viscosity, Dubai Crude which has slightly lower viscosity but is more sour than the Russian blend, Brent Blend from Europe which is light and sweet, and West Texas Intermediate from the United States which is slightly lighter and sweeter than the Brent Blend.

SOURCE: Pomeroy (2014).

within the refined products and must be removed to eliminate potential environmental toxicity. One process used is called the **Claus process**. In the Claus process, the refined product like gasoline is "sweetened" (i.e. the sulfur is removed) by heating it to the point where elemental sulfur falls out of solution and can be collected as a solid. In all, 99.8% of the sulfur can be removed from a hydrocarbon sample using this method (EPA, 1995).

Once refining is complete, the products are ready to be shipped to consumers. These products as well as the initial crude oil resources are transported via tanker and pipelines throughout the world. For instance, gasoline and diesel are shipped to gas stations where individuals purchase fuel for their automobiles. Both gasoline and diesel are used in internal combustion engines where these fuels are combined with oxygen (burned) to form carbon dioxide, water, and considerable amounts of both power and heat. While the internal combustion engine is successful at powering automobiles and trucks, it also results in a number of environmentally harmful by-products such as carbon monoxide, nitrogen oxides, sulfur oxides, unburned hydrocarbons, and carbon dioxide. All of these products contribute to the formation of pollution and greenhouse gases that cause environmental damage including global warming, which will be discussed in the next chapter.

The value of gasoline as well as all of the other products produced from crude oil along with the infrastructure needed for producing, refining, transporting, and storing oil has created the world's largest industry. In total, the oil industry is worth trillions of dollars in the United States alone, and the oil and gas industry accounts for about 570,000 jobs (includes drilling, extraction, and support) (EIA, 2013). Clearly, this is not an industry that will easily be replaced by a renewable resource. However, environmental damage combined with a decreasing supply and continued problems with national security has led to a desire to develop alternative energy resources. One of the promising renewable resources is biofuel, particularly those liquid biofuels like biodiesel that can be used as a drop-in fuel taking advantage of the trillions of US dollars in infrastructure that is already in place in the United States and worldwide. Biofuels are one of the few renewable energy sources being developed that could be used to replace liquid transportation fuels.

The future of energy production and utilization is at a crossroads. This stems from questions of future supplies and costs of fossil fuels, particularly petroleum, continued pressure on national security due to the fact that all fossil fuels are unevenly distributed on the planet, and the increasingly clear environmental damage that burning fossil fuels is having on our planet. Understanding these issues, and their true costs, along with the important history and biochemical nature of coal, natural gas, and petroleum provides a foundation for the remainder of this book. The future of energy in the United States could very well rely on our understanding of this history to set a stage for the rapid and direct development of renewable energy resources in the future.

STUDY QUESTIONS

1. Explain the similarities and differences between the three types of fossil fuels in terms of physical and chemical states.
2. Explain the relationship between endothermic reactions, exothermic reactions, and enthalpy when considering the combustion reaction. Use this relationship to explain why there is not always a linear relationship between increasing molecule size and the release of energy.
3. What factors have the greatest impact on the release of energy? How do they impact energy content in fossil fuels?
4. Explain chemically why food sources have the potential to be used as fuel sources.
5. Briefly describe how fossil fuels are formed and when during the process of diagenesis the formation of oil and natural gas occurs.
6. Briefly explain the differences between the availability of each of the fossil fuels in the future and what sectors of society this availability will most likely impact.
7. Explain how refining crude oil can add to its overall value.

CHAPTER THREE

The Impact of Energy Usage on Climate Change

Fossil fuels have provided the world with an inexpensive source of energy that has become indispensable in powering our way of life; however, the reckless abandon with which society has consumed these resources has left us questioning not only its remaining supply but also its impact on the global environment now and in the future. While you may be familiar with environmental disasters that have brought energy to the forefront of the media including the 2010 Deepwater Horizon oil spill (aka the BP oil spill) in the Gulf of Mexico and the meltdown of the Japanese Fukushima Daiichi Nuclear Power Plant after the tsunami in 2011, many of the long-term environmental impacts these energy resources bring with them go largely unnoticed in the short term. Society widely ignores the impending environmental changes stemming from the heavy consumption of fossil fuels and the associated release of anthropogenic carbon dioxide and other greenhouse gases into the atmosphere. The release of these greenhouse gases is altering the chemistry of the atmosphere at an unprecedented rate, changing the natural cycle of global atmospheric molecules, and accelerating the overall warming trend on the planet at a rate never seen before. As the emission of these gases silently continues to pollute the atmosphere, the climate of the Earth is changing and likely altering the future of this planet forever. The coming chapter will deal with the Earth's atmosphere and the changes that fossil fuel emissions are effecting on the atmosphere's chemistry. You will also learn why and how these anthropogenic greenhouse gases ultimately affect the global climate and what we can likely expect as these changes come into full effect on our planet.

Atmospheric Control of Earth's Energy Budget

Biological life on Earth relies on the presence of a blanket of gases surrounding the Earth known as the atmosphere. The atmosphere plays two major roles in sustaining life: that of provider and protector. The air in the atmosphere is composed of about 78% nitrogen, 21% oxygen, and smaller amounts of other gases including carbon dioxide. These atmospheric gases are critical in critical in photosynthesis and the generation of plant life and in providing the oxygen we breathe. The second role of the atmosphere is that of protector due to the ability of the gaseous molecules within the atmosphere to deflect and absorb much of the intense radiation from the sun. In addition, the absorption of the sun's energy also leads to the transfer of solar radiation into heat energy and results in the warming of the surface of the planet. This warming trend is what buffers us from wild swings in temperature between day and night.

The sun's energy is derived from thermonuclear fusion reactions in which protons are fused to form helium and in the process release high-energy photons. These photons travel out from the core of the sun and through space, eventually coming in contact with the Earth and its atmosphere. Photons are classified based on the **electromagnetic radiation spectrum** where these photons are distinguished by their varying wavelengths (λ), and these wavelengths are based on the energy level of a photon. An example of the spectrum and the relationship between wavelengths and energy levels is shown in figure 3.1. The spectrum is broken into seven distinct radiation categories where photons with longer wavelengths have less energy than photons with shorter wavelengths. These radiation categories include the following: radio waves ($>1 \times 10^{-1}$ meter), microwaves (1×10^{-3} to 1×10^{-1} meter), infrared (7×10^{-7} to 1×10^{-3} meter), visible (4×10^{-7} to 7×10^{-7} meter), ultraviolet (1×10^{-8} to 4×10^{-7} meter), X-rays (1×10^{-11} to 1×10^{-8} meter), and cosmic and γ-rays ($<1 \times 10^{-11}$ meter) (Newman, 2012). Many of these radiation categories commonly function in our everyday lives and may seem familiar including visible radiation that allows for the vision of

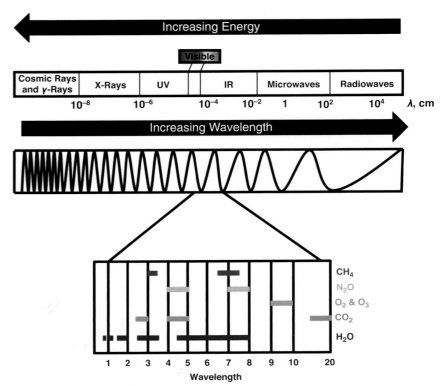

FIGURE 3.1 Diagram of energy and wavelength changes along the electromagnetic radiation spectrum and the major wavelengths at which common greenhouse gases absorb radiation. The top of the diagram shows the common categories of electromagnetic radiation ranging from cosmic and γ-rays with short wavelengths and high energy to radio waves with long wavelengths and lower energy. The bottom section of the diagram shows the wavelengths absorbed by the most common greenhouse gases. Water absorbs the largest range of wavelengths but is not considered an anthropogenic greenhouse gas (data from Grenci and Nese, 2008).

color, ultraviolet wavelengths that cause sunburns, and X-rays that penetrate tissue, making it possible to scan bones. This incoming radiation from the sun also plays an important role in the Earth's energy budget.

The **Earth's energy budget** refers to the balance of incoming solar radiation energy against the outgoing energy from the Earth's surface. Figure 3.2 diagrams the Earth's energy budget. Calculating the budget begins with the constant level of incoming solar radiation of about 340 watts per square meter. This incoming solar radiation is then deflected or absorbed by the atmosphere and clouds, reflected back towards the sun from the Earth's surface, or absorbed directly by the surface of the Earth (Canright, 2011). The second part of the Earth's energy budget is the outgoing radiation in the form of long-wave infrared radiation emitted from the surface of the Earth. The atmosphere absorbs 90% of this outgoing radiation, leaving only 10% lost into space. Most of the outgoing radiation is absorbed and reflected by the greenhouse gas molecules in the atmosphere including water vapor and carbon dioxide. The absorption and reflection of this radiation traps the energy within the atmosphere in a method similar to a blanket trapping heat around your body and results in warmer temperatures on Earth than would be expected based on surrounding planets. This phenomenon also diagrammed in figure 3.2 is known as the **greenhouse effect** (Wong et al., 2014). Naturally, the Earth's energy budget including the greenhouse effect is balanced to maintain the Earth's

temperature; however, as will be discussed later in the chapter, an imbalance can occur when the energy budget is influenced by unnatural factors.

Greenhouse Gases and their Role in Radiative Forcing

The greenhouse effect is a natural process that is important in heating the Earth. In fact, the greenhouse effect is responsible for the balmy average global temperature of 15 degrees Celsius (59 degrees Fahrenheit) present today as compared to the –18 degrees Celsius (–0.4 degrees Fahrenheit) average temperature that the Earth would have without the effect. Without the greenhouse effect, there would be an extraordinary challenge in maintaining life on Earth as we know it now, if not precluding it altogether. It is important to note that the global average temperature is just that—an average—and regional variations in temperatures still occur based on latitude, seasons, and weather patterns; it can be much warmer and much cooler than 15 degrees Celsius in different places around the world.

The atmospheric gases that are responsible for the greenhouse effect are called **greenhouse gases**. Water vapor (H_2O) is the most abundant greenhouse gas present in the atmosphere followed by carbon dioxide (CO_2), methane (CH_4), nitrous oxide (N_2O), and the halocarbons (CFCs, HFCs, HCFCs, SFs). The molecular structure of these greenhouse gases and their current atmospheric concentrations

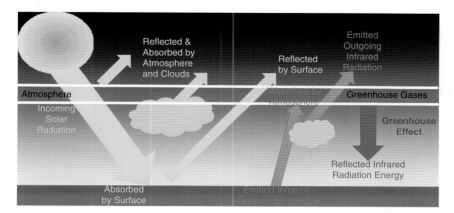

FIGURE 3.2 Diagram of Earth's energy budget and the greenhouse effect. Within the diagram, incoming solar radiation is both absorbed and reflected by the atmosphere, clouds, and the surface of the Earth. Infrared radiation emitted from the surface of the Earth is also reflected back toward the Earth by the atmosphere. This reflection of infrared radiation back and forth between the surface of the Earth and the atmosphere results in the warming of the Earth's atmosphere known as the greenhouse effect (source: Wong et al., 2014).

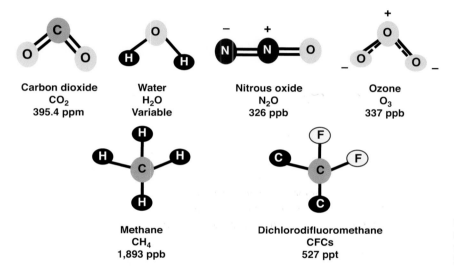

FIGURE 3.3 Chemical structures of common atmospheric greenhouse gas molecules and their atmospheric concentrations (source: Blasing, 2014).

are shown in figure 3.3 (Easterling and Karl, 2012; Wong et al., 2014). Each of the greenhouse gases is responsible for a different percentage of the natural greenhouse effect as a consequence of differences in their atmospheric concentrations and their molecular structures as seen in figure 3.3. Based on the different combinations of atoms, these molecules each absorb energy of a slightly different wavelength. For instance, a carbon–oxygen bond will absorb a different wavelength of energy than an oxygen–hydrogen bond. Figure 3.1 shows an example of the absorption profiles for different greenhouse gases. These profiles show how each gas molecule absorbs energy at slightly different wavelengths (Grenci and Nese, 2008). By absorbing different wavelengths of energy, together these molecules maximize the greenhouse effect and maintain warmer surface temperatures.

While greenhouse gases are the most influential gases at warming the planet, there are also molecules that work in the opposite direction by cooling the planet. Molecules like aerosols can lead to increased cloud formation that disrupts the balance of radiation entering and leaving the atmosphere similar to greenhouse gases but with a cooling effect rather than warming. Naturally, molecules that work to warm the atmosphere and molecules that result in cooling can counteract one another to create a stable temperature on the planet. However, an imbalance of one of these types of molecules will result in a net change in the balance of energy on the Earth. A change in this balance is known as **radiative forcing**. Positive radiative forcing causes the climate to grow warmer, while negative radiative forcing causes the climate to grow cooler. Positive radiative forcing can result from changes in solar irradiance, but it is most commonly attributed to anthropogenic activities that release greenhouse gases. Negative radiative forcing usually results from the release of aerosols either anthropogenic or naturally from volcano eruptions or changes in land use leading to more solar radiation being reflected away from the Earth, thus cooling the temperature. Figure 3.4 illustrates the main positive and negative radiative forcing drivers recognized by the Intergovernmental Panel on Climate

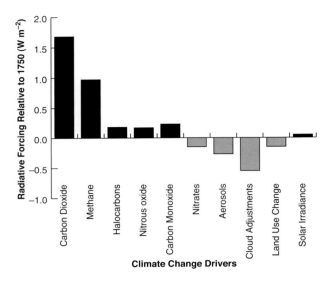

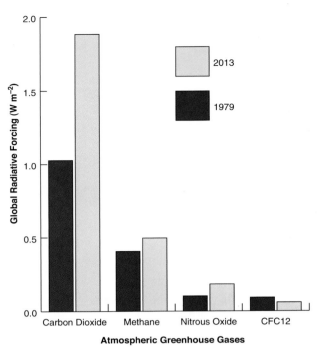

FIGURE 3.4 Radiative forcing of Earth's climate between 1750 and 2011. Forcing shown by black columns represents positive radiative forcing, leading to atmospheric warming. Forcing shown by gray bars represents negative radiative forcing that leads to atmospheric cooling. Positive and negative radiative forcing can balance each other, but since the beginning of the Industrial Revolution, the level of positive forcing has far outpaced that of negative forcing, leading to an overall increase in global average temperature. The high level of positive radiative forcing is almost entirely due to anthropogenic greenhouse gases in the atmosphere including carbon dioxide, methane, halocarbons, and nitrous oxide. Changes in solar irradiation are the only natural driver shown that impacts climate (data from IPCC, 2013).

FIGURE 3.5 Comparison of global radiative forcing for common atmospheric greenhouse gases between 1979 and 2013. With the exception of CFC12, all of the greenhouse gases shown including nitrous oxide, methane, and carbon dioxide have seen increases over the past several decades. Carbon dioxide has shown the largest change at 0.86 watts per square meter (data from Butler, 2014).

Change (IPCC). From the figure, it is clear that carbon dioxide and methane are major drivers of positive radiative forcing (Core Writing Team et al., 2007).

Throughout the Earth's history, both positive and negative radiative forcing agents have generally remained in equilibrium, maintaining a stable average global temperature. Sometimes this equilibrium can change by natural phenomena such as volcanic eruptions, changes in solar irradiance, or the impact of an asteroid, but the recent increase in positive radiative forcing agents cannot be attributed to these natural phenomena (Strom, 2007). The most recent report by the IPCC states that currently radiative forcing is positive as a result of the release of **anthropogenic** or human-made greenhouse gases into the atmosphere, mainly carbon dioxide (IPCC, 2013). Most of the greenhouse gas molecules discussed including carbon dioxide, methane, and nitrous oxide are anthropogenic; however, it should be noted that water vapor is generally not considered anthropogenic as its concentration in the atmosphere is generally not significantly impacted by human activity. Figure 3.5 compares the change in global radiative forcing for the major anthropogenic greenhouse gas molecules over the past 34 years. Carbon dioxide has clearly had the largest impact on radiative forcing, with methane and nitrous oxide also showing increasing impacts.

Based on the IPCC Synthesis Report in 2007, the concentrations of anthropogenic greenhouse gases have seen a tremendous rise over the past 200 years. Figure 3.6 shows the concentrations of common greenhouse gases from the year 1000 to 2014. From the year 0 to 1800, the concentrations of carbon dioxide, methane, and nitrous oxide saw only slight variations; however, after 1800 the concentrations of these gases show a sudden and dramatic increase. For example, in 1800 the concentration of carbon dioxide in the atmosphere ranged from 275 to 285 parts per million (ppm). In 2015, the concentration of carbon dioxide in the atmosphere is almost 400 ppm, an increase of about 30% (Solomon et al., 2007; Tans and Keeling, 2015).

There are skeptics to the idea that humans are causing the Earth to warm by burning fossil fuels. Some of these people say that the increase in the concentration of greenhouse gases, particularly carbon dioxide, is due to natural fluctuation that occurs between glacial and interglacial periods. While a small amount of fluctuation can be seen between these periods, never has such a drastic rise occurred in so short a period of time (Ahlenius, 2007). The scientific consensus attributes this incredible rate at which the concentration of these gases is increasing in the atmosphere to human activities, particularly the burning of fossil fuels. This is supported partially by the rise in greenhouse gases coinciding with the beginning of the Industrial Revo-

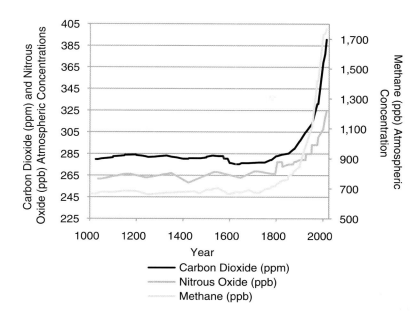

FIGURE 3.6 Change in atmospheric concentrations of carbon dioxide, methane, and nitrous oxide between 1000 and 2014. The concentrations of all of these greenhouse gases remained relatively constant until the middle of the nineteenth century when the Industrial Revolution and the combustion of fossil fuels became widespread. Over the past 160 years, the concentrations of these greenhouse gases has risen both quickly and steadily (data from EPA, 2011; CSIRO Marine and Atmospheric Research and the Australian Bureau of Meterology, 2014).

lution. As discussed in Chapter 2, the Industrial Revolution represents a time period when the discovery and leveraging of abundant fossil fuel reserves transformed society and became the foundation of technological advancements that define us today. The commercial development of coal, natural gas, and petroleum led to breakthroughs in the harnessing of electricity, indoor climate control, and personal automobiles. While these advances in technology brought comfort and easier living—such as with the availability of a larger food supply, the ability to travel great distances quickly, and instantaneous communication with people around the world—these technologies also quietly began changing the environment.

Many and perhaps most of the technologies used today that correspond with the commercialization of fossil fuels during the Industrial Revolution result in the emission of anthropogenic greenhouse gases. Anthropogenic carbon dioxide increases result largely from transportation, heating, cooling, manufacturing, and deforestation; anthropogenic nitrous oxide changes can result from fertilizers and fossil fuel combustion; anthropogenic methane increases may result from agriculture, the use of natural gas, and from landfills; and anthropogenic halocarbons are largely a result of refrigeration (Myhre et al., 2013). These greenhouse gases have increased exponentially over the past 200 years, and each acts as a positive radiative forcing agent on the global environment.

Carbon Dioxide Dominates the Greenhouse Gases

Greenhouse gases play a huge role in maintaining the global temperature; thus as the concentration of these gases change within the atmosphere, the temperature of the

Earth also changes. However, not all greenhouse gas molecules are equally influential in impacting the global temperature. This is due to a phenomenon known as the **band saturation effect** (Archer, 2012). As mentioned earlier, each type of greenhouse gas absorbs radiation at a slightly different wavelength, but the overall level of absorbance is also influenced by the concentration of that gas in the atmosphere. For instance, let us consider a molecule of methane and a molecule of carbon dioxide, both of which can be anthropogenic greenhouse gases that absorb two different wavelengths of radiation. When a molecule of methane is added to the atmosphere, it is 23 times more powerful as a greenhouse gas over a 100 year period than a molecule of carbon dioxide. This is because each molecule of methane turns more radiation into heat and because the concentration of methane is lower than carbon dioxide. The concentration of carbon dioxide in the atmosphere is currently around 400 ppm, while the concentration of methane is currently between 1.75–1.80 ppm (Blasing, 2014; Tans and Keeling, 2015). By adding a single molecule of carbon dioxide to the atmosphere, the impact is not as large as that of methane because there are already more carbon dioxide molecules present in the atmosphere. Due to their lower concentrations in the atmosphere, many of the other greenhouse gases are also considered stronger radiative forcing drivers than carbon dioxide as seen in Table 3.

However, despite methane and other greenhouse gases being stronger forcing drivers when compared molecule to molecule, carbon dioxide as a whole is still the primary cause of global warming today due to the rate at which we are adding carbon dioxide into the atmosphere. The amount of carbon dioxide in the atmosphere is governed largely by the **carbon cycle**. The global carbon cycle is a

TABLE 3
Global warming potential for common atmospheric greenhouse gases

Atmospheric Gas Common Name	Chemical Formula	Lifetime (years)	Global Warming Potential in Given Time Period	
			20 years	100 years
Carbon dioxide	CO_2	5–200	1	1
Methane	CH_4	12	72	25
Nitrous oxide	N_2O	114	289	298
CFC-11	CCl_3F	45	6,730	4,750
HFC-23	CHF_3	270	12,000	14,800
Sulfur hexafluoride	SF_6	3,200	16,300	22,800

NOTE: The table shows the individual gases and their chemical formulas alongside the number of years that an individual gas molecule of that chemical makeup will remain in the atmosphere once released. The last two columns show the potential of an individual greenhouse gas molecule to increase global warming as compared to carbon dioxide. Methane, nitrous oxide, and the other greenhouse gases listed have the potential to increase global warming much faster than carbon dioxide; however, their concentrations are generally much lower and in some cases these gas molecules do not stay in the atmosphere very long.

DATA FROM: Forester et al. (2007).

biogeochemical cycle based on the movement of carbon between the biosphere, atmosphere, ocean, and geological features of the Earth. An example of the cycle can be seen in figure 3.7 (Solomon et al., 2007). This cycle is extremely important because it maintains the balance of carbon within the atmosphere by both sequestering and storing carbon within the biosphere, rocks, and oceans while at the same time releasing carbon back into the atmosphere. Each component of the cycle works in opposition to another component to maintain this balance. For instance, photosynthesis and respiration by plants work in opposition to one another. **Photosynthesis** is the process by which plants use the energy from sunlight to fix carbon dioxide from the air into a useable form of organic carbon such as a sugar. This allows the plants to grow and forms the bottom level of the food chain that supports all life on Earth while simultaneously sequestering and storing carbon in the land or sea. At the same time, respiration typically uses oxygen to break down organic carbon containing molecules and releases carbon dioxide back into the atmosphere. As seen in the figure, these two very important components of the biological carbon cycle are nearly in balance.

The biological carbon cycle is largely responsible for the carbon that is sequestered in the biosphere; however, most carbon is actually sequestered in the rocks and minerals of the Earth's crust as well as within the ocean. The Earth's crust stores an average 100,000,000 petagrams (Pg) of carbon and the ocean about 37,000 Pg, while plants and soils store only about 2,000 Pg (Solomon et al., 2007). Deep ocean carbon is sequestered and stored as a result of chemical reactions occurring at the air–sea surface interface. The exchange of carbon between the biosphere, upper ocean, and the atmosphere occurs relatively quickly,

but the exchange of carbon between the deep ocean and upper ocean occurs much more slowly, on the order of hundreds of years. This difference is significant because it indicates that carbon stored within the deep ocean will remain there for hundreds of years without exchange with the atmosphere. Therefore, this carbon is not impacting the radiative forcing balance in the atmosphere. However, the carbon fluxing into the atmosphere from human activities such as burning fossil fuels and deforestation is upsetting the natural balance of carbon in the atmosphere and leading to an increase in the Earth's average global temperature. Ultimately, rising temperatures also lead to warmer oceans where less carbon is sequestered and stored in the ocean and more carbon remains in the atmosphere. This is compounding the problem and causing atmospheric carbon dioxide levels to rise at even faster rates.

Prior to the Industrial Revolution, the concentration of carbon dioxide in the atmosphere was about 280 ppm largely due to the natural exchange of carbon dioxide on Earth through the carbon cycle; however, the Industrial Revolution has put into motion dramatic changes to that cycle with the combustion of fossil fuels, deforestation and agriculture, and the heavy use of cement. In 1958, Dr. Charles David Keeling from Scripps Institution of Oceanography began monitoring atmospheric carbon dioxide levels from the Mauna Loa Observatory in Hawaii (Keeling Curve, 2012). Prior to this time, carbon dioxide levels in the atmosphere were thought to vary by season but remain constant on average over time; however, with the onset of this study that was soon shown to be incorrect. The results from the Keeling study, which is still ongoing today, have become famously known as the Keeling Curve represented

SHORT-LIVED CLIMATE POLLUTANTS

Travis L. Johnson

Carbon dioxide is the greenhouse gas that accounts for the majority of global warming and it can persist in the atmosphere for hundreds to thousands of years. But, there are other greenhouse gases (GHG) known as short-lived climate pollutants that also contribute to global warming, including methane, black carbon, ozone, and hydrofluorocarbons. These pollutants make up roughly 40% of current global warming, but their lifetimes are much shorter. Black carbon persists in the atmosphere less than two weeks, ozone less than two months, and methane and hydrofluorocarbons less than 15 years. Because of their short lifetimes, we can make significant changes to GHG accumulation relatively quickly if we reduce these emissions. Researchers estimate that decreasing short-lived climate pollutants could reduce global warming by 50% (around 1 degree Celsius) and sea level rise by 25% (around 40 centimeter) by the end of the century.

SOURCE: Ramanathan (2014).

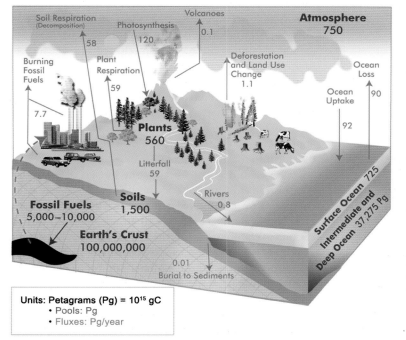

FIGURE 3.7 Diagram of the global carbon cycle. This diagram shows how carbon is exchanged between the land and the ocean as well as how various human activities such as the use of fossil fuels and deforestation impact this exchange. Within the diagram, an average accounting of the amount of carbon in flux through both natural and man-made processes is shown in red, while the average amount of carbon stored in various carbon pools on Earth is shown in blue. For example, 5,000–10,000 petagrams (Pg) of carbon are stored as fossil fuels, but as we extract these fuels and use them in combustion, they add 7.7 Pg to the atmosphere largely in the form of carbon dioxide. This additional carbon in the atmosphere will upset the balance of the natural carbon cycle and ultimately impact the greenhouse effect (image by NASA/GLOBE Program, www.nasa.gov).

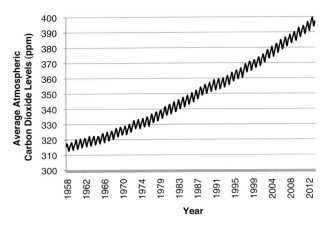

FIGURE 3.8 Change in average atmospheric carbon dioxide concentrations between 1958 and 2013. This graph, known as the "Keeling Curve," is one of the major supporting factors for the rise in anthropogenic carbon dioxide levels and their impact on global warming. Seasonal variations can be seen due to a change in the carbon dioxide levels during different seasons. Carbon dioxide levels are lower in the spring and summer as plants bloom, and higher in the fall and winter as plants degrade in the northern hemisphere (data from Keeling, 1974 and Tans, 2015).

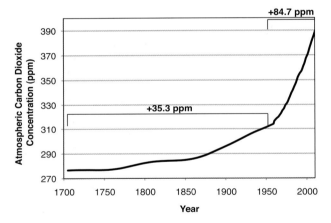

FIGURE 3.9 Change in the atmospheric concentration of carbon dioxide between the years 1700 and 2013. In 250 years, between 1700 and 1950, the concentration of carbon dioxide only increased by 35.3 ppm; however, in only 63 years between 1950 and 2013, the concentration of this greenhouse gas increased by 84.7 ppm. The fast pace of the recent trend in increasing atmospheric carbon dioxide concentrations is a result of the widespread combustion of fossil fuels, leading to the emission of this gas into the atmosphere (data from EPA, 2011).

in figure 3.8. The Keeling Curve shows a continuous steady increase in the level of atmospheric carbon dioxide since 1958. Seasonal variations are seen every year due to more photosynthesis occurring in the spring and summer of the northern hemisphere, thus lowering the atmospheric carbon dioxide levels slightly during that time each year, but the overall trend of increasing CO_2 is readily apparent from this graph.

The Keeling Curve has become an important contributing factor to the argument for the role of carbon dioxide in climate change and global warming over the past few decades. By comparing air samples extracted from Antarctic ice cores with the carbon dioxide readings from the Mauna Loa laboratory, a comparison of carbon dioxide concentration in the atmosphere can be made for the past 300 years. Figure 3.9 shows that from around 1700 to 1960, the concentration of carbon dioxide increased by 35.3 ppm; however, in the past 54 years (1960–2014) this concentration has already increased by 84.7 ppm. The graph draws attention to the fact that not only is the concentration of carbon dioxide in the atmosphere increasing, but the rate of increase itself is also going up and continues to do so, even today (EPA, 2011).

The increased concentration of carbon dioxide in the atmosphere over the last 50 years corresponds precisely with the world's insatiable thirst for fossil fuels. As shown below, combustion reactions involving fossil fuels such as methane result in the formation of carbon dioxide and water:

$$CH_4 + 2O_2 \rightarrow CO_2 + 2H_2O$$

Currently, this carbon dioxide is released directly into the atmosphere faster than carbon-sequestering portions of the carbon cycle can pull the carbon out of the atmosphere due to the tremendous consumption of fossil fuels. As an example, in 1950, the number of vehicles in operation in the United States was 43 million; by 2009, that number was about 250 million vehicles; each of these cars burns gasoline, releasing about 9 kilograms of carbon dioxide into the atmosphere per gallon burned. If every car in 1950 had burned just 1 gallon of gasoline, 391 million kilograms of carbon dioxide would have been released into the atmosphere. But today, when every car burns 1 gallon, the release of carbon dioxide amounts to a stunning 2.2 billion kilograms of carbon dioxide (Davis et al., 2012). Moreover, personal transportation is only one of many applications for fossil fuels; other commercial, industrial, and residential applications all contribute to the overall release of carbon dioxide into the atmosphere.

One way that scientists show the impact of fossil fuel combustion on atmospheric carbon dioxide levels is by comparing carbon isotopic ratios found within atmospheric carbon dioxide molecules. As an atomic element, carbon contains 12 protons within the atomic nucleus; however, a small portion of carbon can also naturally be found that contains variations in this number of protons with either 13 or 14 protons. Atoms with these proton variations are known as isotopes. ^{14}C is known as radiocarbon and will decay over time. Because fossil fuels have been buried for hundreds of millions of years, these fuels contain none of the ^{14}C isotope and this can be traced in the atmosphere. ^{13}C is another carbon isotope present in a very small proportion of total carbon (1.1%). Interestingly, plants have a

JOHN TYNDALL

Travis L. Johnson

Before Charles David Keeling started measuring atmospheric carbon dioxide concentration in the 1950s and subsequently showing its increasing trend, John Tyndall, an Irish physicist, discovered carbon dioxide's radiation-absorbing capabilities nearly 100 years earlier. Tyndall was interested in how radiant energy acted on the molecules of the air, and he invented an instrument that measured the infrared radiation (radiant heat) absorption in atmospheric gases like water vapor, carbon dioxide, nitrogen, and oxygen. He discovered that water vapor and carbon dioxide have powerful heat-trapping abilities and that the majority of the air, nitrogen and oxygen, do not. Tyndall is credited with proving the earth's atmosphere has a greenhouse effect, and upon his discovery he immediately understood the implications for climate change.

SOURCES: Somerville (2014); Tyndall (1872).

preference for ^{12}C over ^{13}C during photosynthesis; therefore, because fossil fuels are derived from ancient plants, they contain a lower abundance of ^{13}C. Comparing the ratio of ^{12}C to ^{13}C between the carbon stores in the ocean and on land today with those of the carbon dioxide molecules in the atmosphere indicates that atmospheric carbon dioxide is composed of higher levels of ^{12}C, an indicator that they are derived from fossil fuels (Shoemaker, 2010).

Carbon Dioxide and Global Warming

As detailed earlier in this chapter, greenhouse gases play an important role in maintaining the Earth's stable average temperature through the greenhouse effect. However, with the rise in anthropogenic greenhouse gases, particularly carbon dioxide, from human activities over the past century, these gases have caused greater positive radiative forcing and thus have begun to increase the average temperature of the Earth's atmosphere. In fact, in the past 100 years, the average global temperature has risen by 1.6 degrees Fahrenheit (0.9 degrees Celsius) and is continuing to rise. Some scientists predict that the ongoing release of anthropogenic greenhouse gases into the atmosphere will cause the average temperature to increase by 2–6 degrees Fahrenheit (1.1–3.3 degrees Celsius) (Samenow et al., 2010). While this may seem like only a small increase in temperature, it could have devastating implications for the balance of life on the planet. For instance, even a small rise in global temperature could potentially decrease plant growth, impacting both the photosynthetic uptake of carbon dioxide and the agricultural supply of food to people around the world. The rising temperature might also impact the Earth's hydrological cycle where warming oceans will lead to melting glaciers and ice sheets, ultimately increasing the input of freshwater into ocean systems. This input of freshwater could increase the sea surface level, thereby flooding millions of miles of coastal land around the world. Thus, small changes in the climate system can have a profound effect on the planet as a whole.

The cause of global warming has been hotly debated for a number of years with theories for why the Earth has warmed ranging from changes in the sun's radiation intensity, volcanic eruptions, asteroids impacting the planet, changes in the thermohaline circulation pattern in the ocean, natural fluctuations to changes in atmospheric greenhouse gas levels. When looking at each of these theories individually, we can assess how it may play a role in global warming. The thermohaline circulation pattern has not changed, and we know there has not been a significant asteroid impact on Earth. Volcanic eruptions typically release ash that acts as a negative radiative forcing driver by reflecting sunlight back into space and thus cooling the Earth's temperature. But no significant volcanic eruptions have occurred that would impact the temperature on the entire planet. Changes in the radiation level being emitted by the sun could cause global warming or cooling, but based on continuous monitoring of the sun's intensity over the past 20 years, this change only accounts for about 3% of the global warming being experienced today (Strom, 2007). Taken together, only one theory fits the measured change in global temperature seen over the past 50 years: anthropogenic greenhouse gas emissions.

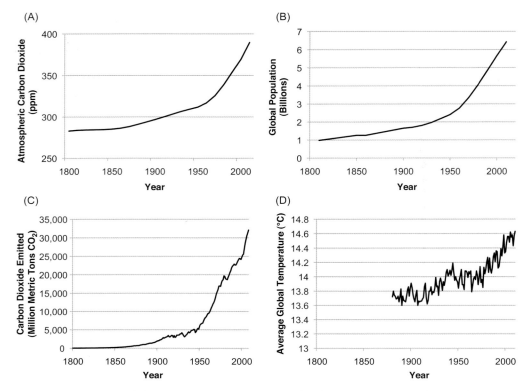

FIGURE 3.10 Comparison of changes in (A) atmospheric carbon dioxide concentrations over the past 200 years with the change in (B) global population, (C) global carbon dioxide emissions, and (D) the global change in land–ocean temperature since the late 1800s. The rise in atmospheric carbon dioxide concentrations beginning around 1900 appears to correspond with a similar pattern of increase in the global population and in the actual emissions of carbon dioxide around the globe. In addition, this time period also shows an increase in the global land–ocean temperature. The similar pattern of increase within graphs A–C is one piece of evidence towards the impact of humans and the use of fossil fuels on the atmospheric carbon dioxide concentrations (data from EPA, 2011; United Nations Secretariat, 1999; NASA, 2013; Boden et al., 2011).

While most individual greenhouse gases are rising, each of these gases can account for only a percentage of the overall global warming due to its band saturation effect and atmospheric concentration. Of the major greenhouse gases, methane accounts for 16%, ozone for 12%, halocarbons for 11%, and nitrous oxide for 5%. These gases combined with the changes in the sun's intensity mentioned previously make up 47% of global warming. Interestingly, the remaining 53% of global warming can be attributed to one greenhouse gas—carbon dioxide (Strom, 2007).

Figure 3.10 compares the trends between atmospheric carbon dioxide levels, population change, global carbon dioxide emissions, and the global land–ocean temperature change. In the first three graphs (A–C), the increasing trend of atmospheric carbon dioxide levels, population change, and global carbon dioxide emissions after 1875 appears to be very similar. The increase in the global land–ocean temperature is not as large or as steady, but there is a clear increase in the temperature that corresponds with the same time period, although slightly delayed. Taken together, these graphs suggest a connection between atmospheric carbon dioxide levels, global

carbon dioxide emissions, and rising global land–ocean temperatures.

While the graphs in figure 3.10 seem to tell a striking story, there is still the possibility that the increase in average global temperature could be a result of natural fluctuations on Earth. Historically, carbon dioxide levels have been known to see great fluctuations leading to periods of warming and cooling such as in the ice ages. Based on geologic precedence, the Earth is naturally experiencing a period of warming; however, the difference between the warming now and in the past is the rate at which the change is occurring and the concentration of carbon dioxide in the atmosphere. The Earth has never seen carbon dioxide levels reach as high in so short a period of time (Ahlenius, 2007). Thus, it seems that the combustion of fossil fuels is the most likely source for the current rise in atmospheric carbon dioxide and the subsequent increase in global temperature.

Global Indicators of Climate Change

Collectively, changes in the Earth's climate due to these rising levels of greenhouse gases is known as **global climate**

change. Global warming is just one of many indicators for global climate change. To understand climate change, you must first understand the difference between two key concepts often confused: climate and weather. **Weather** is the way the atmosphere is behaving in the short term in a specific location, referring to timescales ranging from hours to days to weeks. Weather is typically associated with forecasts that you may see on the news including sunshine, rain, cloud cover, winds, hail, snow, and thunderstorms. **Climate** is the way the atmosphere behaves in the long term, referring to timescales of years, decades, and even centuries. Climate measures are typically averages of precipitation, temperature, or sunshine for a particular place over the course of these long timescales. In general, it is fairly easy to confirm the accuracy of weather changes by simply watching the weather. If a weather forecaster says it is going to rain, and it rains, then the weather forecast was accurate. However, climate is not as easy to confirm for accuracy. Because of the long time spans, scientists must use sophisticated models requiring the input of many environmental factors to predict how the climate will react to a particular change in a single environmental factor. The accuracy of these predictions is very hard to confirm in the short term; however, these predictions can help us to prepare for the future by providing an understanding of key environmental indicators that will likely play a role in global climate change. According to the Environmental Protection Agency's Climate Change Report, these indicators may include global temperature, sea surface temperature, sea level, ocean acidity, Arctic sea ice, glaciers, snow cover, heat waves, drought, precipitation, typhoon intensity, and length of growing season. A few of these climate change indicators will be discussed below (EPA, 2014).

Rising global temperature or global warming is one of the most highly discussed indicators of climate change largely as a result of its influence on other climatic factors. The anthropogenic increase in the concentration of atmospheric carbon dioxide has ultimately resulted in rising temperatures on the Earth. Since the beginning of the twentieth century, the average global temperature has risen by 0.15 degrees Fahrenheit per decade. You may ask why you have not noticed this change in your area. This is because changes in global temperature are not uniform across the globe. While some places are getting hotter, others may not be changing or even getting cooler. Since the 1970s, the average temperature in the northern hemisphere has warmed significantly more than the middle or southern latitudes. This temperature anomaly is apparent when we look at the average temperature of the United States alone. Since the 1970s, the average temperature in the United States has risen by 0.31–0.48 degrees Fahrenheit per decade, which is about twice the global rate. This is a clear example of the nonuniform changes in temperature that will take place across the globe (EPA, 2014).

Changing global temperatures will also bring changes to global precipitation patterns. Increasing precipitation occurs when warming temperatures cause more evaporation; however, this evaporation can only occur in areas that have water to be evaporated such as oceans, lakes, or melting ice. Therefore, predictions suggest that some areas of the planet will receive significantly more precipitation such as those areas near the poles, while other areas will receive much less precipitation including temperate and subtropical regions like the southwest United States. Unfortunately, in areas that receive less precipitation, a shortage of water could lead to drought, a condition that can greatly affect agriculture productivity and water availability as seen recently in California. In the past decade, 20–70% of US land area experienced drought conditions at any given time (EPA, 2014).

While heat waves and droughts can have significant negative effects, increased precipitation can be equally problematic. Drought occurs when the rate of evaporation is higher than the rate of precipitation; however, the evaporated water must still go somewhere. Therefore, other areas will experience higher levels and more intense precipitation that may lead to increased flooding. For instance, since 1901, global precipitation has increased at a rate of about 2.2%, while the rate in the United States has increased by about 5.5% (EPA, 2014). This increase is not uniform across the United States as indicated in the juxtaposition of news stories of severe flooding and land erosion alongside record-setting droughts.

The ocean plays a critical role in the balance of energy on the Earth. Oceans cover 70% of the planet, giving them ample space to store enormous amounts of energy from the sun. The ocean not only stores this energy, but is also instrumental in moving energy around the world via ocean currents. Ocean currents bring warm water from near the equator to the poles, all the while bringing cool water toward the equator. These currents and the movement of this energy in the form of warm and cool waters have a huge impact on the climates of the surrounding landmasses and could be impacted by changing ocean temperatures.

Rising global temperatures will affect the temperature of the ocean, particularly the sea surface temperature. The sea surface temperature is the temperature of the water at the ocean's surface and is an important indicator of the health of the ocean. There are normal variations in this temperature with the warmest being in the tropics and the coolest being in the poles, but overall the average temperature of the sea surface is rising. Since 1901, the sea surface temperature has risen at an average rate of 0.13 degrees Celsius per decade (EPA, 2014). While small, this increase in sea surface temperature can still affect ocean circulation patterns, can alter marine ecosystems, such as killing coral reefs and increasing harmful algae blooms, and can change the evaporation and precipitation rates around the world. One extreme example of the effects of changing sea surface temperature can be seen when examining the intensity of tropical storms. Hurricanes, tropical storms, and cyclones build energy from warm ocean surface waters, and this fact explains why they are most often associated with the warmer summer months. As the global sea surface temperature rises, more energy trapped in warmer tropical waters leads to increases in storm intensity, and this increase has been observed over the last few decades.

Sea surface temperature plays an important role in climatic changes because it is in direct contact with the atmosphere, but the energy content or heat content of the ocean as a whole is also an indicator of climate change. Water has a higher heat capacity than air and the oceans can absorb a larger amount of energy than the atmosphere without substantially changing temperature. This is important for the energy balance of the Earth because if the oceans were to stop absorbing this heat, the temperature of the Earth's surface would increase much faster. Increasing atmospheric carbon dioxide levels and thus increasing global warming can affect the overall temperature of the ocean as well as the sea surface temperature. While the sea surface is warming, the deep ocean can also be absorbing energy and its temperature can also rise. This deep water temperature change could certainly affect the ecological diversity of the ocean. In addition, because warmer water takes up slightly more space than cooler water due to its higher energy level, the oceans will expand as their temperatures increase, a phenomenon known as thermal expansion. This expansion plays a role in the sea level rise we see now and will continue to see in the future.

In the twentieth century, global sea level has risen by about 0.06 inches per year, and like other climatic indicators, this rate of change has doubled in the last 40 years to 0.11–0.12 inches per year (EPA, 2014). Rising sea level has the potential to cause erosion and flooding, forcing people to move from their homes and changing the ecosystems in coastal regions, particularly those of saltwater marshes critical for the reproductive cycles of many fishes. It is estimated that by 2080, 240 million people will be at risk from global sea level rise, including some entire countries (Coley, 2008).

While thermal expansion is one of the causes for sea level rise, another cause is the melting of ice and snow due to rising global atmospheric temperatures. Snow and ice make up the solid form of water on Earth and they are important for the global climate system due to their reflective capabilities. In general, the reflection of the sun's radiation by snow and ice is one of the factors allowing these areas to maintain temperatures low enough for ice to form. Because snow and ice exist in a condition very close to their own melting temperatures, increases in global temperature could have profound effects on the quantity of snow and ice present on the planet. Changes in the levels of snow and ice will play a role in many different systems including sea level rise, effecting ocean currents due to changes in water density as saltwater is replaced with freshwater from melting ice, and will likely impact the biodiversity in polar regions and other regions around the world.

One example of changing levels of snow and ice on Earth can be seen in Arctic sea ice. Sea ice typically forms in the winter and covers much of the Arctic. When temperatures warm during the summer months, the ice melts and total sea ice is reduced to a minimum point for the year, called a sea ice minimum. This ebb and flow of ice is a normal process, but in the past several decades, the minimum level of sea ice for a given year has been decreasing. In 2013, it was predicted that the minimum sea ice level had decreased by 700,000 square miles compared to the 1979–2000 historical average. This is equivalent to the loss of an area of ice equivalent to twice the size of Texas (EPA, 2014). These lower sea ice minimums are another climate change indicator. Conversely, there is also a sea ice maximum, the time when the largest area of the Arctic sea is frozen, and this usually occurs in late winter or early spring. The smallest sea ice maximum recorded was measured in the spring of 2015, another indication of how global warming is impacting the environment (Viñas, 2015).

Rising global temperatures can also affect ice found on land. Almost one-fourth of the land in the northern hemisphere is considered permafrost or land that stays below freezing for more than 2 years. Permafrost is important because it is a major storage reserve for carbon in the form of peat and methane. As temperatures rise and permafrost begins to melt, this carbon will be unlocked and methane will begin to seep from the soils. As discussed earlier, methane is 23 times more powerful than carbon dioxide as a greenhouse gas, and an increase in the concentration of this gas in the atmosphere could have a huge impact on the rate of global warming.

Temperature is not the only climatic indicator concerning the atmosphere or the ocean; the actual chemistry of the ocean is changing as well. As you learned in the carbon cycle, the ocean absorbs a significant amount of carbon dioxide, and as the atmospheric levels increase, the ocean continues to absorb more of that carbon dioxide. As the ocean absorbs more carbon dioxide, it is chemically converted into carbonic acid, bicarbonate, and carbonate. This change in carbon chemistry also comes with an overall change in the acidity of the ocean. As more carbon dioxide is absorbed, the ocean's pH is decreased, becoming more acidic. The acidity of the ocean plays an important role in the physiology of many organisms including corals and phytoplankton. These organisms rely on the presence of calcium carbonate to form their anatomical structures such as their shells; however, the decreasing pH of the ocean is literally dissolving calcium carbonate and making it unavailable for these organisms. Thus, the phytoplankton that rely on the use of calcium carbonate and play a crucial role in the production of atmospheric oxygen through photosynthesis could be greatly impacted by changes in ocean chemistry and lead to further changes in atmospheric chemistry.

These are just some of the environmental indicators being used to monitor changes in the global climate. Continued research and time will allow for a better understanding of the intensity of these changes in the future and how ecosystems will adapt to these changes.

Changing the Future

The indicators of global climate change discussed in previous paragraphs will undoubtedly have an impact on ecosystems and the many organisms they support. One of the organisms impacted will certainly be humans. Even at the current global average temperature increase of about 1.65 degrees Celsius, humans are going to feel the impact on food production, human health, and the economy (Strom, 2007). Scientists predict that for every 1 degree Celsius rise in temperature, grain production will be reduced by 10%, sending millions of people into starvation. At the same time, diseases known only to the tropics have the potential to spread to temperate regions as the climates warm. These diseases could cause human outbreaks affecting the health of people around the world, not just in tropical regions. Finally, the economy will suffer for countries all around the world. Here in the United States, it will likely be felt through loss of productivity in agriculture, but also in changes in insurance premiums due to the increasing intensity and probability of weather-related disasters in various regions around the country. Based on predictions in *Hot House* by Robert Strom, it is estimated that cutting greenhouse gases might cost 1% of the global gross domestic product, but the cost to the economies of the world if global climate change is left unchecked could be nearly 20% of the global gross domestic product (Strom, 2007). The saying "an ounce of prevention is worth a pound of cure" also holds for the world economy under climate change.

With little time to adapt and a continuing stream of carbon dioxide entering the atmosphere from fossil-fuel-driven human activities, a rise in global temperature appears unstoppable. The question remains: how high will the temperature reach and at what point will people start curbing carbon dioxide emissions to prevent further environmental catastrophes? The concentration of atmospheric carbon dioxide today is about 400 ppm and the global temperature has risen by 1.65 degrees Celsius over the last 100 years. It is expected that an atmospheric carbon dioxide concentration approaching 450 ppm will lead to a 2 degrees Celsius increase in temperature, while 550 ppm will see a 3 degrees Celsius increase in temperature (Meehl et al., 2007).

To avoid these increases in global temperature and prevent further environmental harm from climate change, it is crucial that steps be taken immediately to avert the worst outcomes.

Currently, the two largest emitters of anthropogenic carbon dioxide are China and the United States, making up about 42% of the world's emissions (EIA, 2012). China recently moved ahead of the United States as the largest carbon emitter, with other developing nations such as India and Brazil showing signs of increase as well. But the answers to climate change problems start with us; it starts with the changes we make in our households, our communities, our industries, our states, and ultimately our country to create a society that is sustainable for generations still to come. To do this, we must take a critical look at our society and the impacts that fossil fuels have on our lives. We must ask the question: what can we do to change the future?

STUDY QUESTIONS

1. Briefly describe the Earth's energy budget, the greenhouse effect, and the role the atmosphere plays in maintaining the Earth's average overall surface temperature.
2. Explain the role of greenhouse gases (both anthropogenic and non-anthropogenic) and radiative forcing in the atmosphere.
3. Which anthropogenic greenhouse gas is most prevalent? Does this greenhouse gas have the greatest warming potential? Why or why not?
4. Explain the role that the carbon cycle plays in global warming.
5. Discuss the potential reasons for global climate change. Which one is the most likely to be the reason for the current trend in global warming? Why?
6. What is global climate change? Why is it associated with greenhouse gases? Explain some of the major environmental impacts expected due to global climate change.

CHAPTER FOUR

Methods for Reducing Our Fossil Fuel Usage:
Renewable Energy Sources and Uses

Over the past century, society's expanding consumption of energy has mainly focused on energy derived from fossil fuels. Fossil fuels are abundant and relatively cheap resources, and they are remarkably energy dense fuels. However, for the greater part of human history, people have actually relied mostly on other natural resources for energy. For instance, humans first used wood and dung to produce fire for warmth and cooking, the ancient Egyptians used boats with sails powered by the wind to travel the Nile River, and the Romans used running water to turn waterwheels for irrigation and for grinding grain (Nersesian, 2010). These natural resources are still widely available and used in a similar manner around the world today, but their consumption is significantly less due to the prevalence and efficiency of fossil fuels for energy production.

Renewable and Sustainable Energy Resources for the Future

How are wood, wind, and water different from fossil fuels? All are essentially natural energy resources derived from the Earth, and all can be used as a type of fuel, but a key difference between some natural resources and others comes down to two simple words, **renewable** and **sustainable**. These terms are commonly used interchangeably, but their meanings are not precisely the same: a renewable resource is defined as a resource that can be exploited from the Earth and then replaced within a short period of time, while a sustainable resource refers to a resource that is utilized in such a way that the use does not impact its availability for future generations. Consider the wind. Humans have used wind power for many centuries, yet there is no shortage of wind today. Wind is created from the heat of the sun on Earth; it is constantly produced and the Earth's generation of wind is not directly impacted by human activity. Thus, wind is both renewable and sustainable. Fossil fuels, while

naturally created, cannot be sustained in the face of rapidly increasing consumption and therefore cannot be guaranteed to be an available resource in the future. In addition, as you learned previously, fossil fuels take millions of years to develop, meaning these resources cannot be replaced in a short period of time; therefore, fossil fuels are both nonsustainable and nonrenewable. The development of an ideal energy resource to replace fossil fuels will need to be both renewable and sustainable to support all generations in the future and maintain environmental quality.

All of the energy that is consumed today, regardless of whether it is from a finite or infinite resource, comes from only one of three **primary energy sources**: the sun, the moon, or the core of the Earth. For example, fossil fuels, which are a product of millions of years of decayed plant matter, are ultimately derived from the sun since plants capture solar energy through photosynthesis. As other renewable energy sources like water, wind, waves, and tides are discussed, it is important to remember that they can all be tracked back to one of these three primary energy sources.

Two of these primary energy sources, the sun and the core of the Earth, generate their energy through **thermonuclear reactions**. Thermonuclear reactions involve the combination or dissociation of atomic nuclei, either of which results in the release of large amounts of energy. These reactions are classified as either nuclear **fission** reactions when the nuclei dissociate (i.e., break apart) or nuclear **fusion** reactions when two atomic nuclei combine (i.e., fuse). Figure 4.1 shows examples of nuclear fission and fusion reactions. Nuclear fission reactions usually involve the breaking apart of larger atomic elements into smaller elements. An example of a nuclear fission reaction is when uranium is split into two smaller atoms like krypton and barium as shown. Also diagrammed in the figure, nuclear fusion reactions involve the combination of very small

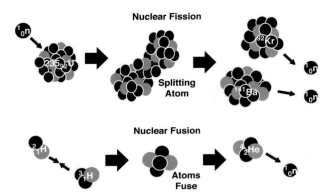

Nuclear Fission

Splitting Atom

Nuclear Fusion

Atoms Fuse

FIGURE 4.1 Diagram of thermonuclear fission and fusion reactions. In this example, nuclear fission occurs when a neutron (n) collides with a Uranium (U) atom, causing it to split into one atom of krypton (Kr) and one atom of barium (Ba). This results in the release of two neutrons that can continue the cycle by colliding with other atoms of uranium. Nuclear fusion occurs when two atoms of hydrogen (H) collide with one another and fuse to generate one atom of helium (He) and release one neutron. Both reactions produce large quantities of energy due to the loss of neutrons during the reactions.

atomic elements like the combination of two hydrogen atoms to form a helium atom. Whether the reaction is a result of fission or fusion, the changing atomic structure releases a huge amount of energy. This is obvious if you consider the colossal amount of energy produced through thermonuclear fusion by both the sun and the core of the Earth. Nuclear fusion reactions in the core of the sun produce the energy that lights and warms Earth externally, while fusion within the core of the Earth heats the Earth from the inside out. The sun alone is estimated to produce a level of energy equivalent to 3.8×10^{26} joules per second. To put this into perspective, the United States consumed only about 1.00×10^{23} joules total during 2014. Therefore, the sun produces an amount of energy in one second that is more than 1,000 times the amount of energy consumed in the United States in one year (Barbier, 2012; EIA, 2015). While these reactions occur naturally, scientists and engineers have also been able to reproduce thermonuclear fission reactions to generate nuclear power.

Nuclear Energy: An Alternative Energy Resource, but Is It Really Renewable?

When we think of thermonuclear reactions, it is probably not the sun or center of the Earth that comes to mind but rather nuclear energy production. Nuclear energy is a form of energy that results from mass-to-energy conversion within the nucleus of an atom. Because atoms use a lot of energy to hold the protons, neutrons, and electrons inside and around their nuclei, the splitting or combining of individual atomic nuclei can release an immense quantity of energy, generally in the form of heat. As a reminder, nuclear fission reactions occur when large atoms such as uranium or plutonium are split into smaller products such as kryp-

ton and barium. During this reaction, a neutron is released from the splitting nucleus. This neutron can then impact another atomic nucleus, causing it to split. The splitting process results in a chain reaction leading to the continuous splitting of atomic nuclei and release of extremely large amounts of energy (Bodansky, 2005). Nuclear fission is the type of reaction usually involved in the production of commercial nuclear energy today in nuclear power plants.

In 2013, nuclear energy provided about 4.4% of the world's total energy consumption and 8.3% of US primary energy. Despite a fairly low percentage of total energy, this nuclear energy consumption is significant because it represents the second-highest consumed nonfossil-fuel-based energy resource with only hydropower having a higher consumption at 6.7% of the world's primary energy (British Petroleum, 2014). In addition, as a nonfossil fuel, no carbon dioxide is released into the atmosphere from nuclear reactions.

The utilization of nuclear energy today has its origins in the concept of the atom initially proposed by Greek philosophers in about 370 BC. However, it was not until early in the twentieth century that physicists began to fully grasp the quantities of energy contained within an atomic nucleus. In 1911, Ernest Rutherford, a British scientist, discovered that an atom has a nucleus and wrote about the heat produced as radium decayed. In 1934, a physicist named Enrico Fermi demonstrated the interactions between neutrons and nuclei that led to the discovery of nuclear fission. This discovery was quickly followed by the first self-sustaining nuclear reaction in 1942 focused mostly on the use of this technology for military purposes including nuclear weapons. The obvious risks associated with the use of this technology for weapons development led to the establishment of the Atomic Energy Commission in 1946. Their goal was to create a nuclear energy program for peaceful civilian use. In 1951, the world's first electricity-generating nuclear power plant came on line (Bodansky, 2005; WNA, 2014).

Energy generated from nuclear reactions today is largely a product of the fission of uranium. Uranium is an atomic element that naturally occurs in most rocks and in seawater. Uranium nuclei are naturally found as two different radioactive isotopes, ^{238}U (99.3%) and ^{235}U (0.7%). The difference between these two isotopes is based on a variation in the number of neutrons and protons present in the nucleus where ^{238}U has three more neutrons than ^{235}U. ^{235}U is the most utilized of the uranium isotopes for regular nuclear fission reactions. When the uranium atom captures a stray neutron, it will split into isotopes such as krypton (^{92}Kr) and barium (^{142}Ba). Although this is a volatile reaction, scientists have learned to sustain and control the reaction allowing for the energy to be captured and used in the production of electricity (Mudd, 2011).

This ability to control and harness energy from nuclear reactions is largely based on the design of a nuclear reactor. If nuclear reactions containing uranium were allowed to take place without control mechanisms, there would eventually be a nuclear meltdown, releasing radioactivity and heat without capturing any energy. Instead, the nuclear reactor is specifically designed to control the fission reaction. The uranium is formed into pellets and then placed in rods, about the diameter of a dime, called fuel rods. The rods are attached together and submerged in water within the center of the reactor known as the nuclear core. The water is crucial in keeping the rods cool during the reaction and avoiding an uncontrolled chain reaction resulting in a nuclear meltdown; however, the water alone is not enough to control the nuclear reaction. In order to truly control a nuclear reaction, scientists use control rods made of materials designed to absorb excess neutrons. These control rods are raised or lowered to control the number of neutrons that are coming in contact with the fuel rods. In this way, if the control rod is lifted, then the uranium rods are exposed to more neutrons and more energy is produced in the form of heat, but as the control rods are lowered, some of the neutrons are blocked, lowering the amount of energy released. When a nuclear reactor is shut down, usually in case of an emergency or maintenance, these control rods are completely lowered to stop the reaction (Nersesian, 2010).

The energy from a nuclear reaction is usually in the form of heat. This heat energy can be utilized to heat water and generate steam. The steam is then used in a traditional steam-turbine-driven power plant designed to produce electrical power as described in Chapter 2. Once the steam has been utilized in power generation, it cannot simply be released into the environment due to the detrimental environmental effects that would result from the direct release of the extremely hot water. To avoid environmental damage, large distinctly shaped cylindrical towers, often associated with nuclear power plants and called cooling towers, are assembled. These cooling towers allow the gaseous steam to condense into cooler water before being released into the environment.

Nuclear energy has become an alternative to fossil fuel energy due to its release of large amounts of energy with much lower greenhouse gas emissions. Reactor grade uranium has an energy density of 3.7×10^6 megajoules per kilogram, enough to power a 100 watt lightbulb for 1,171 years. When compared to the energy density of coal (32.5 megajoules per kilogram) or crude oil (41.9 megajoules per kilogram), the difference in magnitude is immediately recognizable. The same amount of coal or crude oil could only power that same lightbulb for 3–5 days (Touran, 2012). The potential of nuclear energy has resulted in the development of many nuclear power plants. In 2015, there were 438 operating nuclear reactors in 30 countries, producing 378,870 megawatts of energy, an additional 69 reactors under construction, and 184 reactors ordered or planned in the future (WNA, 2015). As mentioned previously, these nuclear reactors supply about 4.4% of the world's total primary energy (British Petroleum, 2014). Figure 4.2 compares

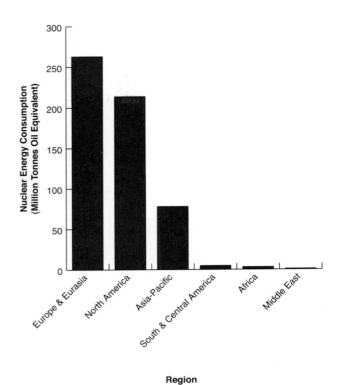

FIGURE 4.2 Comparison of nuclear energy consumption by region in 2013. Europe and Eurasia are the leading consumers of nuclear energy at 263 million tonnes oil equivalent (mtoe), followed by North America at 213.7 mtoe and Asia-Pacific consuming 77.8 mtoe. Within these regions, the United States, France, and South Korea are the highest individual nuclear energy consuming countries, respectively. Other regions including South and Central America, Africa, and the Middle East are minor contributors to global nuclear energy consumption (data from British Petroleum, 2014).

nuclear energy consumption in the various regions of the world where Europe, Eurasia, and North America clearly consume the most nuclear energy. Within these regions, the United States is the leading consumer of nuclear energy, accounting for about 33% of the world's nuclear energy consumption (British Petroleum, 2014). Most of the nuclear power plants located in the United States have been in operation for a few decades, and they continue to become more efficient and produce more power. In the United States, 99 nuclear reactors produce 98,756 megawatts of total nuclear energy including 790.2 billion kilowatt-hours of electricity. This is about 19.4% of the total electric power generated in the United States (EIA, 2015; WNA, 2015).

Despite its sometimes-confusing classification as a sustainable energy resource due to its potential to produce power for a very long period of time without carbon dioxide emissions, nuclear energy is not a renewable energy source by the traditional definition. Nuclear energy requires uranium or some other radioactive atomic element found within the Earth. These sources are finite and must be mined. Once these resources run out, nuclear power will no longer be possible. Currently, scientists estimate there to be enough uranium to provide nuclear power to the

NEXT-GENERATION NUCLEAR REACTORS

Travis L. Johnson

Nuclear fission reactors have been designed and built since the mid-twentieth century and nuclear technologies have advanced through the years. The oldest reactor designs are called Generation I and were built in the 1950s and 1960s as prototypes. Most current nuclear reactors are Generation II, which were built from the late 1960s to the late 1990s and are based on the light water reactor (LWR) concept. In this design, the reactor core is immersed in water and the heat from the reactor is transferred to a secondary water system that transforms into steam and turns a turbine generator. This reactor type has high power densities and requires high safety redundancy to minimize loss of coolant (water) in the reactor, which could lead to a meltdown. Generation III reactors are newer designs built in the 2000s and Generation III+ reactors are improved versions that will be built in the 2020s and 2030s. These are based on LWR, but also include passive safety systems such as natural convection air discharge, outside cooling air intake, internal condensation, and natural recirculation. These designs are safer and less complex than Generation II reactors and make it less likely for a meltdown to occur. Looking forward, the next-generation reactors would be Generation IV and nuclear fusion reactors. Generation IV would take the passive safety design to the extreme, making a meltdown physically impossible as well as improving the economics of the reactor and reducing the amount of long-lived waste products. Finally, scientists are researching fusion reactors, which would never have meltdowns, would not produce long-lived waste, and would have life spans of up to 100 years rather than the 40–60 year life spans of current reactors, but fusion reactors appear to still be several decades away.

SOURCE: Tynan (2014).

existing nuclear power plants for only the next 90 years. These uranium reserves are located largely in mines found in Australia, Kazakhstan, Russia, and Canada (WNA, 2014). So why would this resource sometimes be classified with the other truly sustainable resources? One of the answers lies in the development and implementation of a second type of nuclear reactor known as the **breeder reactor**. In a breeder reactor, more fissile material is produced than consumed. In these reactors, nuclear material undergoes fission reactions to produce energy and also generates new material capable of additional rounds of fission, thereby continuing the energy generation cycle. In the end, this continuous reaction is predicted to be able to provide the substrate for energy production for a much longer period of time, possibly 1,000 years (Nersesian, 2010; Sevior et al., 2010).

While nuclear energy is an important source of power and one that could likely replace coal- and natural-gas-generated electricity easier and faster than some of the other truly renewable energy resources, nuclear power does have some significant problems including public safety, vulnerabilities to terrorism, and significant issues with long-term storage of the radioactive waste. In order to continue developing nuclear technology and to expand the utilization of nuclear power, it will be necessary for the entire global community to address each of these concerns.

Geothermal Energy

Nuclear energy used in nuclear power plants is not truly a sustainable resource due to the reliance upon a finite supply of uranium from the Earth. However, another energy source that does have the potential to be sustainable is derived directly from the Earth's core in the form of **geothermal energy**. Geothermal energy is the result of heat transferred from the core of the Earth to its surface. The energy is initiated from radioactive isotopic decay and other potential nuclear reactions at the core that heat the surrounding rock and turn it into magma. The heated magma then moves up through the various layers of the Earth until it gets closer to the surface, becoming cooler within every layer. As magma gets closer to the surface, it heats the land and water at the surface (Isherwood, 2011). Sometimes hot magma is released from the Earth through volcanoes and at other times it is trapped within the Earth and acts to heat its surroundings. One common way to experience how magma heats the Earth is to enjoy a natural springs thermal bath. These warm waters are naturally created when heat energy from the core of the Earth acts to warm waters trapped within the Earth before they seep to the surface. Sometimes these waters can be warmed by magma to such an extent that steam is created. If this highly energetic steam is funneled toward a small exit within the Earth, it can create an explosion of water vapor known as a geyser.

The warming of water trapped just below the surface of the Earth is actually the exact way we use geothermal energy as an energy resource. In some simple cases, piping is laid down under a home or building, allowing water to be pumped underground where it is warmed by the Earth, and then brought back up to be used to heat a home or building. Other larger-scale commercial power stations use steam generated from within the Earth to produce electricity. Some power stations can rely directly on natural water trapped deep in the Earth as steam. These stations simply place a pipe deep into the Earth, creating an exit for the pressurized steam. This high-energy steam will shoot to the surface with plenty of velocity to spin a turbine and generate electricity. Once the pressure of the natural water supply becomes too low, water can be injected into the Earth to replenish the water supply and rebuild the pressure. This technique was successfully employed by the largest natural steam field in the world north of San Francisco known as The Geysers. Here, they used processed wastewater from local communities to replenish the natural water supply allowing for continued generation of energy from this field (Nersesian, 2010).

The creation of pressurized steam generally requires temperatures above 300 degrees Fahrenheit; however, there are geothermal resources closer to the surface that are below this temperature and can still be used to produce power. In these situations, a heat exchange occurs where the heat from the geothermal water is transferred to a second fluid with a lower boiling point such as isopentane. The vapor from this second fluid can then be used to drive the power-generating process (Nersesian, 2010).

The most active areas for geothermal energy are those areas where hot rocks are close to the surface, which often occurs where tectonic plates come together. These areas are also associated with volcanoes and earthquakes. The Ring of Fire, located around the rim of the Pacific Ocean, is one of these geothermally active regions. Figure 4.3A shows a map of geothermally active regions in the United States. The darker red and orange areas, representing many regions in the western United States, are areas where the prospect of geothermal power generation is favorable. In 2013, the United States produced a total of 16.5 million watts of power using geothermal energy, an amount equivalent to 0.4% of total US power generation (EIA, 2015). Nearly 80% of this geothermal power is generated in the state of California, with the remainder produced largely in Nevada, Hawaii, Idaho, and Utah as shown in figure 4.3B (Roberts, 2014).

The generation of power utilizing geothermal energy is a sustainable manner of power production that could play a continued and larger role in replacing coal and natural gas for electricity generation. While certainly less detrimental than fossil-fuel-based energy resources, the consumption of geothermal energy also has environmental consequences. Geothermal power plants can release both hydrogen sulfide and carbon dioxide, but generally at much lower levels than coal or natural gas power plants. By using a scrubbing system to remove contaminants from the steam, hydrogen sulfide can be limited to about 3% of that released by fossil fuel power plants. In addition, carbon dioxide from a geothermal power plant is less than 4% of that released by the dirtier fossil fuel plants (Kagel et al., 2007). Another potential environmental consequence to consider related to geothermal energy is the locations of good steam fields. In some cases, these steam fields may be located in regions that are designated national parks like Yellowstone National Park. The development of these natural regions to fulfill our desire for energy could result in a significant level of ecological destruction. With the continued development of geothermal energy, it will be necessary to figure out ways to use the heat from the Earth without ruining delicate ecosystems at the Earth's surface.

Solar Energy

Perhaps the most powerful primary energy resource available on Earth comes from the sun. The sun continuously produces an average of 340.4 watts per square meter of energy, and 163.3 watts per square meter of this energy arrives at the surface of the Earth (Canright, 2011). The sun produces this energy through nuclear fusion reactions that are released in the form of photons. These photons travel toward the outer boundaries of the sun and into space and eventually come in contact with the Earth. Photons that are not absorbed by the Earth's atmosphere impact the surface of the planet and can either be absorbed by vegetation and inorganic materials on the surface or be reflected back into the atmosphere. The continued reflection of energy between the surface of the Earth and the Earth's atmosphere results in the warming of the planet due to the greenhouse effect as discussed in the previous chapter. Considering solar energy in terms of individual photons may not reveal the impact of energy from the sun on the Earth, but in fact, one hour's worth of solar energy striking the Earth is greater than all of the energy consumed by the world's population in one year (Lewis and Nocera, 2006). This means that the sun is easily capable of sustaining all of the Earth's energy needs if the proper technologies were developed to harness this energy.

One of the most common of these technologies available today is solar power. Solar power can be classified into one of three subcategories: passive, active, and photovoltaic.

(A)

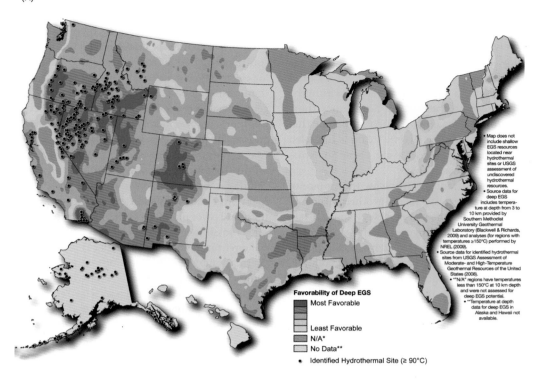

Favorability of Deep EGS

Most Favorable

Least Favorable
N/A*
No Data**

• Identified Hydrothermal Site (≥ 90°C)

- Map does not include shallow EGS resources located near hydrothermal sites or USGS assessment of undiscovered hydrothermal resources.
- Source data for deep EGS includes temperature at depth from 3 to 10 km provided by Southern Methodist University Geothermal Laboratory (Blackwell & Richards, 2009) and analyses (for regions with temperatures ≥150°C) performed by NREL (2009).
- Source data for identified hydrothermal sites from USGS Assessment of Moderate- and High-Temperature Geothermal Resources of the United States (2008).
- **N/A** regions have temperatures less than 150°C at 10 km depth and were not assessed for deep EGS potential.
- ***Temperature at depth data for deep EGS in Alaska and Hawaii not available.

(B)

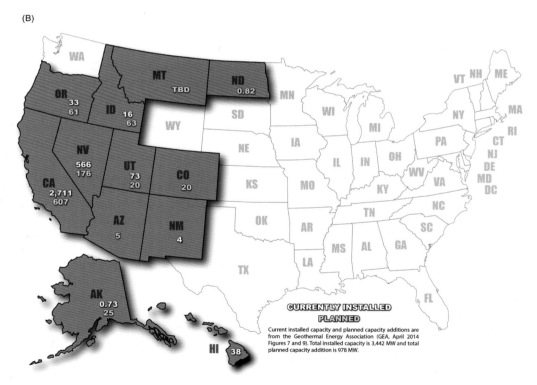

CURRENTLY INSTALLED
PLANNED

Current installed capacity and planned capacity additions are from the Geothermal Energy Association (GEA, April 2014 Figures 7 and 9). Total installed capacity is 3,442 MW and total planned capacity addition is 978 MW.

FIGURE 4.3 Geothermal resources in the United States. (A) Map of potential geothermal resources within the United States, showing locations of identified hydrothermal sites and favorability of deep enhanced geothermal systems (EGS). Areas shown in red and dark orange, largely in the western portion of the country, represent areas favorable for the development and implementation of geothermal systems. (B) Map of current and planned geothermal power generation capacity (in megawatts) in each state. Installed capacity shown in white and planned capacity shown in yellow (images by Roberts, 2009, 2014).

FIGURE 4.4 Photograph of parabolic solar trough at Harper Lake in California. These troughs reflect the sunlight to a central line where the increased intensity of the light and heat can be used to heat a fluid that can be used for many purposes including as an energy resource (image by "Parabolic trough at Harper Lake in California" by Z22, Own work. Licensed under CC BY-SA 3.0 via Wikimedia Commons, http://commons.wikimedia.org/wiki/File:Parabolic_trough_at_Harper_Lake_in_California.jpg#/media/File:Parabolic_trough_at_Harper_Lake_in_California.jpg).

Passive solar energy occurs when a structure such as a house, apartment, or dorm room absorbs heat from the sun and that heat is used to replace heat usually obtained by methods derived from fossil fuels such as a space heater. Passive solar energy is the easiest type of solar energy to use because it does not require any mechanical equipment but rather takes advantage of the inherent properties built into a structure. These properties may include south-facing windows or insulation. To better understand this concept, imagine a car left parked in the sun with the windows rolled up for a period of time. Within a matter of minutes, the interior will have become warm if not outright hot. The capture of thermal energy inside the car is an example of passive solar heating. The same thing can occur in a home where glass windows, particularly those facing south, can trap heat inside and thereby warm the room. Once the room gets too warm, the window can be opened allowing the air to flow out and be replaced by cooler air from the outside. In this way, passive solar energy can be used for heating without the need for fossil-fuel-derived energy supplementation.

Active solar energy is the next logical step beyond passive solar energy, where instead of relying on the structure inherent in a building itself, mechanical structures are built to actively absorb and collect the sunlight. The heat from this sunlight can then be used for a specific purpose like heating water or generating steam. Thermal solar energy uses solar collectors to collect sunlight in ideal locations, which is then used to heat water (Gabbard, 2011). The simplest of solar collectors are those often associated with heating water for individual home use, generally either a swimming pool or a shower. These solar collectors are made of rubber or plastic and designed in a tube-like structure to allow water to flow through. As the water flows through, it is warmed by the sun then returned to its original source. In a swimming pool, for instance, it will raise the temperature of the water within the pool, thus creating a heated swimming pool without the use of large amounts of electricity.

More complicated solar collectors are designed to collect enough solar energy to actually produce electricity from heated water and other materials (Philibert et al., 2010). These solar collectors come in three main forms: a solar dish, a solar trough, and a solar power tower. An example of a solar trough collector is shown in figure 4.4. All of these solar collectors are designed so that they reflect sunlight to a designated point, either the apex of a solar dish or tower,

SOLAR PHOTOVOLTAIC TECHNOLOGIES

Travis L. Johnson

The sun is an enormously important source of energy for earth, and one way we are able to harness its power is by converting the sun's photons directly into electrons using photovoltaic (PV) cells. This power comes in the form of sunlight, which is also called solar irradiance, and it has different components at the earth's ground level. One component is the direct normal irradiance that is the radiation from the sun coming directly from the sun's rays. If one pointed a tube straight at the sun and tracked it as it moved through the sky, one would be capturing the sun's direct normal irradiance. Another component is the diffuse irradiance, which is the radiation that has been scattered by atmospheric molecules and distributed throughout the sky.

The most common form of PV technology is the flat panel PV array, which can be found on watches, calculators, and residential and commercial roofs. This technology uses both diffuse and direct irradiances and converts photons to electrons at around 20% efficiency. Another PV technology is concentrated PV, which consists of a mirror system that concentrates direct irradiance into the PV cell. A first mirror is shaped like a dish and can capture more area of direct irradiance and reflect it to another smaller mirror in the middle of the dish that then reflects the sunlight into a multi-junction cell. The multi-junction cell can capture more wavelengths of light, converting 35–40% of the photons to electrons, making it more efficient than flat panel PV cell. Concentrated PV cells are typically about 100 times more expensive than a flat panel PV, but they are able to concentrate the solar radiation by 300–1,000 times.

SOURCES: Coimbra (2014); 3Tier by Vaisala (2015).

or at the midline of a trough. These solar collectors will track the sun as it moves through the sky and reflect as much radiation energy as possible on to these individual points. In many cases, these points are directly associated with flowing water or other materials in a tube similar to the simpler thermal solar collectors. This water can be heated to such an extent that steam is formed, and just like with coal power generation, this solar-generated steam can be used to generate electricity (Nersesian, 2010).

Finally, photovoltaics represents another solar technology developed to take advantage of solar energy. Photovoltaics means "electricity from light" and is different than both passive and active solar power because a photovoltaic cell can skip the step of heating a material to produce steam that is used to generate electricity, and instead directly produce electricity from sunlight through the **photovoltaic effect** (Knier, 2011). The photovoltaic effect occurs when the semiconducting material of a photovoltaic solar cell absorbs photons of energy from the sun. Eventually the semiconducting material absorbs enough photons that it forces electrons to be ejected from atoms within the material. Due to the negative charge of electrons, these electrons are attracted to the front of the solar cell, which is coated in positive charges. The separation of the negatively charged electrons from the positively charged atomic nuclei creates a voltage potential. Once the front of the solar cell is connected to the back via an electric conductor, this voltage potential caused by the attraction between the positive charges and the negative charges will directly create a flow of electricity (Knier, 2002).

You may be familiar with photovoltaic technology if you have ever used a solar calculator that does not require a battery. While this is a small example, combining many photovoltaic cells can produce enough electricity to power a home or building. Sometimes this electricity exceeds what is needed by the individual structures and can be sent onto the electric grid. Photovoltaics are beneficial in that they reduce carbon dioxide emissions by replacing fossil fuel power-generating technologies like coal; however, they also have consequences for the environment. Photovoltaic cells have a lifetime of about 25 years and contain toxic materials such as lead, mercury, and cadmium. These are finite materials that can result in toxic emissions during their initial mining and preparation, and must be dealt

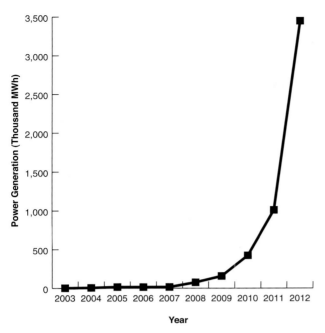

FIGURE 4.5 US net power generation from solar photovoltaic technology from 2003 to 2012. The expansion of photovoltaic power generation began in the United States in 2007 and has rapidly increased each year (data from EIA, 2015).

with properly for disposal (Alsema et al., 2006). In order to expand photovoltaic technology for the replacement of coal-generated power, it will be necessary to find ways to safely dispose of these materials and in some cases find methods to recycle these finite resources.

The demand for solar photovoltaic power generation is growing. Figure 4.5 shows the net power generation from solar photovoltaic utilization in the United States since 2003. Beginning in 2008, solar photovoltaic power generation quickly began rising with net power generation of 76,000 megawatt-hours in 2008 and 8,327,000 megawatt-hours in 2013, an increase of 10,856%. This compares to a much slower increase in solar thermal power generation from 788,000 megawatt-hours in 2008 to only 926,000 megawatt-hours in 2013, an increase of just 17.5%. Despite the increase in power generation from photovoltaics, total solar technology is still only used to provide 2.3% of total power generation in the United States (EIA, 2015). The widespread use of photovoltaic technology is limited due to the initial economic burden or cost to the consumer. This is largely based on the initial cost of installation of solar cells. Photovoltaic solar cells cost about $4–10 per watt of installed capacity compared to a coal power plant of $3 per watt of installed capacity (Gabbard, 2011). In addition, the intensity of sunlight is not equivalent across the entire United States. NREL has shown that areas in the southwestern United States can produce more than 6.0 kilowatt-hours per square meter per day from a photovoltaic solar collector; however, regions in the northeast may only be able to produce less than 4.0 kilowatt-hours per square meter per

day on average (Roberts, 2009). This difference in solar irradiance between these regions represents a potential differential in price and availability of power. Homes in the northeast may need to continue supplementing power generated from fossil fuels to meet their needs, while homes in the southwest that can produce more power than needed can get paid to put their extra power into the electric grid. While the installation costs today may keep some people from installing this technology, the continued decrease in the price of the technology, coupled with the ever-increasing price of fossil fuel electricity, will likely help push more people to use photovoltaics. Developing cheap and efficient methods to store photovoltaic power and creating a more ecologically sustainable process for solar cell generation would also allow photovoltaics to play a much larger role in the future of renewable energy.

Wind Energy

Wind is another renewable energy source that is ultimately derived from the sun. At first this may not seem intuitive, but the regional heating of the Earth by the sun generates wind. Warm air will rise, leading to a void that then pulls the cooler air in to replace the rising warm air. As this air rushes in, wind is created (Olanrewaju, 2011). This process is evident on a global scale by studying hemispheric wind patterns. As depicted in figure 4.6, in the tropics near the equator, the sun is intense and heats the air, resulting in warm, moist tropical air. This air rises into the atmosphere near the equator in what is known as the intertropical convergence zone. However, this rising warm air must also be replaced by cooler air located closer to the Earth's surface in the subtropical regions. Subtropical cool air drops below the tropical warm air to create the Easterlies or Trade winds often found in the tropics and subtropics.

Wind can also be created on a much smaller regional scale as seen with ocean front climates. Daily sea breezes felt when visiting the beach are the result of variations in air temperatures. During the day, the sun warms the land more quickly than the water. This warm air over the land will rise and the cooler ocean air will flow in, creating the afternoon onshore sea breezes. However, at night the process is reversed as the air above the land cools much more quickly, while the air over the ocean remains warm, resulting in an offshore flow. Whether on a global, regional, or local scale, wind is ultimately a manifestation of solar energy.

Wind as an energy resource is a very old concept dating back to the ancient Egyptians using sail power to navigate the Nile River. Civilizations have used windmills for the last 4,000 years for everything from grinding grain to pumping water. In the United States, the development of wind turbines resulted from the need to provide rural farms with electricity. Today, wind power has become an important sustainable resource for energy generation due to the rising prices of fossil fuels (Olanrewaju, 2011).

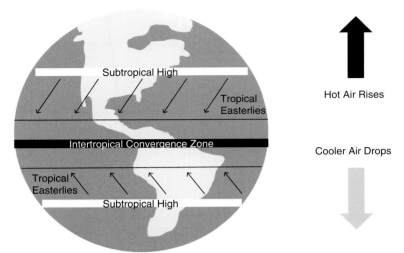

FIGURE 4.6 Basic diagram of hemispheric wind patterns. Warm, moist air in the tropics rises within the intertropical convergence zone, allows cooler air from the subtropics to be pulled under the warm air, and creates a global wind pattern known as the trade winds.

Wind power is generated using a wind turbine, the modern-day windmill. Wind turbines are fairly simple devices designed to capture the kinetic energy of the wind with long blades as shown in figure 4.7. These turbines function when the wind turns the blades that are connected by a shaft to a gear box (equivalent to a turbine) that powers the generator and produces electricity. This electricity travels to the ground and can be used by individual buildings in rural areas or connected to the power grid to supply electricity over a much wider area (Olanrewaju, 2011). As you can see, this process is similar to the normal steam-generated power production process but without the need for fossil fuels. The largest wind turbines are built to tower over the land, sometimes at the height of nearly a 20 story building, in order to obtain higher wind speeds and less turbulent winds. These turbines can be very efficient and often have a rotating diameter nearly the length of a football field (BLM, 2012).

The best places to take advantage of wind energy are locations where wind turbines or groups of wind turbines (known as wind farms) can be placed on the tops of hills, across open plains, along shorelines, or within mountain gaps where wind funneling can occur. In general, wind speed increases with altitude, making regions at higher altitudes with these characteristics even more valuable (Nersesian, 2010).

Wind power has become the second most utilized renewable resource for power generation in the United States providing 4.1% of total power generation and 32.1% of renewable power generation. Figure 4.8 shows the growth of power generation from wind between 2002 and 2012. Like solar photovoltaics, wind energy production in the United States has also seen a huge increase in the past decade, growing from 10,354,000 megawatt-hours in 2002 to 167,665,000 megawatt-hours in 2013, with the largest increase in production occurring steadily since 2007. This amount of electricity is enough to power over 10 million households (EIA, 2015).

Wind energy is one of the cleanest and most environmentally friendly forms of renewable energy. Wind turbines do not release harmful emissions, they do not consume water, and they have a low land use footprint; however, just like any energy source, wind turbines also come with trade-offs. One of the greatest weaknesses of wind power is that wind does not blow all of the time, and in many cases when the wind does blow, it does not correspond to peak electricity usage periods. Finding a way to store power generated from wind turbines will be critical in the future development and implementation of this technology on a wider scale. Some people are also concerned that wind turbines are not aesthetically pleasing and that they are a threat to wildlife, particularly migrating birds. Steps are being taken to improve wind turbines and the control of wind turbines to avoid disturbing bird migration patterns, but more birds are killed yearly by pesticides (estimated 72 million due to pesticides application in agriculture) than by wind turbines (33,000 bird deaths) (FWS, 2002).

Wind farms are also being built offshore where wind is often found to be more constant; however, these turbines must have added features to allow them to withstand the harsher oceanic elements such as wave action and powerful storms. Although wind turbines may not have a huge land footprint, they are still unrealistic for large-scale use in urban areas; thus, new urban wind turbines are also being developed with designs appropriate to urban environments, such as vertical wind turbine shafts placed on the sides of buildings. These modern wind turbines are effective in capturing wind in the urban environment while also architecturally designed to flow with the surrounding urban neighborhoods.

Hydropower

In 2013, hydropower represented the renewable energy resource that resulted in the highest percentage of total US power generation at 6.6% and represented 51.5% of renewable power generation (EIA, 2015). Just as wind is an

FIGURE 4.7 Example of wind turbines. The long blades are capable of capturing the kinetic energy of the wind and transferring this energy to the compartment at the base of the blades that houses a rotor. Spinning the rotor can be used to generate electricity. This electricity is transferred back to the ground down the pole where it can enter the power grid (image by ©iStockphoto.com/globestock).

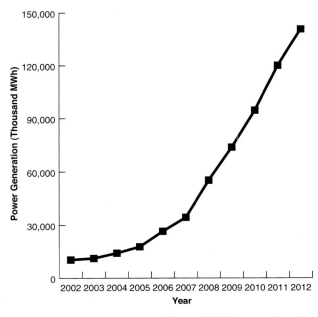

FIGURE 4.8 US net power generation from wind. The generation of wind power in the United States has seen a steady increase since 2002 with the most rapid increase after 2007 similar to the rapid increase of solar photovoltaic usage (data from EIA, 2015).

indirect renewable energy resource from the sun, so is hydropower through the **hydrological cycle**. Hydropower is often generated in a dam. Dams have a long history of providing water for domestic use, irrigation, and in controlling flooding. Humans have historically used moving water to turn wheels for lifting water, grinding grain, and a number of other tasks. Today, we have combined the beneficial aspects of dams and waterwheels to create hydroelectric power plants (Nersesian, 2010).

Hydropower is simply power produced from the kinetic energy of moving water. The potential of water as an energy resource is best understood by reviewing the Earth's hydrological cycle (Perlman, 2012). As shown in figure 4.9, we can begin tracking the cycle as warmth from the sun causes water to evaporate, particularly from large bodies of water like the ocean. This warm moist air travels into the atmosphere and is transported towards shore and eventually over land. Once over land, the air begins to condense into clouds and water vapor. Eventually, the air is so heavy that precipitation begins to fall often times in association with mountains as snow or rain. The precipitation falls to the Earth and travels into the ground by infiltration becoming groundwater or travels on the surface of the ground as runoff into streams and rivers. These rivers flow back towards the ocean potentially delayed for a short period of time within inland lakes. Overall, this cycle, driven by the evaporation of water by the sun, is what provides the water that makes rain and snow, fills rivers and lakes, and irrigates crops. This same cycle can be used to produce hydropower.

The production of hydropower is only as efficient as the Earth's hydrologic cycle. In most cases, hydropower is generated by blocking a river's flow with a dam. This creates an unnatural lake behind the dam where water can accumulate to a significant depth. This depth and the natural force of gravity are what generate the kinetic energy needed for the generation of hydropower. Figure 4.10 shows a basic diagram of how water flowing through a dam can generate potential and kinetic energy to drive the production of hydropower. Within the dam, there is a chute or chutes known as penstocks. On the side of the dam closest to the lake the opening of the penstock is at the top of the dam, but on the downstream side of the dam, the exit of the penstock is located at its bottom. This engineering sets up a drop in elevation within the dam. As the water flows down the penstock, it builds up energy and before exiting the penstock encounters a turbine. The kinetic energy within the flow of the water spins the turbine that is attached to a generator and produces electricity.

The largest hydroelectric power station associated with a dam in the United States is Grand Coulee Dam on the Columbia River in Washington state with a total capacity of 6,809 megawatts. Other large power stations that contribute to the total production of hydropower in the United States include the Robert Moses Niagara hydropower station in New York, Bath County Pumped Storage Station in Virginia, and the Chief Joseph Dam and The Dalles hydropower

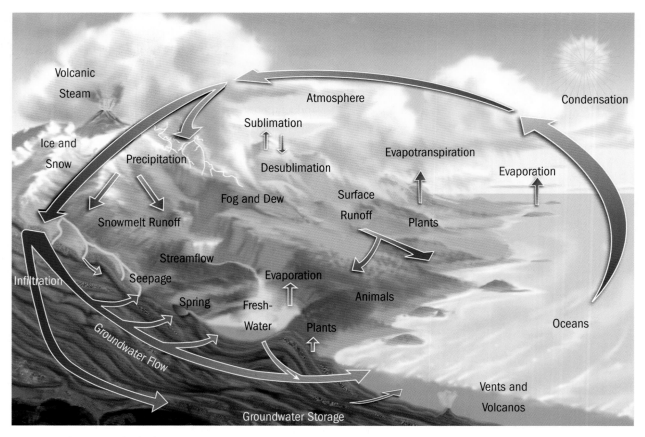

FIGURE 4.9 The water cycle also known as the hydrological cycle diagrams how water is continuously recycled on the Earth. Arbitrarily beginning with the ocean, water is evaporated due to the warmth of the sun. This moist air in the atmosphere travels over land and condenses to form precipitation, usually snow, ice, or rain. This precipitated water then either infiltrates the land and flows as groundwater back toward the ocean or flows as surface water runoff into streams, rivers, and lakes that ultimately lead back to the ocean (image by Evans and Perlman, 2014).

stations located in Washington and Oregon, respectively. One of the most widely known hydroelectric power stations in the United States is known as the Hoover Dam, shown in figure 4.11. This dam, originally built as the Boulder dam, is over 700 feet tall and over 650 feet wide at its base. The dam is located along the Colorado River where it blocks the flow of the river to create Lake Mead. The Hoover Dam produces 1,039.4 megawatts of electricity, which is provided to the surrounding states of Arizona, Nevada, and California. The dams in the United States are not the largest or most prolific hydropower stations in the world. The largest power station was recently built in China called the Three Gorges Dam. This dam has an installed capacity of 22,500 megawatts. In fact, five of the world's seven largest power plants are hydroelectric power plants, while the other two are nuclear power plants (Global Energy Observatory, 2014).

Hydroelectric power is responsible for an estimated 6.7% of total primary energy production around the world today (British Petroleum, 2014). Figure 4.12 shows how hydroelectric power is utilized in different regions of the world. Nearly 60% of all hydroelectric power consumed in the world occurs in the Asia-Pacific and Europe and Eurasia

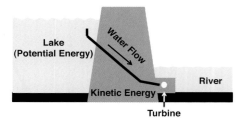

FIGURE 4.10 Basic diagram of a dam showing the buildup of water on one side representing the potential energy of water and the flow of this water through a penstock to create kinetic energy that will spin a turbine at the bottom to generate power.

regions, with China representing the single country with the greatest consumption of hydropower due to the use of the Three Gorges Dam (British Petroleum, 2014).

Hydroelectric power plants have a long lifespan and minimal operational costs after the very large capital cost initially invested in building a dam. This low operational cost provides a significant economic advantage for hydropower

FIGURE 4.11 Photographs of (A) Hoover Dam and the (B) hydroelectric power-generating facility within the dam (images by Adam and Carla Jones, 2011)

A)

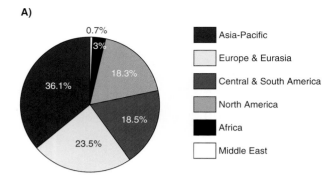

- Asia-Pacific
- Europe & Eurasia
- Central & South America
- North America
- Africa
- Middle East

B)

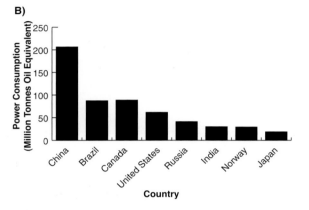

FIGURE 4.12 Hydroelectric power consumption around the world in 2013.

A A graphic comparison of hydroelectric power consumption by region in 2013. The Asia-Pacific and Europe and Eurasia regions consumed 59.6% of all hydroelectric power, while areas such as Africa and the Middle East played a very small role in hydroelectric power.

B In 2013, the single largest consumer of hydroelectric power was China at about 206.3 million tonnes oil equivalent equal to approximately 2.4 billion megawatt-hours. Brazil, Canada, and the United States were the next highest consumers of hydroelectric power.

DATA FROM British Petroleum (2014).

as a renewable resource for energy production (Casalenuovo, 2011). However, despite this advantage, its production of virtually emissions-free power, and its already important role as a renewable resource for power generation in the United States, it seems unlikely that conventional large-scale hydroelectric power capacity can be significantly and sustainably expanded in the coming years. One of the largest issues with the expansion of hydroelectric power in the United States is that most of the locations where large dams could be built already contain hydroelectric power stations. Therefore, it may be necessary to take advantage of microscale hydroelectric stations that rely on the flow of natural rivers and streams rather than on the creation of potential energy through the building of a traditional dam. Hydroelectric power stations, particularly the building of dams, also have important environmental repercussions. When a dam is built, it blocks the flow of

water to create a large lake, thereby flooding all of the surrounding land and having a severe impact on the ecosystem of that area. Also, within the river itself, the ecosystem suffers due to a change in the flow of the water, the blockage of fish migrations, and/or alterations in local fish populations (Casalenuovo, 2011). Overall, hydroelectric power is a clean source of energy that could help in replacing electricity generation from coal and natural gas; however, its expansion in the United States will likely depend largely on the development of more small-scale, environmentally sustainable water turbines that are designed to provide electricity for smaller surrounding communities.

Ocean Energy

Seventy one percent of the Earth is covered by water. Of these water sources, 97% are saltwater, mainly oceans, and only 3% are freshwater systems like rivers or lakes (Perlman, 2012). When discussing renewable energy technologies, we generally associate freshwater systems with hydroelectric power; however, this leaves 97% of the planet's water resources unaccounted for in terms of energy including the oceans that are capable of absorbing and transporting large amounts of energy. Therefore, it is necessary to look at the ocean as a source of renewable energy. Currently, research and development are focused on a limited number of potential commercial-scale technologies for ocean energy including **wave energy**, **ocean thermal energy conversion** *(OTEC)*, and **tidal energy**.

Wave energy is also derived from the sun as a primary energy source. Waves are generally created from wind, and as was discussed earlier in this chapter, wind is created from the heating of the air by the sun (Nersesian, 2010). Waves are built up as the wind blows along the surface of the water, creating friction. This friction first produces a ripple, building up with time to create much larger waves. Ultimately, the size of a wave depends on the wind speed, the length of time the wind is blowing, and the distance over which the wind pushes the wave.

If you have ever been to the beach and played in the surf, then you have experienced the up and down motion of waves. It is this motion that ultimately generates the kinetic energy that can be used to produce power. This energy can be captured by placing a buoy or other device similar to that shown in figure 4.13 within the path of the wave. When the water moves up and down, it pressurizes the air within the buoy. As the air moves in and out of the buoy, it drives a piston or spins a turbine that can generate power. Almost all devices being developed to generate power from wave energy are designed around the idea of using pressurized air to spin a turbine. In the end, scientists envision creating a farm of these buoys in the ocean where each buoy is connected to a central line that leads to land, carrying the generated electricity into coastal communities (Nersesian, 2010).

The most wave energy is found in coastal areas in Scotland, Canada, Africa, the United States, and Australia. In the

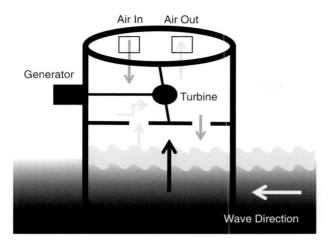

Air In   Air Out

Generator

Turbine

Wave Direction

FIGURE 4.13 Diagram of wave power generation within a buoy. The air moving in and out of the buoy from the top would spin a turbine connected to a small generator used to produce power. While one single buoy may only be able to generate a small amount of power, a large collection of these buoys has the potential to produce greater quantities of power (source: California Energy Commission, 2012).

United States, the west coast, particularly the southwestern Alaska coast, has the greatest potential for wave power generation. The Electric Power Research Institute estimates that the Alaskan coast could produce 620 terawatt-hours per year of power with wave energy. This is enough power to supply about 58 million homes annually in a state with only about 308,000 housing units (US Census Bureau, 2013; BOEM, 2014). Although wave energy technologies are not commercialized, they are continuing to be developed due to their potential for power generation.

The second renewable energy technology from the ocean being developed but currently not commercialized is Ocean Thermal Energy Conversion (OTEC). As was discussed earlier, the ocean takes up a massive percentage of the Earth and basically represents a giant solar energy collector and storage system. The energy absorbed and stored in the ocean can be used to produce power using OTEC.

OTEC is based on the natural formation of thermoclines in the ocean. A thermocline is a transition layer that separates the surface waters from the deep waters of the ocean. In general, the surface waters and deep waters individually maintain fairly constant temperatures; however, the thermocline layer is the transition layer with temperatures changing based on depth. The oceanic thermocline typically forms between 200 and 1,000 meters in depth and separates the warm surface waters from the cold deep waters. OTEC uses this thermal gradient to produce a power cycle. As long as the temperature in the deep ocean water is at least 20 degrees Celsius (36 degrees Fahrenheit) lower than the temperature at the surface, power can be produced. This temperature extreme is generally only found in tropical and subtropical oceanic areas closer to the equator, with the largest extremes seen in the western Pacific Ocean (Burman and Walker, 2009).

A closed OTEC system uses a material with a low boiling point that when exposed to the warm surface temperatures of the ocean will turn into vapor. This vapor is then pushed through a turbine that drives a generator. This vapor is then pushed down the pipes through the thermocline and into the colder, deeper layer of the ocean. The cooler temperatures cause the vapor to condense back into liquid and the cycle will repeat itself. The system can also be designed as an open system working in basically the same manner except that water is used under pressure to create the vapor, and then once the water is condensed by cooler ocean water, it creates desalinated water (Burman and Walker, 2009).

The final ocean energy technology does not directly rely on the energy from the sun as its primary energy source but rather the interaction of the Earth with the moon and, to a small extent, the sun to create tides. Tides form as large bodies of water rise and fall due to the gravitational forces as the Earth rotates. Most places have two high tides and two low tides each day. The timing of tides is predictable and consistent due to the alignment of the sun, the moon, and the Earth (Nersesian, 2010). When the sun and the moon are directly opposite from one another with the Earth in the middle, the positions on the Earth located on this linear trajectory will have the largest tides. The moon is closer to the Earth than the sun, so its gravitational forces are a little stronger and create a slightly higher tide on the side of the Earth closest to the moon. As all of that water moves across the Earth, it contains kinetic energy.

In about 20 places around the world, the ebb and flow of the changing tide is over 16 feet, the minimum difference needed to capture the potential of tidal power (Burman and Walker, 2009). Although tidal energy has been used for many centuries on a smaller scale, scientists are now trying to develop ways to use tidal energy on a more widely available commercial scale. One possible method is to build a tidal barrage or dam. In this scenario, the tide will come in and flood a basin located behind the dam. As the tide flows out, the difference between the height of the flooded basin and the level of the water on the ocean side of the dam can be significant enough to allow the water to flow out via penstocks and generate electricity like at a hydroelectric power plant. Unlike traditional hydroelectric power plants, the potential energy from tides is only available for about 10 hours a day during tidal changes. Thus, power generation is more limited. Another method being considered is a method that would take advantage of the tide coming both in and out where the tidal water would be funneled through channels packed with water turbines. The turbines would be designed to switch directions depending on whether the flow of the tide was going inland or out towards the sea. There are a few tidal plants in existence around the world. The largest is located in La Rance, France, and is capable of producing about 240 megawatts of power from a tidal difference of 26 feet (Nersesian, 2010).

Biomass

Up to this point, numerous renewable resources have been discussed that use the sun as their primary energy source, and all of these resources have been developed primarily to assist in the replacement of coal- and natural-gas-generated electricity. While the replacement of this fossil-fuel-generated electricity with renewable energy is essential for the planet, there is also a very real need to create renewable liquid fuels for the transportation sector. Biofuels created from biomass could be used for electricity generation, but could also play a critical role in the production of liquid transportation fuels. Biomass is grown as a result of the absorption of energy from the sun by the process of photosynthesis in plants. Photosynthesis evolved millions of years ago first in cyanobacteria, then in algae, and finally in plants, as a means to harvest energy from the sun and convert it into stored chemical energy. This biological process is also largely responsible for oxygenating the atmosphere and is the most important primary metabolic function on the planet.

For thousands of years, people have used biomass as an energy resource, mostly in the form of burning wood or dung, for heating and cooking. Burning biomass is still a fairly common method for producing heat around the world, but in the United States and Europe burning wood amounts to a very limited part of the total energy utilization (<1%) (EIA, 2015). However, in other parts of the world wood is used extensively to produce heat and light and for cooking. Wood represents a finite supply at any given time in many places in the world, and without careful and sustainable forestry management, the use of wood can result in significant environmental damage from deforestation as well as the release of atmospheric pollutants in smoke.

Burning wood is just one of many sources of energy derived from biomass. Biomass can also be used to produce a variety of different biofuels including bioethanol from sources like corn and sugarcane; biodiesel from soybeans, rapeseed, and algae; biohydrogen from algae; and biogas from biomass waste. The remainder of this book will focus largely on these sources of biofuels and their importance in the future of renewable energy.

Renewable Energy Utilization

Renewable energy sources like water, wind, and wood are the foundation from which energy was developed over thousands of years; however, despite the many sources of renewable energy that are available, renewable sources make up a small fraction of energy consumption around the world today. Figure 4.14 shows the global utilization of energy by source where fossil fuels make up 86.7% of consumption and non-hydropower-based renewable sources make up only 2.2% (EIA, 2015). With all the developments in renewable energy technologies over the past century and the known environmental damage by fossil fuels, why do

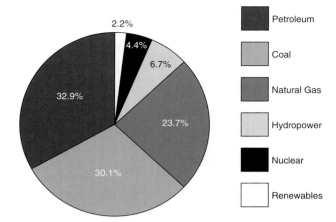

FIGURE 4.14 Global primary energy consumption by source in 2013 showing that the fossil fuels (petroleum, coal, and natural gas) make up 86.7% of global energy consumption, while renewable energy including hydropower makes up only 8.9% (data from EIA, 2015).

these sources not play a more significant role in energy utilization around the world? One of the main reasons these renewable energy technologies have not been able to develop a stronger foothold in energy consumption is their lower power density compared to fossil fuels. Overall, the lower energy density for renewable energy technologies often means that more of the resource must be harvested to generate enough energy to be competitive with fossil fuel resources. This drawback ultimately means that, in most cases, renewable energy technologies are more expensive and not economically competitive with fossil fuels; however, with dwindling fossil fuel supplies and concerns over their environmental impact, the price of fossil fuels may eventually overtake many of these renewable resource technologies and make them competitive in the global energy market.

In the United States, renewable energy (including hydropower) makes up about 10% of total energy consumption (EIA, 2015). As shown in figure 4.15, water, wood, biofuels and wind were all significant contributors to total renewable energy in the United States in 2013. In thinking about the utilization of these various renewable resources, approximately 78% of them are used primarily for power generation, yet 36% of energy consumption in the United States in 2013 came from petroleum largely used in the creation of transportation fuels (EIA, 2015).

Petroleum as an energy resource in the transportation sector is unique not only because it has an extremely high energy density, but also because it is easily transported and stored due to its liquid form. With the transportation sector making up 28% of energy consumption, it is important that a renewable energy technology be developed that can replace this liquid fuel source (EIA, 2015). Most of the renewable energy technologies discussed above may prove to be excellent replacements for coal- and natural-gas-

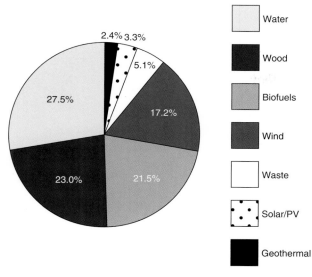

Water

Wood

Biofuels

Wind

Waste

Solar/PV

Geothermal

FIGURE 4.15 US renewable energy consumption in 2013 showing that water, wood, biofuels, and wind all contribute significantly to total renewable consumption, while waste, solar including photovoltaics, and geothermal are only minor contributors to US renewable energy (data from EIA, 2015).

derived power, but they would be unfeasible for the replacement of liquid transportation fuels. However, over the past several decades, it has become increasingly evident that a potential source for the replacement of liquid petroleum fuels may be the original source of those very fossil fuels: plants and algae. Rather than waiting millions of years to allow biomass to be buried to create fossil fuels, biomass energy technologies are now being explored and commercialized in order to use these photosynthetic resources directly for the production of liquid fuels.

The use of biofuels as a source of transportation fuel is certainly not a new concept; in fact, the first combustion engines actually ran off of a biofuel, peanut oil. Today, the most important biofuel in the United States is corn-derived ethanol that is used as an additive in gasoline. This came about in response to the issue of rising petroleum prices, along with concerns about the environment and energy security. To address these issues, the US government developed the Renewable Fuel Standard in 2007 that calls for the replacement of 36 billion gallons of petroleum-based fuels with renewable biofuel sources by 2022. Fourteen billion of

those gallons can come from corn ethanol, while the rest must come from non-corn ethanol-based sources (EPA, 2014).

This chapter has given you an overview of the various forms of renewable energy resources available on the planet. The latter parts of this book will focus exclusively on the various biomass sources for the production of biofuels and their commercialization. Because the production of biomass will result from agricultural processes developed over thousands of years, it is important to have a basic understanding of how the use of industrial agricultural knowledge can help both maintain a robust food supply and expand the opportunity to use plants and algae as a source of biofuels. Accordingly, the next chapter will introduce the fundamentals of industrial agriculture.

STUDY QUESTIONS

1. Define renewable and sustainable. Explain why fossil fuels are neither renewable nor sustainable. Is nuclear energy renewable and sustainable? Why or why not? Give some examples of truly renewable and sustainable energy resources.
2. Briefly explain the importance of the design of the nuclear reactor in the use of nuclear fission reactions for the production of electricity.
3. Briefly describe the energy resources that we can utilize from each of the primary energy sources. How can these resources be utilized best in society (electricity, transportation, etc.)?
4. Explain the three main types of solar power generation and how they differ in using the sun's energy to generate power.
5. Briefly discuss how water on the planet can be used to generate energy.
6. Explain why renewable energy technologies do not play a more dominant role in global energy consumption.
7. Explain why biofuels may have a different niche in the renewable energy future than many of the other renewable resources.

CHAPTER FIVE

Linking Food and Fuel:
The Impact of Industrial Agriculture

Plants have the capability of converting solar energy and carbon dioxide into chemical energy and oxygen through the process of photosynthesis. Because of this, they are directly responsible for our ability to live on this planet. Not only does photosynthesis allow for the oxygen we breathe and the growth of the food we eat, but it is also the source of most of our energy resources including those ancient sources like fossil fuels. Photosynthesis represents a very powerful method for converting solar energy into chemical energy, only in a much shorter timescale than fossil fuels. We call energy generated from today's photosynthesis bioenergy. As mentioned in earlier chapters, bioenergy is simply the utilization of the photosynthetic process to transfer electromagnetic energy into chemical energy stored in the form of biomass. Once produced, biomass can then be used in a variety of methods to produce food and fuel. While plant growth is a renewable process, it is deeply dependent on the balance of many environmental factors to maintain the proper conditions that allow for efficient and productive growth of these organisms. Striking this balance is the fundamental concern of modern agriculture. Agricultural breakthroughs over the last century have allowed for an unprecedented expansion of crop production and hence an enormous expansion of the human population. Accordingly, this chapter will address not only the science of agriculture, but also how energy impacts agriculture, and how agriculture has changed the world and could change the future of the energy landscape.

For many, it may not be immediately obvious what agriculture has to do with renewable energy, but in reality, the answer is simple: biofuels, the focus of the remainder of this book, are derived from the same photosynthetic biomass used in the traditional agricultural production of food. Therefore, the development of commercial-scale biofuels will greatly depend on knowledge gained from agriculture, particularly modern industrial agriculture prac-tices and those practices implemented during the Green Revolution. Many of these practices have been developed in the agricultural field to promote increased efficiency in food production to be able to feed a rapidly increasing world population that has now surpassed 7 billion. These practices, including the use of fertilizers, production machinery, irrigation, and food and feed transportation, rely heavily on the use of fossil-fuel-based energy resources; thus, there is also now an inseparable link between the price of food and the price of fuel. For many years, when oil was inexpensive, the energy costs for agriculture, although important, were not the primary factor in the price of food; those were labor costs. As the cost of energy has increased, it has become a larger and larger part of agriculture costs. Today energy is a dominant cost in food production, and hence energy cost and food costs are now tightly linked (Woods et al., 2010). Finally, it is becoming ever more crucial to maintain and even advance agricultural productivity in order to increase efficiency as arable land and resources become scarcer and hence more expensive. These advancements in the efficiency of food production will also need to coincide with establishing a sustainable biofuel industry.

Chemical Energy: Linking Food and Fuel

The words food and fuel likely conjure up very different ideas. When we imagine food, we think of everything from bread to ice cream, but when considering fuel, we usually imagine the pungent liquid we put into our gas tanks or the smelly diesel that keeps our trucks running on the highways. Although fuel is not exactly synonymous with food, food is a form of fuel. It is the form of fuel that provides the nutrition to power the metabolic processes in the human body. So while humans require fuel in the form of food to survive, humans have also come to rely on another type of fuel: fossil

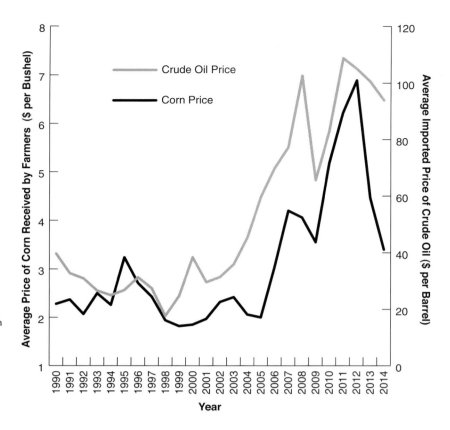

FIGURE 5.1 Graphical comparison of the average annual price of corn (black line) with the average annual real price of imported petroleum (gray line) between 1990 and 2014. The price change of these two commodities clearly follows the same pattern, suggesting a link between the price of food and the price of fuel (data from USDA, 2014a; EIA, 2014b).

fuel. Food and fossil fuels are both different forms of chemical energy, which as a reminder can be defined as the energy released by a chemical reaction to do work. We can eat a pizza and use that energy to run a marathon or we can put a gallon of gas in our gas tank and use that energy to drive the same distance; both rely on chemical reactions that do work. In this most basic form it is easy to see that food and fuel are really two sides of the same coin; both are simply chemical energy. Another way to look at the two sides of this energy coin is to examine the average American energy consumption. The average American consumes about 2,637 Calories of food per day and about 31,500 Calories, or 1 gallon, of gasoline per day (UM, 2011; EIA, 2014a). As we will see, food and fuel are closely linked in many ways due to modern agricultural production processes today being completely dependent upon fossil energy utilization to remain productive.

Modern agricultural practices, along with the Green Revolution, led to huge advances in agriculture productivity. However, these advances came with a pretty significant energy bill. Consequently, a link has been established between the price of food and the price of fuel. This link is apparent when we compare the price of corn with the price of crude oil as shown in figure 5.1. Between 2002 and 2012, the increase in the price of crude oil is matched by a rapid increase in the price of corn. It is not until 2013 that we begin to see both the price of crude oil and the price of corn begin to drop. The link between food and petroleum is largely derived from the need for crude oil resources to power large industrial farming machinery, for food processing and packaging, and for the transportation of food around the globe. In addition, nitrogen-based fertilizers are also derived from fossil fuels like coal and natural gas through processes that require high temperature and pressure. In modern industrial agriculture, these fertilizers are critical to the productivity of most crop plants and further link the cost of fossil fuels with the cost of food production.

Recently, the link between food and fuel prices was strengthened by the creation of the Renewable Fuel Standard in the United States, where now, not only do we use crude oil as a resource in the production and distribution of our food, but we also use food sources as a legally mandated replacement for the crude oil products themselves. These fuel replacements largely come from corn-derived bioethanol, and to a much smaller degree from biodiesel derived from soybeans. With food and fuel so inextricably linked, it is not surprising that the prices of these two commodities tend to rise and fall together. Thus, any increase in the price of fuel is likely to impact the price of almost all types of food.

In a time when the price of fuel remains at historically high levels, it is critical that we develop new sustainable methods to maintain and even improve agricultural productivity. While it may be possible to survive without fossil fuels, it is impossible to survive without food. It is not only a matter of producing enough food to feed more than 7 billion people; it is also a matter of producing this food at a price affordable to everyone. On the modern farm, the ability to produce cheap food has become wholly dependent on the availability of cheap energy. As the price of fossil fuels rise, and as biofuels grow in demand as an alternative to crude oil, it is unknown how the world will strike a balance between food and fuel. But one thing is for certain: the ability to produce biofuels while maintaining high-level production of affordable food will test the agriculture techniques developed over the last century. To find a path forward that will allow us to produce food at affordable prices in a time when oil prices seem unlikely to return to the levels of past decades, we will have to renew our efforts to keep the Green Revolution alive and continue to deploy the advancements made in agricultural biotechnology.

The Agricultural Revolution

Agriculture is the cultivation of animals and plants through the distribution of natural resources like nutrients and water. Agriculture provides abundant sources of food and fiber including foods like cereals, fruits, vegetables, and meat, and also common fibers such as cotton, wool, hemp, silk, and flax. But beyond these, agriculture also yields raw building materials like lumber and bamboo and the feedstocks to produce biofuels like methane, ethanol, and biodiesel. Modern industrial agriculture has transformed human society by granting individuals the free time to devote to pursuits outside of acquiring food. The efficiency of modern agriculture has allowed humans to become specialists in specific trades and develop societies into immense booming metropolises. At the time of writing this book, less than 1% of the population in the United States claim farming as their primary occupation and less than 2% even live on farms (EPA, 2013). Today, very few farmers are required to provide all of the food for the booming population.

It should be noted that humans are not the only animals that practice agriculture. In fact, leaf cutter ants have been agricultural specialists for millions of years, predating the development of human agriculture. Leaf cutter ants are the dominant herbivores in the tropics, capable of harvesting huge quantities of forest vegetation (15–20%), which they interestingly do not eat. It is routine to see a line of these tiny ants carrying large pieces of leaves along the forest floor. In a classic example of biological symbiosis, they bring this foliage back to their burrow where the harvested leaves are used to cultivate a fungus. The fungus is the actual food source for the ants, while the leaves provide nutrition for the fungus. Remarkably, the ants actually go a step further in their agricultural methods. These ants have a natural pesticide on their bodies that when transferred to the fungus prevents unwanted bacteria from growing and taking over (O'Brien and Walton, 2010). Thus, these leaf cutter ants have not only figured out how to "farm" their fungus, but also developed a method to "protect" their crop. Both domestication and crop protection are key fea-

tures in modern agriculture developed by humans millions of years after these ants had developed similar practices.

Human agriculture emerged in about nine areas on four continents including within the Fertile Crescent and China on the continent of Asia, in West Africa and Ethiopia on the continent of Africa, in the Amazon and Andes in South America, and within both the eastern United States and midwest on the North American continent (Diamond, 2002). The first step toward the revolution of agriculture was the domestication of plants and animals. **Domestication** is the process of taking a wild organism and optimizing it to grow and thrive within a semi-controlled and confined environment. Plant and animal domestication is one of the most important developments in all of human history because it allowed humans who had always been hunters and gatherers to settle in a defined location. While hunting and gathering could provide food, it often resulted in the expenditure of a significant amount of energy because people had to be on the move to keep up with the cycles of plant growth and animal migration. It also required that almost all people participated in hunting or gathering to secure enough food to sustain their own bodies. With the domestication of plants and animals, this changed. Humans could now plant their food close to their homes, reducing the need to travel long distances to collect enough food. In turn, people also realized that using their own hands to do all of the work was not as efficient as utilizing large mammalian herbivores to help in the process. Thus, animals were also domesticated to help supplement the physical labor requirements. However, the domestication process was not and is not easy. Many species of both plants and animals are difficult to control. Today of the 148 species of large mammalian herbivores, only 14 are domesticated, and of the 200,000 species of higher plants, only 100 species yield large-scale agricultural products (Diamond, 2002). Despite what may seem like small numbers, these domesticated plants and animals have become the mainstay of agricultural production in society today.

Following domestication, the next major development in agriculture occurred when humans in Egypt and Mesopotamia learned to supplement natural rain levels in their fields by storage and transportation of water from other areas. This process is known as **irrigation**. Before the advent of an irrigation system, regular natural flooding and rain served as its precursor. This natural flooding is best known in areas of Mesopotamia and The Fertile Crescent where the Tigris and Euphrates rivers regularly flooded and irrigated the land (Cowen, 1999). About 3,000 years ago, the first primitive irrigation system was developed and is known as the Karez irrigation system. The Karez irrigation system worked through a network of canals in a mountain that collected rainwater and then funneled it down to fields below. In this way, water could be shuttled via a primitive aqueduct from the mountains to the fields (Khan and Nawaz, 1995). From this first irrigation system, irrigation developed into one of the most important features for the advancement of agriculture, so much so that today 275 million hectares or 679.5 million acres of land worldwide are irrigated, representing 20% of cultivated land and accounting for 40% of global food production (WWAP, 2012).

Agriculture was further advanced about 2,000 years ago when the Romans realized that by alternating the crops grown on a single plot, higher productivity for the crops could be achieved. This method became known as the "food, feed, and fallow" rotation, and is still practiced today in many crop rotation strategies. Under this rotation, the Romans would plant a crop used for human consumption like wheat, then in the next year a crop designed mostly to create feed for the domesticated animals, and finally in the third year no crop at all was planted, allowing the land to lie fallow. This ordered planting of specific crops on the same field is known as **crop rotation** (Science Encyclopedia, 2012). Crop rotation is important because it minimizes soil erosion, reduces pest and disease within the crop, allows for better nutrient balance, and increases the overall crop yield. Today, it is common to see a four-crop rotation that is very similar to this initial approach. The rotation begins with wheat or a similar food crop, followed by a feed crop such as barley in the second year. Then, the field may be planted with a legume or some other nitrogen-fixing-type plant to help increase nitrogen levels in the soil, and finally in the fourth year, the land may lay fallow to conserve nutrients and build up moisture within the soil in preparation for the process to begin again. Crop rotation can have a huge benefit on yields year after year.

Overall, these three major advances in agriculture—domestication, irrigation, and crop rotation—allowed the population of the world to go from about 1 million people, located mainly in and around the world's fertile lands, to a population of 1 billion people by 1850. But it was not until the commercialization of fossil fuels that agriculture really began to expand into what is known today as industrial agriculture, allowing the world's population to dramatically expand again.

Industrial Agriculture: Advancing Agriculture with Fossil Fuels

Modern agriculture began in the 1700s with the advent of the seed drill in 1701, the plough in 1730, and the thresher in 1786. These pieces of equipment revolutionized and commercialized the farm industry, allowing for much greater individual efficiency, and hence the numbers of farmers began to decrease. At the same time, the productivity of crop production increased. In 1868, the steam tractor was introduced and slowly began to supplant traditional horse-drawn plows and harvesters. The tractor was so successful that while in 1830 it took 250 labor hours to produce 100 bushels of wheat, by 1955 it took only 6 labor hours (Spielmaker, 2006). This change in efficiency was largely due to the replacement of traditional domesticated

animals such as horses and mules with fossil-fuel-powered farm equipment.

Another important change in traditional farming brought about by the introduction of fossil fuels was the development and use of commercial **fertilizers**. Three key chemical fertilizers, nitrogen, phosphorus, and potassium, have become mainstays for agricultural productivity throughout the last 60 years.

Phosphorus is a key element for agriculture, and is an essential elemental necessity for all life. In plants, phosphorus functions in the storage and transfer of energy throughout the plant as well as promotes root and seed development. However, phosphorous is often a limiting nutrient for crop productivity, and must be mined and processed into fertilizer-based nutrient supplements. Today, phosphorus is becoming less readily available as global phosphorus production is believed to be declining (Van Kauwenbergh, 2010; Wood et al., 2012). Figure 5.2 shows the reserve-to-production ratios for phosphate rock in countries with significant reserves. Just as in fossil fuels, the reserve-to-production ratio equals the number of years remaining for phosphate based on current production levels in that country. Countries in North Africa, including Algeria and Morocco, and those in the Middle East, including Syria and Iraq, have very large phosphate supplies and a smaller level of agricultural demand, leading to a much longer availability of phosphate resources. However, many countries with higher productivities in agriculture, including the United States and China, also have large supplies of phosphate, but their demand is so high that the number of years this resource will be available is limited to 37 for the United States and 39 for China. Luckily, assuming global cooperation, world reserves for phosphorus are likely to last about 300 years (Jasinski, 2014). However, this assumption could be premature considering that many of the countries controlling the long-term supplies of phosphate are potentially unstable or will require their entire supply of phosphate to support agricultural needs for their own populations. For many countries including the United States, this leads to two significant problems: cost and supply security. Because there are no substitutes for phosphorus in agriculture, once phosphorus supplies are exhausted or become too expensive, it will be necessary to develop ways to recover and reuse phosphorus to avoid massive decreases in global agricultural productivity.

Potassium is another important component of common fertilizers. Potassium promotes stem and root growth as well as protein synthesis (Johnston, 2003). As you can see in figure 5.3, the reserve-to-production ratio for potassium is highest in Belarus and Brazil with available supplies lasting almost 700 years for each of these countries, significantly longer than any other country. Based on the production levels for 2014, the predicted global potassium reserve will last about 183 years (Jasinski, 2014).

Nitrogen is another crucial nutrient for plant growth and thus an important component of fertilizer; fortunately,

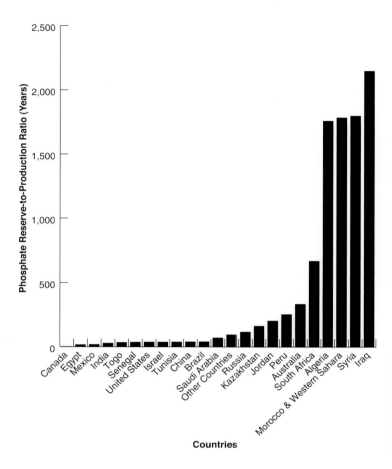

FIGURE 5.2 Comparison of the reserve-to-production ratios for phosphate in major phosphate-producing countries, 2014. The reserve-to-production ratio equals the number of years phosphate will be available at current production levels. The graph clearly shows that countries in North Africa and the Middle East will have the longest availability of phosphate resources (data from USGS, 2014a).

nitrogen fertilizers can be manufactured. The air is full of nitrogen, but this nitrogen is a gas that is unusable by plants. In order for a plant to be capable of using nitrogen, it must be in the form of reduced nitrogen. Nitrogen gas can be converted to reduced nitrogen by nitrogen-fixing bacteria associated with some plants as plant symbionts. These bacteria can convert the inorganic nitrogen gas into a useable organic form of reduced nitrogen such as ammonia. Because the nitrogen-fixing bacteria are not associated with every crop plant, modern agriculture often requires supplementation with reduced forms of nitrogen to maintain high productivity of these crops.

Nitrogen is not seen as a limiting resource for crops around the world due to the development of the **Haber–Bosch process**. This process generally requires fossil-fuel-based energy, usually natural gas or coal, to synthesize ammonia from nitrogen gas using hydrogen, high pressure, high temperature, and an iron catalyst. In Europe, it is estimated that about 34 million tonnes of nitrogen are produced through the Haber–Bosch process each year. Globally, about 61% of all nitrogen is produced through this process. Fertilizer production as a whole utilizes about 1.2% of the world's energy (Jenkinson, 2001; Tomlinson, 2012). Thus, while nitrogen may not be limited, coal and natural gas are finite resources and the environmental impacts of coal and natural gas consumption could

worsen with an increasing demand for nitrogen-based fertilizers. The dependence of agriculture on artificial fertilizers will likely lead to a doubling of the nitrogen requirement by 2050, resulting in the use of significantly more fossil fuels and thus the release of more carbon dioxide into the atmosphere due to fertilizer production (Jenkinson, 2001).

Advances in industrial agriculture have dramatically increased the productivity of food production. While the inputs of land and human work have remained fairly constant, agricultural output and productivity as well as the inputs required in the form of energy resources have increased steadily. Not only have the quantities of inputs such as fertilizers and other crop-protecting agents like pesticides increased over time, but the price of these irreplaceable agricultural inputs has also increased. Essential fertilizer nutrients have seen a steady increase in price largely since the beginning of the twenty-first century. For example, the price of phosphate increased from about 30 cents per pound to about 90 cents per pound between 2000 and 2008 (Huang, 2008). This increase in price can be explained by both a diminishing supply and an increase in the cost in fuel. Interestingly, many of these important agricultural products have values disproportionate to their actual use. In other words, while a particular fertilizer may only be worth so much due to its demand in a particular year, the

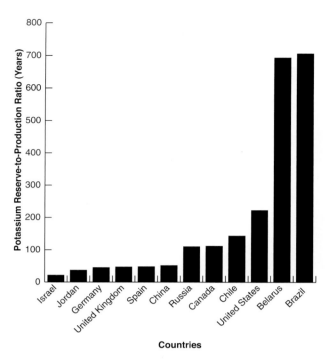

FIGURE 5.3 Comparison of the reserve-to-production ratios for potassium, 2014. Belarus and Brazil have supplies predicted to last almost 700 years each at current production levels. This is significantly longer than any other country. Based on reserve-to-production ratios for 2014, the global supply of potassium will last about 183 years (data from USGS, 2014b).

reality is that as this fertilizer becomes more limited in the future, the price will likely increase. Thus, the cost seen today is also based partially on the expected cost in the future.

With a steady rise in world population over the next century, the ability to provide adequate and affordable food for all will depend largely on the balance between the availability of natural resources like arable land, fertilizers, and irrigation water and the growth in agricultural productivity. Currently, regardless of production levels, the price of food is closely correlated with the price of fuel; as the price of fuel increases, the price of food also increases. This is not only due to the rising costs of agricultural inputs like fertilizers, but also due to the fuel used during planting and harvesting, the energy required to process food, the fuel used during the transport of food, and, in more recent years, the amount of food being used to actually generate fuel, largely in the form of corn- and sugarcane-derived ethanol.

The Green Revolution

Advances in the use of irrigation, fertilizers, and large fossil-fuel-powered farming equipment introduced during the industrial revolution allowed for significant advances in productivity relative to traditional agricultural techniques.

These techniques resulted in a tremendous increase in agricultural outputs, leading to a population increase from 1.0 billion in 1850 to 2.5 billion in 1950, but, since 1950, the population has exploded to over 7 billion (US Census Bureau, 2014). While the techniques developed for industrial agriculture were sufficient to match agricultural productivity to the population increase between 1850 and 1950, these techniques alone were not sufficient to match agricultural output with the population explosion since 1950. In fact, starting in the 1940s, it became obvious that another type of agricultural revolution would be necessary to maintain increases in agricultural productivity equal to the rate of population growth. This time the revolution was seated in the biology and genetics of the crop plants themselves and came to be known as the **Green Revolution**.

The Green Revolution is defined as the period of time beginning in the 1940s when agricultural practices underwent significant evolution. It all began in 1943 when Norman Borlaug in Mexico introduced new techniques to produce higher-yielding and pest-resistant crop varieties, for which he won the Nobel Peace Prize in 1970 (AgBioWorld, 2011). These revolutionary techniques stemmed from a better understanding of plant biology, particularly genetics. It was known that for any individual crop some plants would do better than others and yield more of the desired product. From this idea, Borlaug developed the concept that something within the genetics of these individual plants must be providing a competitive advantage over the other plants in a particular environment. Thus, he isolated these superior plants from the plant population and used them as a breeding stock to develop improved crops with desirable characteristics and an increased overall yield. Some of the genetic and physiological traits that were desired included the production of more leaves to increased solar energy capture, reduced plant height to limit resources going to the production of the stalk, a shortened growth duration to allow for the planting of more than one crop per year, and the identification of genes so that plants could survive under various abiotic stresses including drought tolerance and heat (Hazell, 2002). Overall, breeding and selecting of high-yielding crops had a huge impact on agricultural productivity.

Along with breeding and selection, farmers quickly realized that no matter how high yielding a particular crop may be, it was still susceptible to a large array of pests. This can be particularly problematic for crops that have been bred to be very similar genetically. Without genetic diversity, a single pest or disease can easily wipe out an entire crop. In order to combat this problem, new tools for crop protection were developed that included herbicides, insecticides, and fungicides. Crop protection was not an altogether new idea in agriculture; in fact, salt had been used as a form of crop protection for hundreds of years. The idea of actually producing a host of chemicals to use as crop protection, however, was new and largely stemmed from the ability to chemically synthesize these molecules using petroleum-based chemical feedstocks. The first generation of synthetic chemical crop-protecting agents included arsenic and hydrogen cyanide. These agents were discontinued due to their extreme toxicity toward many forms of life. Next came the development of second-generation crop-protecting agents known as dichlorodiphenyltrichloroethane (DDT), but these had negative environmental impacts and ultimately resulted in the development of the less toxic synthetics including organophosphate insecticides and acidic herbicides used today (Zadoks and Waibel, 2000).

Finally, continued improvements in plant genetics and a better understanding of the role of DNA and genes in plant biology during the Green Revolution led to advances in plant biotechnology. Through genetic engineering, genes are moved between organisms that are not sexually compatible, allowing the creation of novel plant varieties with desirable characteristics. These **transgenic** crop plants were generally both higher yielding and more tolerant to disease or herbicide use. There are five desirable characteristics generally sought after in transgenic crops including (1) herbicide tolerance; (2) disease and insect tolerance; (3) improvements in the quality of a product; (4) tolerance to stress such as drought, temperature, and salt; and (5) growth and productivity enhancements like increasing crop yields. The first example of a major biotechnological advance came in the early 1990s, when a protein found in the bacterium *Bacillus thuringiensis* known as Cry protein was shown to inhibit insect growth by getting them to stop eating plants. By inserting this gene into corn or cotton known as "Bt corn" and "Bt cotton," these crops were shown to have greatly reduced damage from insects (UMT, 2012). Another important biotechnological advance came with the discovery of the EPSP synthase gene from the bacterium *Agrobacterium*. When engineered into plants, this gene was shown to give resistance to glyphosate, a weed herbicide, thus, allowing the herbicide to be sprayed on the plants without negatively impacting growth but destroying the surrounding weeds (Funke et al., 2006).

The impact of the Green Revolution on corn production is evident in figure 5.4 where there is a clear and rapid increase in corn production beginning in 1950 and continuing today. The development of hybrid plant species and industrial processes such as crop protection followed closely by advances in molecular science increased the yield of corn from around 2,206 million bushels per year in 1940 to over 14,000 million bushels by the year 2014 (USDA, 2014a). This amazing advancement and increase in productivity during the latter half of the twentieth century carried over to many other staple crops around the world, providing food for world population growth.

The Green Revolution was not limited to increases in plant crop productivity; it also spread into the food animal produc-

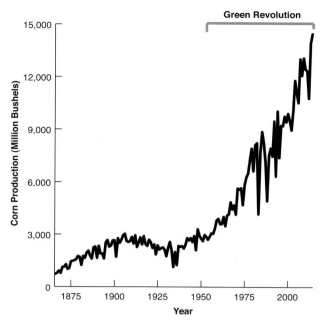

FIGURE 5.4 The rapid increase in corn production shown as millions of bushels between 1950 and 2014 indicates the impact of the Green Revolution on the production of this important crop (data from USDA Economic Research Services, 2012).

tion markets. Through genetic selection and improvements in nutrition and feed utilization, animals paralleled the productivity increases in crops. For example, chickens currently grow to a market weight in six to seven weeks, a period of time threefold less than the average maturation age of a commercial chicken 50 years ago (Perry et al., 1999). This same concept can be seen repeatedly with many food animal products. It is interesting to note that in agricultural markets like beef, the ability to produce more beef likely influences the overall environmental impact of this industry as well. Beef production is currently a significant contributor to greenhouse gases, accounting for 14–22% of all greenhouse gas emissions according to the United Nations Food and Agriculture Organization (Fiala, 2009). The high emission levels for beef production are largely a result of removing cattle from the open range and providing their nutrition through agriculturally produced feeds made with grains, both of which require an energy-intensive production process.

Food Production, Processing, and Transportation

Up until 2000, the trend in food prices was mainly downward, but with the rising cost of fossil fuels over the last decade, the trend has reversed itself and the price of food has begun to increase. This rise in price cannot be accounted for solely by increasing demand from a growing population; rather, it reflects costs associated with the production, processing, and transporting of food.

AGRICULTURAL INTENSIFICATION AND CLIMATE CHANGE

Travis L. Johnson

The Green Revolution was crucial to creating food security for the growing number of people around the world, but it is important to note that increases in agricultural production actually contribute to climate change through soil emissions, enteric fermentation (cow burps and flatulence), manure, rice paddy emissions, and land use change (deforestation), all together producing roughly 25% of greenhouse gas emissions. It would seem that the Green Revolution may have negatively impacted the environment, except research has shown that the most detrimental part of agriculture is land use change, which is using land for agriculture that would otherwise be carbon-sequestering forests. Because of the productivity increases from the Green Revolution, instead of having to use more land to meet the demand for growing new crops, crop improvements through agricultural biotechnology provided more harvestable food per acre. So in effect, the Green Revolution actually prevented a massive amount of deforestation by intensifying agriculture and using less land. It is estimated that 161 gigatons of carbon have been saved because of this, or roughly 34% of all carbon emissions since the Industrial Revolution.

SOURCE: Burney (2014).

In today's society, access to cheap locally produced foods has become more limited as small-scale local production does not afford the economies of scale of industrial mass production, making locally produced foods more expensive. Most grocery store chains in the developed world have the majority of their shelf space dedicated to large industrial food production and processing companies, as these foods are in greater demand due to their cheap price tags and easy preparation. Processed foods have made life easy with the ability to buy a microwaveable meal in a box or soup in a can, but the reality is that all of this processed food comes with an added energy requirement. In the state of California, food processing is the third largest industrial energy user in the state. The food processing industry in California generates over $50 billion in gross annual revenues, but it comes with the cost of consuming 600 million therms of natural gas and 2.7 billion kilowatt-hours of electricity, used in everything from processing machines to refrigerated warehouses (Amon et al., 2008). The packaging of processed foods also results in a significant amount of consumer waste.

Perhaps the biggest change in the last decade in terms of food price and availability is the massive expansion of food transportation. It is now possible to get almost any type of food at any time of year regardless of whether that food is in season. Take a moment and consider the last time your local grocery did not have tomatoes or bell peppers in stock. The answer is probably never, yet both of these items are only grown seasonally between June and October in the United States. The reason these items are always available is because many food products are now being shipped all around the world to provide consumers with the convenience of having any food in their grocery store at any time of year. As an example, during the North American summer months, the agricultural production in the United States is high and a lot of this fresh produce is shipped to other countries including those in South America, where it is winter. Then in the winter months for the United States when productivity of fruits and vegetables is low, it is summer in South America so productivity is high, and the transportation of food is reversed, resulting in a large array of produce being shipped from those South American countries back to the north. This massive international shipping cooperation feeds importation and exportation businesses for agricultural products.

The US Department of Agriculture Economic Research Service compares the import and export of agricultural commodities. Interestingly, exactly the same types of agricultural commodities including livestock, poultry, dairy, vegetables, fruits, nuts, grains, and many other edible products are both imported and exported in the United States. For example, the United States spends $6,211 million

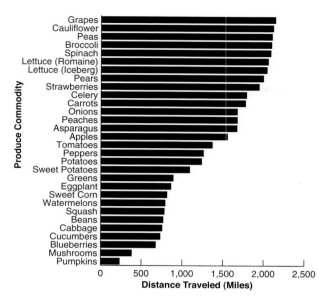

FIGURE 5.5 Comparison of average distances that food travels within the continental United States with a final destination of Chicago, Illinois (data from Piroq and Van Pelt, 2002)

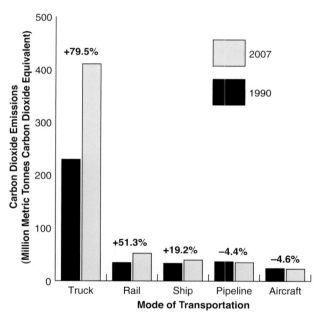

FIGURE 5.6 Comparison of carbon dioxide emissions from US domestic freight transport between 1990 and 2007. Values shown above columns indicate the percent change in emissions during this time period (data from US Department of Transportation, 2009).

importing vegetables and receives $9,881 million exporting vegetables, many of these being the same vegetables (USDA, 2014b). The importing and exporting of food products is a big business driven by a consumer mindset of having a large variety of food available year-round. This luxury in food availability is not the case for many developing countries or those with smaller economies as evidenced by dif-

ferences seen in the weekly food budget for the average citizen. According to Time Magazine, the average American family spends about $341 a week on groceries; however, in countries such as Bhutan or Chad, the average family spends less than $5 a week on food (Menzel and D'Aluisio, 2007). Thus, these countries simply cannot afford most of the fresh or processed foods we are accustomed to in the United States, Europe, and Japan.

This wide-scale global transport of food to economically advantaged countries also comes with a high cost in fuel consumption and greenhouse gas emissions. The relationship between energy consumption and food transportation has recently become known as a "food mile." By determining where a particular food or textile item was originally grown or produced, the consumer can determine how many food miles were actually "used" to get that item from the original location to the consumer's plate. Figure 5.5 shows a comparison of the average number of truck miles traveled by conventional fresh produce between the farm and the table in the continental United States. Here it is apparent that many of the different types of produce that we consume on a regular basis like grapes, broccoli, cauliflower, and peas travel distances over 2,000 miles to reach a destination like Chicago, Illinois, a distance equivalent to almost the entire continental United States traveling east to west (Piroq and Van Pelt, 2002). As mentioned earlier, some of these same types of produce travel even further if they originate outside the continental United States, requiring travel by plane or ship before being imported. Clearly, these long travel distances result in the use of large quantities of fossil fuels, especially if this produce is going by air. By knowing the distance and method of travel for an individual item, the consumer can have a much better knowledge of the energy and environmental impact of that food item and choose more sustainable options.

Domestically, most shipping is done by truck, but rail, ship, and air are also used to transport products around the world. Due to commercial trucking, emissions from domestic cargo transports have risen twice as fast as the emissions attributed to personal transportation as seen in figure 5.6. With a 79.5% increase in carbon dioxide emissions between 1990 and 2007 resulting in over 400 million metric tonnes of carbon dioxide equivalents being emitted into the atmosphere from commercial trucking alone in 2007, it is clear that shipping by truck has a significant environmental impact due to the emissions resulting from the consumption of diesel fuel (US Department of Transportation, 2009).

While domestic transport is important, the availability of fresh produce year-round often requires this produce to be shipped long distances from international locations. Imported items are usually shipped by air or sea. The type of transport chosen to ship produce and other products around the world can have a large impact on the amount of carbon emitted during cargo transport. For instance, shipping by plane can result in over 500 grams of carbon dioxide emitted per ton of cargo per kilometer shipped

SAUDI ARABIA'S ABANDONED AGRICULTURAL INDEPENDENCE

Travis L. Johnson

Saudi Arabia's land is mostly the arid Arabian Desert, and its minimal rainfall has historically prevented it from major food production. In the late 1930s, massive oil reserves were discovered in Saudi Arabia, and it eventually became one of the largest oil producers and exporters in the world, increasing its economic prosperity and potential. With the nation's new wealth, the government turned its attention to food independence and began investing in homegrown agriculture. By the late 1970s and 1980s, the government's Ministry of Agriculture and Water had begun surveying and pumping nonrenewable groundwater to irrigate wheat farms. The government subsidized the technology, equipment, seeds, fertilizers, engineers, and the farm workers for this endeavor, guaranteeing farmers an exorbitant 3,500 riyals ($933.50) per ton of wheat. By 1992, Saudi Arabia had increased its wheat production by nearly 30 times to 4.1 million tons, making it the world's sixth-largest wheat-producing country. They were producing enough wheat to feed themselves as well as export to the Soviet Union and Syria. In order to do this, they increased their wheat-producing land by nearly 14 times and overall irrigated land to 1 million hectares. It is estimated that between 1980 and 1999, 300 billion cubic meters of water—two-thirds coming from nonrenewable sources—were used for irrigation. This program was not sustainable, as Saudi Arabia was producing crops in an area not meant for agriculture using money made from fossil fuels and irrigation from nonrenewable water. Eventually, the Saudi government scaled back their agricultural subsidies due to a decrease in oil prices during the 1980s. By the end of 1996, 76% of the wheat-growing land had been abandoned and production had dropped by 70%. The government downsized their purchasing of domestic wheat and in 2008 announced that it planned to entirely rely on imports by 2016. It is estimated that Saudi Arabia spent $84 billion between 1984 and 2000 on this agricultural project, which does not include indirect subsidies such as farm fuel and electricity or bureaucratic administrative costs. Saudi Arabia's failed attempt at food independence serves as a sober reminder of why sustainability is essential for food and energy production.

SOURCES: Elhadj (2008); Karam (2008).

---

compared to between 10 and 100 grams for trucks, and less than 10 grams for trains and cargo ships (Penner et al., 1999).

Worldwide food transportation has also had a huge impact on the distribution of global populations. Today, immense population centers exist in areas that have no precedence for supporting agriculture. Just as an example, the Middle East has a growing population despite having very few agricultural resources. Many of these countries rely heavily on income from fossil fuel exports to purchase and import foods to meet the nutritional needs of their population. What will happen to these populations when these fossil fuel resources run out?

Balancing Food and Fuel

The agricultural revolution is not nearly over. In the next 40 years, the global population is expected to increase by about 40%, reaching 9.3 billion people. Over this same period, the Food and Agricultural Organization estimates that food and feed production will need to increase by 70% (Hofstrand, 2012). This huge increase in agricultural productivity must also be balanced with a growing biofuel industry that will likely place added strain on the availability of natural resources for food and feed production. The changing biofuels industry combined with the increasing cost of fossil fuels, shrinking agricultural resources like land and water, and the adverse effects of climate change on agricultural productivity are likely to result in a significant challenge for humankind when attempting to meet future agricultural demands.

Understanding and balancing the biofuel industry with agricultural food and feed production is the first step to facing this challenge. In the United States, biofuels are largely produced from two major food resources: corn for the pro-

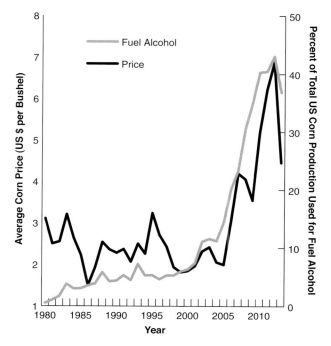

FIGURE 5.7 Comparison between the percent of US corn production used for the generation of fuel alcohol and the average price of corn, 1980–2013. The increase in the price of corn between 2000 and 2012 coincides with a large increase in the use of corn for fuel alcohol production (data from USDA, 2012).

duction of bioethanol and soybeans for the production of biodiesel. At first glance, biofuels may seem like a great solution to the energy crisis, and in particular as a replacement for liquid transportation fuels, but a closer study will reveal that the ability of biofuels to play a role in our energy future without disrupting food and feed production highly depends on the source of the biomass used in biofuel production. Figure 5.7 compares the percent of US corn production used for the production of fuel alcohols like ethanol to the price of corn. Beginning in about 2002, as a larger percentage of corn was used to produce fuel, the price of corn also went up. This price increase in corn carries over into the food market for consumers all over the world. The use of food to produce fuel has the potential to increase food costs by decreasing food availability, and with 2.4 billion people living on less than US$2 per day, increasing the cost of food even slightly can have very severe consequences (World Bank, 2014).

A Sustainable Agricultural Future

Regardless of how agricultural resources are ultimately used for both food and fuel, it will be essential to tap into the vast, accumulated knowledge of best agricultural practices developed by farmers over millennia to increase overall agricultural productivity. It will also be critical to employ those practices in a manner that is sustainable for the environment. To do this, agriculture will need to become

less dependent on fossil fuels. This can occur by identifying new traits bred or engineered into crops to reduce the need for inputs while also increasing yields. Food will need to be transported more efficiently and perhaps over shorter distances, and the waste of food resources will need to be eliminated to maximize the efficiency of food production.

Another important consideration in the future of sustainable agriculture will be the types of crops grown in our changing climate. With average temperature increases, changing rainfall patterns, rising atmospheric CO_2 levels, and growing climatic variability and extreme weather events, the ability to grow specific crops in certain regions will likely change. It will be important to predict how these changes will affect agriculture and make the necessary adjustments in crops to allow for more stable agricultural production going forward.

We have the opportunity to learn from the past and make decisions about agricultural practices for both food and fuel production that will result in a healthy environment. We must understand that food prices are tied to energy prices and that food processing and transportation adds significantly to the energy input for food products. When we plan our food consumption, we must consider that different foods have different carbon footprints. For instance, produce like fresh vegetables have a carbon footprint only a few percent of that of beef (UM, 2014). In addition, we must consider population distribution around the world and how it will be impacted by food availability and the availability of natural resources like water. Finally, we must consider the impact that turning food into fuel will have on the availability, and perhaps more importantly the cost, of food for people around the world. As coming chapters will elucidate, finding a nonfood biomass resource that can be grown on nonarable land, using non-freshwater sources is a central concern for the success of bioenergy in the future.

STUDY QUESTIONS

1. Why might we correlate fluctuating prices of food with fuel?
2. Briefly explain the impact of agricultural advancements on food production as well as the potential limitations of these in the future.
3. Describe the "Green Revolution" and its importance for modern agriculture.
4. Explain why food transportation impacts both the price and the emissions associated with food production.
5. Why is it important to find a balance between food and fuel when considering biofuels in the future?

CHAPTER SIX

The Past and Present of Bioethanol: Corn, Sugarcane, and Cellulosics

As a reminder, biomass is a biological material that is derived from a living organism, a recently living organism, or a metabolic by-product of a living organism (EIA, 2014). Corn is an example of a living source of biomass, chopped wood is an example of a recently living source of biomass, and methane gas emitted from cows is an example of a metabolic by-product of a living organism. All of these biomass sources and many more either are directly biofuels themselves or can be used in the production of biofuels. One of the most common biofuels used today is bioethanol, the focus of this chapter. To understand how bioethanol and other biomass sources are used to produce fuel, it is important to first understand the photosynthetic process that results in the massive quantities of biomass found on Earth.

Introduction to Photosynthesis

Primary production is the biological conversion of chemical energy into biomass and can occur through either **photosynthesis** or **chemosynthesis**. Photosynthesis is the conversion of carbon dioxide into biomass using energy from the sun, whereas chemosynthesis is the conversion of either carbon dioxide or methane into biomass using inorganic molecules such as hydrogen sulfide as a source of energy. In discussing biofuels, this book will focus on photosynthesis.

As can be seen by the equation below, photosynthesis uses the energy from sunlight to drive the fixation of six molecules of carbon dioxide (CO_2) and six molecules of water (H_2O) to form six molecules of oxygen (O_2) and one larger carbon-based organic sugar known as glucose ($C_6H_{12}O_6$). Glucose is used in both storing energy for the plant and providing a biological building block for the production of important plant metabolites such as larger carbohydrates, proteins, and lipids.

$$6CO_2 + 6H_2O + Sunlight \rightarrow C_6H_{12}O_6 + 6O_2$$

Photosynthesis is most often associated with plants but is also an important metabolic process in algae and cyanobacteria. While this process is uniform across these three very different organisms, the housing of the photosynthetic machinery varies based on cell type. In eukaryotic plants and algae, the chloroplast or plastid is a separate organelle that houses all of the machinery needed for photosynthesis. Cyanobacteria are prokaryotes, meaning they do not have separate organelles; therefore, their photosynthetic machinery is spread throughout the cell. This may be better understood by comparing the three different cell types visually in figure 6.1. Here it is apparent that the more simplistic cyanobacterium cell is much less structured and organized when compared to the more complex algae and plant cells. Understanding the differences between these cells is important as this can play an important role in the biotechnological methods that can be used to improve photosynthesis and ultimately biofuel production in each of these organisms. For the remainder of this section, we will discuss photosynthesis as it occurs in the chloroplasts of eukaryotic plant cells, but the basic process of photosynthesis is very similar in all three of these organisms.

Within the chloroplast organelle, one of the most important components of the photosynthetic machinery is the chlorophyll molecule, as it is responsible for capturing sunlight. As shown in the basic diagram of the plant chloroplast in the inset of figure 6.1, the chlorophyll molecules are packed into thylakoid membranes that are then combined into stacks called granum. The thylakoids are often extended to connect with one another via the lamella. The soluble portion of the chloroplast is called the stroma, and this is where key photosynthetic enzymes and the chloroplast DNA are located. A double layer membrane that ultimately makes up the chloroplast boundary then surrounds the stroma.

Photosynthesis consists of two sets of reactions: the light reactions and the dark reactions. While the light reactions

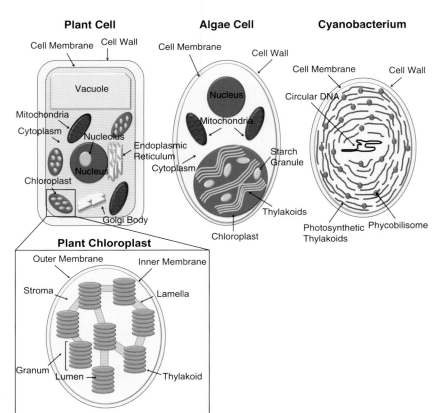

Plant Cell

Cell Membrane — Cell Wall

Vacuole

Mitochondria

Cytoplasm

Nucleolus

Nucleus

Endoplasmic Reticulum

Chloroplast

Golgi Body

Algae Cell

Cell Membrane

Cell Wall

Nucleus

Mitochondria

Starch Granule

Cytoplasm

Thylakoids

Chloroplast

Cyanobacterium

Cell Membrane

Cell Wall

Circular DNA

Photosynthetic Thylakoids

Phycobilisome

Plant Chloroplast

Outer Membrane

Inner Membrane

Stroma

Lamella

Granum

Lumen

Thylakoid

FIGURE 6.1 Basic examples of photosynthetic cells including a plant cell, algae cell, and a cyanobacterium cell. Notice that in the plant and algae cells the chloroplast is a membrane-bound organelle shown in green, but in the cyanobacterium cell the thylakoids are dispersed throughout the cell and much less organized. Inset: magnified example of a plant chloroplast showing the thylakoids forming stacks known as granum and all housed within the double layer membrane of the chloroplast organelle.

as their name suggests are dependent on the availability of sunlight, the dark reactions are actually light-independent reactions that can occur either in the light or in the dark.

We will begin by discussing the light reactions. The light reactions occur in two separate photosystems: photosystem I (PSI) and photosystem II (PSII). Both of these photosystems are embedded in the thylakoids and each contains chlorophyll molecules responsible for the initial absorption of sunlight. These photosystems work together to form an electron transport chain (ETC) along the membrane of the thylakoid as shown in figure 6.2. The goal of the photosynthetic ETC is to transfer the energy of photons from sunlight into the formation of an energy storage molecule known as ATP (adenosine triphosphate) and to produce molecules with a reducing potential known as NADPH (nicotinamide adenine dinucleotide phosphate). As you can see in the figure, sunlight drives PSII to strip electrons from water generating hydrogen protons in the thylakoid lumen to produce oxygen that ultimately oxygenates the planet. The electrons then move through various molecules and proteins along the ETC until eventually being used to reduce NADP+ to create NADPH. As the electrons cycle through the ETC, more and more protons build up in the thylakoid lumen creating a hydrogen proton concentration gradient. Because molecules want to move from areas of high concentration to those of lower concentration, the protons stream through the embedded proton transporter

known as ATP synthase and enter the stroma while at the same time generating the energy molecule ATP. In the end, the light reactions produce oxygen, NADPH, and ATP in preparation for the next set of reactions in photosynthesis, the dark reactions also known as the Calvin cycle.

The Calvin cycle is responsible for the conversion of carbon dioxide (CO_2) gas from the air into glyceraldehyde 3-phosphate (G3P), a building block for sugars and other organic molecules as shown in figure 6.3. This cycle was named after Melvin Calvin, James Bassham, and Andrew Benson at the University of California Berkeley who discovered the importance of these "dark" reactions in the fixation of carbon (Taiz and Zeiger, 2010). While the light reactions occur largely in the thylakoids, the Calvin cycle occurs within the stroma of the chloroplast. The Calvin cycle occurs in three phases where a carbon substrate is fixed, reduced, and regenerated. The cycle begins by using the most abundant enzyme on the planet, ribulose bisphosphate carboxylase (RuBisCO), to catalyze the capture of a molecule of carbon dioxide gas and combine it with the 5-carbon ribulose bisphosphate, generating a 6-carbon intermediate that is quickly split to create the 3-carbon 3-phosphoglycerate. This is the carbon fixation step of the cycle. Next, the 3-phosphoglycerate must be reduced to G3P in a two-step process requiring the ATP and NADPH generated during the light reactions. Some of the G3P can then be used as a building block for sugars and other

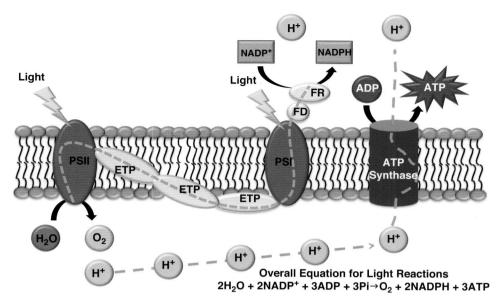

Overall Equation for Light Reactions
$$2H_2O + 2NADP^+ + 3ADP + 3Pi \rightarrow O_2 + 2NADPH + 3ATP$$

FIGURE 6.2 Diagram of the light reactions occurring during photosynthesis in the chloroplast. Diagram shows how electrons are used to reduce NADP⁺ to NADPH (orange dotted line) and how protons derived from water are shuttled through the membrane protein ATP synthase to convert ADP into ATP (blue-green dotted line). Both of these substrates are important in the fixation of carbon during the Calvin–Bensen cycle.

ABBREVIATIONS: PSII – photosystem II; ETP – electron transport proteins; PSI – photosystem I; FD – ferredoxin; FR – ferredoxin reductase.

SOURCE: Campbell and Reece (2002).

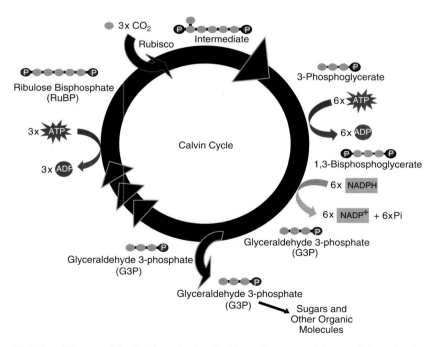

FIGURE 6.3 Diagram of the Calvin cycle showing the cyclic nature of the use of six molecules of carbon dioxide to create glyceraldehyde 3-phosphate (G3P) utilizing NADPH and ATP from the light reactions. The G3P can then be used to create sugars and other organic molecules in the photosynthetic organism (source: Campbell and Reece, 2002).

organic components of the plant cell. However, because this is a cycle, a portion of the G3P must continue through a multistep regeneration process to create the original ribulose bisphosphate. In the end, three molecules of carbon dioxide introduced into the Calvin cycle are used to create six molecules of G3P. Five of these G3P molecules are used in the regeneration of ribulose-1,5-bisphosphate and one G3P molecule can combine with another G3P molecule previously produced to generate glucose (C_3HO_3P), a sugar. The overall Calvin cycle reaction is summarized below:

$$3CO_2 + 9ATP + 6NADPH + 6H^+ \rightarrow$$
$$C_3H_6O_3P + 9ADP + 8Pi + 6NADP^+ + 3H_2O$$

A basic understanding of the two-step process of photosynthesis is a critical foundation to understanding the formation of biomass and other metabolic products derived from biomass, including those used to produce biofuels. But this knowledge goes a step further when you recall that all fossil fuels (coal, petroleum, and natural gas) are products of biomass. Without this intricate combination of reactions ultimately driven by sunlight, there would be no biomass and thus no fossil fuels or biofuels.

Introduction to Biofuels

There are four main types of biofuels including bioalcohols, biodiesels, biochar, and biogas. These different types of biofuels are separated from one another based on the types of molecules that they contain and the biological processes needed to produce these molecules. A common bioalcohol is bioethanol produced from carbohydrates including sugars, starches, and cellulose in plants. Biodiesels are similar to petrodiesel molecules but are typically made from triacylglycerides and fatty acids found in plants. Biochar is any biomass material used directly for combustion such as wood or dung. Finally, biogas can be produced either directly or indirectly as biomethane or biohydrogen by a number of different organisms most commonly employing anaerobic digestion. Over the next four chapters, each of these types of biofuels and their biomass substrates will be discussed.

Biomass, including wood and dung, has been used as a direct source of combustible energy since the discovery and use of fire in the early Stone Age about 1 million years ago (Choi, 2012). Wood is an example of biomass produced directly from plant material and dung is an example of a metabolic by-product generated by living organisms. Since their initial use, combustible biomass resources have continued to play an important role in heating and cooking. Even today these primitive biofuels still represent one of the most utilized renewable energy resources in the world (UN, 2013). Here in the United States, as shown in figure 6.4 about a quarter of all renewable energy consumption is derived from wood biomass combustion and 13% is a result of ethanol.

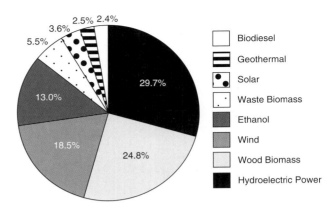

FIGURE 6.4 Renewable energy consumption in the United States in 2013. Each renewable energy resource is shown as a percentage of the total 8.62 quadrillion BTU of renewable energy consumed. Wood biomass makes up about 25% of all renewable energy consumption, while ethanol represents 13% of this consumption (data from EIA, 2014b).

While there is a long history of using biomass materials, the wide-scale use of liquid and gaseous biofuels came about more recently. First-generation liquid biofuels including bioethanol and biodiesel were largely developed as a result of the large-scale availability of agricultural products and their ability to be blended with fuels that work with the internal combustion engine found in most vehicles. Both have continued to play a role in renewable energy consumption since their initial usage in the late 1800s, but only in the last half century has a renewed interest in the research and development of novel sources of biofuels taken off. This renewed interest has spawned second- and third-generation biofuels as an answer to sustained concern over the tension between products that can act as both food and fuel. The remainder of this chapter will turn to bioethanol specifically and its role as a renewable liquid fuel with biodiesel, the other first-generation biofuel, discussed in the next chapter.

Ethanol's History in Recreation and Energy

Bioethanol, more commonly referred to as ethanol, is a clear, colorless volatile liquid that like all alcohols contains a characteristic hydroxyl group (–OH). The chemical structures for ethanol and other common alcohols including methanol, propanol, and butanol are shown in figure 6.5. Here, you can see the characteristic –OH group on the right end of each of these alcohols. The first known use of ethanol is believed to be by the Chinese in 9000 BC, as identified by residues found on pottery. Ethanol was one of the first drugs used by humans and has a long history due to its intoxicating effects. Ethanol's true value as a source of fuel in the United States was realized in 1896 when Henry Ford built the first automobile to run on 100% ethanol. In 1906, a $2 per gallon tax on ethanol, created during the Civil War, was removed, allowing for the affordable larger-scale

BIOMASS FOR THE BOTTOM BILLION

Travis L. Johnson

About 2.4 billion people use biomass for their heating and cooking needs. These are primarily the poorest global communities living in developing nations, and they rely on biomass because they either do not have access to electricity and other fuels or cannot afford them. Although biomass like wood, crop residue, and animal dung are all renewable, they each have drawbacks that cause environmental and health damage, and impede social mobility. First, harvesting wood for fuel leads to deforestation. This reduces the amount of trees and degrades local ecosystems. Second, harvesting and transporting biomass takes a large amount of time—time that could otherwise be used for education, business, or leisure. Third, when biomass is burned for cooking and lighting, particles from the smoke are inhaled and can cause acute lower respiratory damage, resulting in infection and pneumonia. This is the second largest cause of death for African children, killing about 1 million a year. The burned biomass also creates a short-lived climate pollutant called black carbon (soot), which is the most potent greenhouse gas per unit mass of any substance and is the second largest contributor to global warming behind carbon dioxide. Researchers are currently looking at ways to develop and distribute affordable clean stoves for cooking in these communities, which will decrease the emissions of black carbon, help reduce pollution, reduce negative health impacts, and significantly improve the lives of people in the developing world.

SOURCE: McCord (2014); Ramanathan (2014).

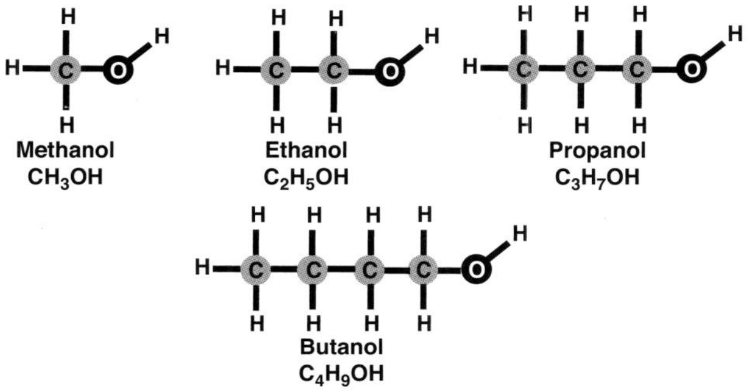

FIGURE 6.5 Basic chemical structures of ethanol (C_2H_5OH) and other common alcohols including methanol (CH_3OH), propanol (C_3H_7OH), and butanol (C_4H_9OH).

production and commercialization of ethanol as a source of fuel. However, the use of ethanol was short-lived due to the onset of Prohibition in 1919, resulting in a severe limitation in the availability of this fuel type. When prohibition came to an end 14 years later, interest in ethanol as a liquid fuel additive once again emerged with a product called "gasohol." However, its success was once again brief due to a production shortage following World War II when grain crops used to produce ethanol were needed for export to support war victims overseas (Ethanol History, 2011).

The most recent recurrence of ethanol as a liquid transportation fuel in the United States resulted from the need for a fuel additive to enhance engine performance, concerns about the effects of fossil fuel combustion on the environment, and concerns about the availability of petroleum-based fuels in the future. In 1973, the newly created US Environmental Protection Agency (EPA) eliminated the use of lead as a fuel additive due to its environmental toxicity and replaced it with methyl tertiary butyl ether (MTBE). MTBE was later found to be toxic and so ethanol was used to replace MTBE as a gasoline additive to enhance fuel performance. In the late 1970s and 1980s, the government exempted ethanol-blended gasoline from the federal fuels tax in order to increase its use during the international oil crisis that limited the supply of oil coming into the United States (MLR, 2009; Ethanol History, 2011).

However, as the oil crisis was resolved and petroleum prices dropped, the ethanol market in the United States once again slumped, forcing half of the ethanol plants to close. Ethanol was reinvigorated yet again by the US Congress in 1990 with the passage of the Clean Air Act, which

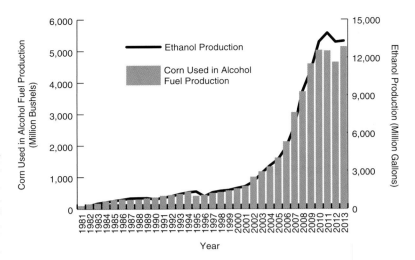

FIGURE 6.6 Comparison of the production of ethanol in the United States with the use of corn for the production of fuel alcohols between 1981 and 2013. The graph shows that the amount of corn used for fuel alcohol production changes at a pace equivalent with the increase seen in the level of ethanol production (data from EIA, 2014c; USDA, 2014).

required oxygen, found in ethanol, to be blended with gasoline to reduce carbon monoxide emissions, a major source of air pollution at the time, by creating a better burning fuel (MLR, 2009). Today, the blending of ethanol with gasoline continues due to its appeal as an inexpensive fuel additive that both increases engine performance and cuts down on carbon monoxide pollution. In addition, the US Federal Government has signed into law federal fuel mandates as of 2007 that require an increased use of renewable fuels. This Renewable Fuel Standard (RFS) requires that at least 36 billion gallons of renewable fuel be blended with transportation fuel by 2022. Ethanol, whether from corn or other biomass sources, plays an important role in meeting this standard today, and will continue to do so in the future (EPA, 2014).

Ethanol as a Fuel Source Today

As mentioned in the previous section, ethanol's main energy-related appeal is its ability to be added to gasoline to raise the performance or octane level of the fuel. The octane rating of fuel represents how much a fuel can be compressed before it spontaneously ignites. By increasing the octane level of gasoline through the addition of ethanol, there is less chance of premature burning before ignition, thus reducing engine knocking and increasing the efficiency of the engine and its performance. However, there are some drawbacks of ethanol as a transportation fuel including its lower energy density when compared to gasoline. By adding ethanol to gasoline, the mileage of a gallon of gas decreases. For instance, a gallon of gasoline has an energy density of about 116,090 BTU per gallon, while a gallon of ethanol has an energy density of about 76,330 BTU per gallon or about two-thirds that of gasoline (DOE, 2012). Thus, a car that can travel 25 miles on a gallon of gasoline would go only 17.6 miles on a gallon of ethanol.

Another drawback to the use of ethanol is that standard car engines cannot be run on 100% ethanol due to compression limitation differences between the two fuels. Most normal combustion engines found in cars today can be supplemented with only up to 10–15% ethanol without altering the construction or operation of the engine (Milnes et al., 2010).

In the United States, ethanol is produced primarily from corn. As figure 6.6 illustrates, there is a direct relationship between the amount of corn used in the production of fuel alcohols and the level of ethanol production. The use of corn and the production of ethanol gained huge momentum beginning in 2005 largely due to the establishment of the first iteration of the RFS. The original RFS mandated that 7.5 billion gallons of renewable fuel be blended into gasoline by 2012, but as technology advanced, this was later revised to the much higher 36 billion gallons by 2022 (EPA, 2014). With ethanol being one of the most highly developed and used biofuels in the United States, this mandate meant that ethanol producers obtained market certainty, leading to ethanol refineries multiplying throughout the country. But concerns about the price of grain and the impact of the use of corn for ethanol production on the availability of food resulted in the revised mandate, created in 2007, focusing on commercializing advanced biofuels that are not solely dependent on the use of food crops for fuel production (i.e. second- and third-generation biofuels such as cellulosic ethanol). These advanced sources of ethanol will be discussed in greater detail later in this chapter.

Most of the bioethanol produced around the world currently comes from one of two sources: corn or sugarcane. For many years, sugarcane dominated the ethanol production market, largely coming from Brazil. Even today, Brazil has a very successful alternative fuels program, replacing nearly 40% of its gasoline with sugarcane-derived ethanol (Sugar-

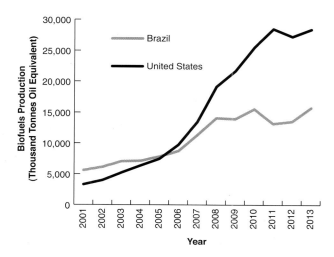

FIGURE 6.7 Comparison of biofuel production levels in the United States and Brazil between 2001 and 2013. The production of biofuels in the United States, largely dominated by the corn-based ethanol industry, surpasses Brazil's biofuel production, largely dominated by the sugarcane-based ethanol industry, in 2005. The production of biofuels in the United States has continued to increase at a much faster rate than the production of biofuels in Brazil (data from BP, 2014).

Cane.org, 2014). However, in 2005 with the establishment of the RFS, the production level of ethanol from corn in the United States quickly overcame the production of biofuels in Brazil, as depicted in figure 6.7. By 2009 US ethanol production used 3.8 billion bushels of corn to produce 10.6 billion gallons of ethanol, and also produce 30.5 million metric tons of distillers grains and corn gluten used in livestock feed. This number rose to 13.3 billion gallons of ethanol in 2013 (ACGF, 2011; RFA, 2014). While the RFS mandate has boosted the US biofuels industry, it has also been complicated by changes in specific requirements within the mandate when considering corn- and sugarcane-derived ethanol. As mentioned previously, this mandate requires that increasing amounts of ethanol be obtained from advanced sources every year going forward. Sugarcane-derived ethanol is considered an advanced source of ethanol; thus, corn-based ethanol producers in the United States are actually exporting their ethanol to Brazil and then importing sugarcane-based ethanol. This allows these ethanol producers to technically be meeting a higher level of advanced biofuel production. This import and export process has the potential to negatively impact the development of advanced biofuels within the United States (Colman, 2014). It will be up to the US government to set future standards to help balance this complicated ethanol trade business.

Ethanol Production

Corn and sugarcane remain the feedstocks of choice for ethanol production due to their high sugar content. Ethanol is made from the **fermentation** of these sugars, usu-

ally either glucose or sucrose. In the case of sugarcane, the juice found within the sugarcane called "garapa" contains 10–15% pure sucrose, which can be used directly in the fermentation process (Basso et al., 2011). Using corn, however, is a little more complicated because the sugars are locked up in starch within the protective kernel of the corn. Starch is a long chain of glucose molecules that must be broken down into individual glucose units in order to be used during fermentation. As diagrammed in figure 6.8, starch can be broken down through an acid hydrolysis reaction where individual glucose units shown in black are cleaved from the longer starch chain shown in gray. Starch can also be broken down through an enzymatic reaction using the enzyme amylase. While available commercially, amylase is also found naturally and is crucial to the breakdown of starchy foods by the human metabolism.

Once individual glucose sugar units are separated, fermentation can be used for the production of ethanol. Fermentation is a metabolic process where biological molecules like glucose are converted into other metabolic products including alcohols. For the production of bioethanol, yeast is used for the fermentation process diagrammed in figure 6.9. Yeast uses glucose and sucrose to generate the intermediate metabolite pyruvic acid through a metabolic process known as glycolysis. Pyruvic acid can then be reduced to form ethanol. Overall, the fermentation process has an average efficiency of about 52% where theoretically 1.1 grams of simple sugars like glucose can yield 0.57 grams of ethanol (Borglum, 1980).

While fermentation may seem rather simple, the commercialization and production of such large quantities of ethanol have required a massive scale-up of these processes and required scientists, farmers, and engineers to make the process from beginning to end as efficient as possible in order to reduce the overall cost. In the United States, ethanol production begins with the harvesting of corn from the fields and the removal of any contaminating substances like dirt, cobs, or broken corn kernels from the mix through sifting. Following this initial cleaning, the corn must be prepared and the starch within the kernel made available for fermentation. This occurs in one of two ways: **wet milling** or **dry grinding**. The difference between these two techniques is largely based on a step known as **steeping** where the corn kernels are soaked in a mixture of water and dilute acid to break down their tough outer shells, allowing better access to the sugars inside. During wet milling, the steeping process occurs immediately following the initial cleaning. Once steeped, a grinding process is used to disassociate the softened outer shell of the corn kernel from the other metabolites including the carbohydrates, proteins, oils, and minerals. Water can then be added to this mixture and centrifuged to separate the starch from the other components. Once separated, the starch undergoes three sequential processes of **liquefaction**, **saccharifaction**, and fermentation as described below (Mosier and Ileleji, 2006).

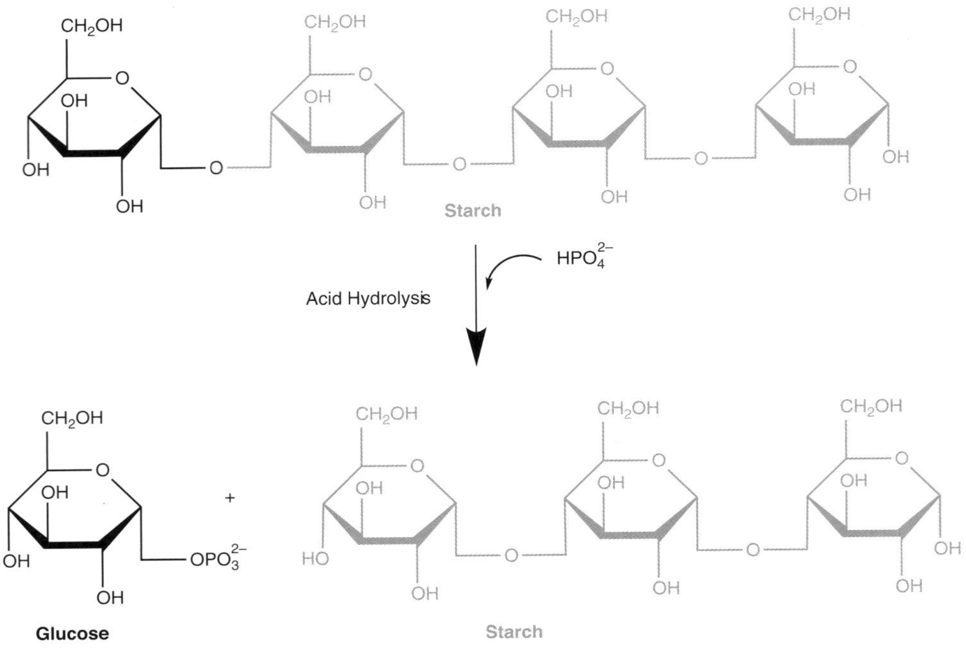

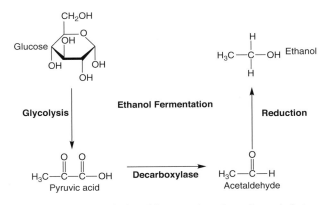

FIGURE 6.8 Diagram of the acid hydrolysis of starch (gray) into individual glucose molecules (black) using an acid catalyst (HPO_4^{2-}).

FIGURE 6.9 Diagram of ethanol fermentation where glucose is first converted into the intermediate pyruvic acid through a process called glycolysis. The pyruvic acid is then decarboxylated and reduced to form the final product ethanol.

By comparison, about 70% of US ethanol is made using the dry grind process rather than the wet milling process. In dry grinding, there is no steeping step. The corn kernels are submitted to a coarse grinding and the starch separated from the other components by sifting before undergoing the liquefaction, saccharification, and fermentation processes. In all, the corn kernel consists of about 84.1% carbohydrates, 9.5% protein, 4.3% oil, and 1.4% minerals, and each of these components can be used to produce different products that add value to the ethanol fermentation process (Mosier and Ileleji., 2006). For example, corn fiber can be used as a basis for corn feed and the protein components

can be used to produce gluten meal, both important agricultural products.

Whether the corn kernel is prepared by the dry grind or wet mill method, the processes of liquefaction, saccharification, and fermentation are very similar. During liquefaction, the starch separated during wet milling or the coarse corn kernel material created during the dry grind are mixed with water containing ammonia and lime to create slurry. This slurry is heated and the enzyme amylase is added to cleave the starch molecules into smaller units. The ammonia and lime in the liquefaction process create an environment with a pH of about 6.5 optimal for the enzymatic reaction, but the next process, saccharification, requires a pH of only 4.5. To achieve this lower pH, sulfuric acid is added to the slurry followed by a second enzyme known as beta-amylase. The role of beta-amylase is to break down the smaller pieces of starch into individual glucose units as is diagramed in figure 6.8. Once the saccharification process is complete, the slurry is called mash and is ready for fermentation (Borglum, 1980; Mosier and Ileleji, 2006).

Fermentation begins with the addition of microorganisms, usually a yeast species like *Saccharomyces cerevisiae*, to the slurry. This is the same type of yeast that is used for brewing beer and baking bread and its one job is to convert the sugar glucose into ethanol. From the fermentation chemical equation below, you can see that one molecule of glucose produces two molecules of ethanol and two molecules of carbon dioxide while utilizing and producing other metabolic components:

$$C_6H_{12}O_6 + 2 \text{ ADP} + 2 \text{ NAD}^+ + 2\text{Pi} + 2\text{NADH} + 2\text{H}^+ \rightarrow$$
(Glucose)

$$2C_2H_5OH + 2 \text{ ATP} + 2\text{NADH} + 2\text{H}_2\text{O} + 2CO_2$$
(Ethanol)   *(Carbon dioxide)*

In most cases, the final ethanol fermentation step results in a mixture containing about 8–12% ethanol (Basso et al., 2011). To generate a purer ethanol, a final separation step, known as distillation, must occur. This distillation step is the same as that described in Chapter 2 for purifying different hydrocarbons from crude oil. During distillation for ethanol production, the fermented liquid portion of the mash is heated. Ethanol boils at a lower temperature than water; therefore, ethanol will boil and evaporate before water. As ethanol evaporates, it will condense within a cold trap and drain into a secondary container. In the end, this process results in 92–95% ethanol and can be even further purified by using molecular sieves, an absorption material, to absorb the water that comes over during distillation, resulting in a final product that is 99% pure ethanol (Mosier and Ileleji, 2006).

While the above processes focus on the production of ethanol from corn, the method of creating ethanol from sugarcane is not significantly different. Sugarcane contains 10–15% sucrose as a cane juice, which can be separated from the other parts of the plant like the stems and leaves known as bagasse. The cane juice is filtered and pasteurized prior to a precipitation step to create a molasses that is very high in sugars. The sugars, particularly sucrose, can be fermented by yeast in the same manner as described above for corn (Basso et al., 2011). One of the differences between the preparation of ethanol from corn and sugarcane comes in the distillation process. In corn-based ethanol production, distillation is typically driven by heat from a fossil fuel source, largely either coal or natural gas, while in the case of sugarcane-based ethanol, distillation is usually driven by the burning of the bagasse. This difference plays a large role in the relative ecological footprints from ethanol produced from corn versus sugarcane because corn-based ethanol production results in higher carbon dioxide emissions due to the use of fossil fuels.

The Ethanol Dilemma in the United States

In the end, whether the feedstock is sugarcane or corn, the result is the same: ethanol. Most ethanol used in the United States is derived from corn because not only is corn a prolific starch producer, but it is also the most widely produced feed grain in the United States. In 2013, US farmers harvested about 87.7 million acres of corn (Capehart, 2014). Yet, the abundance of corn production does not wholly account for the strong US ethanol market. The ethanol market remains highly supported by federal policy, not only through federal gasoline mandates, but also through agricultural support programs dating back to the 1930s.

Although only 1% of the population works on a farm, the United States still relies heavily on agriculture for the pro-

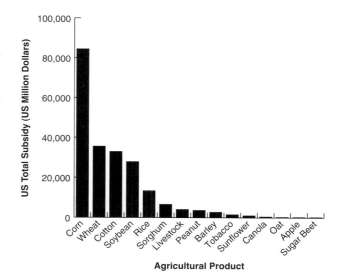

FIGURE 6.10 Graphical comparison of total US farm subsidies for individual agricultural products between 1995 and 2012. Corn is the most highly subsidized agricultural product over this time period (data from EWG, 2012).

duction of food both to support domestic consumption and as an export commodity (EPA, 2013). The value of agriculture is obvious when you consider the way the government subsidizes agricultural commodities. Figure 6.10 compares total US farm subsidies across different crops between 1995 and 2011. This graph clearly shows that although the United States subsidizes many different agricultural commodities, corn has easily received the largest subsidized support over this time period. According to the Environmental Working Group, the corn industry receives a staggering $2–4 billion in government support every year (EWG, 2012). Many of these subsidies were put into place primarily because of the role corn plays as a food source; however, the amount of the corn used to produce fuel alcohols has risen significantly in recent years, as was shown in figure 6.6, reducing the overall value of the subsidy toward food production. In the end, subsidizing corn as an agricultural food product indirectly subsidizes the production of ethanol, a controversial part of current farm bills debated by the US government.

The increase in the use of corn for ethanol production results in less corn available for food and feed, raising the price of food and placing enormous pressure on the food market. Figure 6.11 shows how, as the production of corn-based ethanol increased, the price of corn also rose between 2005 and 2012. If the ethanol market drives up the cost of corn, then the cost of food also rises, which can ultimately result in more people around the world being pushed into poverty. The market for corn just in the United States is also very large as it is a major source of feed for animals and an important source of food both as a grain and through its by-products like high-fructose corn syrup. Fast food chains in the United States rely heavily on this cheap source of food; in fact, almost every food item sold at some fast food

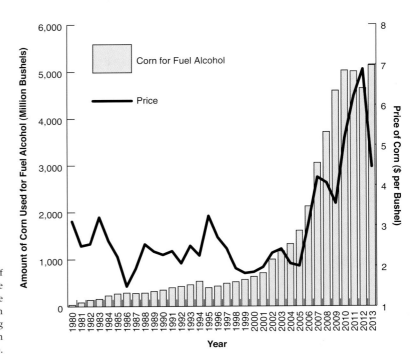

FIGURE 6.11 Graphical comparison of the amount of corn used in the production of fuel alcohol with the price of corn between 1980 and 2013. The increase in the price of corn largely follows the same pattern as the steep increase in the amount of corn being used in the production of fuel alcohol between 2005 and 2012 (data from USDA, 2014).

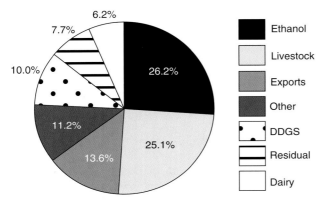

FIGURE 6.12 Comparison of end products for corn use in the United States in 2013. Ethanol and livestock make up over half of the end products utilizing corn. DDGS – distiller's dried grains with solubles (data from IowaCorn, 2014).

chains can be traced back to having an association with corn. Amazingly, about 70% of all processed food found in American grocery stores contains some corn-derived product. As mentioned, corn is also important as an animal feed with 93% of the beef used in burgers being associated with a corn-based diet (Pollan, 2007; Biello, 2008; Jahren and Schubert, 2010). The competition for corn as a source of both food and fuel is apparent in figure 6.12. As of 2013, ethanol and livestock are split almost equally in their use of over 50% of the corn harvested in the United States. The remaining 50% of the corn is used for other products including dairy production, exporting, and dried distillers grains and solubles (DDGS), a by-product of the corn mash that is used as a livestock feed.

The tension between food and fuel vis-à-vis corn-based ethanol is only one of the many reasons why corn is not the ideal candidate for further expansion of ethanol production. Given the role corn plays in the food supply, it is likely that ethanol from corn could never replace more than about 10% of the total gasoline demand in the United States (Cooper, 2012). Furthermore, corn-based ethanol also has some serious environmental concerns that make it less desirable as a renewable source of fuel. Corn is grown largely in the Midwest, and in order to obtain the efficiency of plant growth needed to meet current demand, farmers must use large amounts of fertilizer and water. Typically, 785 gallons of freshwater are used in irrigation per gallon of ethanol produced (Aden, 2007). With climate change already stressing the freshwater supply, it seems unlikely that the use of this much water to produce a fuel source is sustainable. In addition, the fertilizers, such as nitrogen, used to provide the nutrients needed by corn to grow properly have their own deleterious environmental impacts. Runoff from fertilizers leaking down through the soil into the water supply results in **eutrophication**. Eutrophication and other environmental impacts of biofuel production are discussed in more detail in Chapter 12. Even though corn-based ethanol is termed a biofuel and sounds like it should be an environmentally benign source of fuel relative to fossil fuels, it is marked with many problems that could be greatly exacerbated with any further expansion of the industry.

A benefit of biofuels as fuel sources should be a reduction in carbon emissions over the life cycle of these products compared to fossil fuels. One of the central points of biofuels is to lower the carbon dioxide levels in the atmosphere

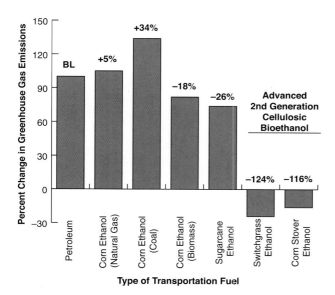

FIGURE 6.13 Comparison of greenhouse gas emissions for transportation fuels. Each alternative fuel is compared to petroleum fuels as a baseline (BL). The percent change in emissions over a 30 year period based on current alternative fuel generation methods is shown above each column. Switchgrass and corn stover represent advanced alternative cellulosic biomass sources that could help sequester carbon dioxide from the atmosphere during bioethanol production (data from EPA, 2009).

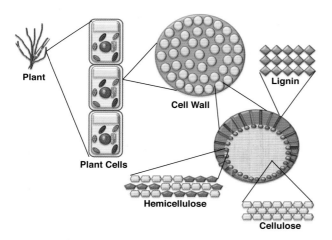

FIGURE 6.14 Diagram of plant cell wall components including cellulose, hemicellulose, and lignin. Cellulose and hemicellulose are structural components that have the potential to be used as alternative biomass sources for bioethanol production. Lignin is not used for bioethanol production but is an impediment to the production of this cellulosic ethanol; thus, its presence in a potential biomass source must be considered.

and decrease the risks associated with global warming. Ideally, a biofuel such as ethanol will be **carbon neutral**, where, although burning the fuel may emit carbon dioxide, the growth of the plant used to produce the biofuel should offset this emission by sequestering an equivalent amount of inorganic carbon. However, the process of creating ethanol from corn is not a carbon neutral process, mainly because a significant amount of inputted energy, usually derived from fossil fuels including coal and natural gas, is needed to produce and distill the ethanol. Figure 6.13 compares the greenhouse gas emissions of alternative fuels including corn-based ethanol with emissions from petroleum-based fuels. The graph shows that these emissions are dependent on the source of the energy used to produce ethanol. While using natural gas and coal for the input energy increases the greenhouse gas emissions compared to petroleum by 5% and 34%, respectively, burning the excess biomass from corn or sugarcane to obtain this energy decreases the emissions by 18% and 26%, respectively. Sugarcane therefore represents a lower emission biofuel. However, the tropical and humid conditions needed for the growth of sugarcane limits its utilization in the United States and Europe, thus the use of corn. The best source of ethanol in terms of emissions may be another alternative or advanced source derived from cellulosic biomass feedstocks.

Cellulosic Ethanol as a Fuel Source

As described, first-generation bioethanol is produced from the small dimeric sucrose sugars found in sugarcane juice and the long glucose chains of starch found in the kernels of corn; however, plants have other components that can also be considered for the production of bioethanol. For many years, additional structural components composed of sugar units and other biomolecular polymers have been ignored for their potential in biofuels development. But recent shifts in societal, political, and economic views toward corn-based ethanol and questions as to the ability of this resource to meet future demand have resulted in a renewed interest in these other plant molecules.

The main structural components of plants can be summarized into three classes of biopolymers: cellulose, hemicellulose, and lignin; and these molecules will be the focus of the last part of this chapter. These three biopolymer components important in the structural stability of a plant cell are illustrated in figure 6.14.

Lignin is a polymer of cyclical units of molecules that function in giving a plant strength and stiffness. Lignin is not used for the production of bioethanol; however, due to its strength and resistance, lignin is considered an impediment to unlocking and accessing the other structural components of a plant for biofuel production. Therefore, many avenues of research are being conducted to find biomass sources low in lignin content as well as methods to break down lignin in an effort to create better bioavailability of other plant biopolymers.

Hemicellulose is the second most abundant biopolymer in plants. It is not particularly well defined, but it is considered to be the structural component of plants containing 5- and 6-carbon monosaccharide (single sugar) units. As will be discussed in Chapter 10, there is the potential to use hemicellulose in the production of biofuels, particularly alcohols, but biotechnological methods will need

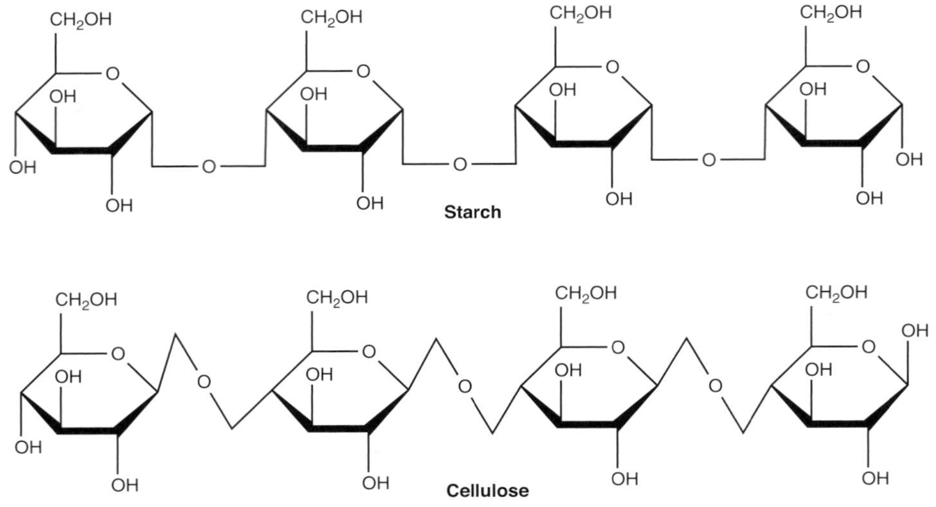

FIGURE 6.15 Chemical structures of starch and cellulose. Starch contains an alpha linkage between each glucose unit, while cellulose contains a beta linkage between each glucose unit.

to be advanced in finding organisms that can efficiently make alcohols from the mixture of 5- and 6-carbon sugars found in this molecule.

As depicted in figure 6.14, the main structural component of plant cell walls is **cellulose**. Cellulose is one of the most abundant biomolecules on Earth, and plants can make hundreds of billions of tons of this material every year. Cellulose is made of a long chain of glucose molecules. While at first glance it may seem that cellulose and starch have very similar structures, they have one significant molecular difference, the formation of the bond between each glucose unit. This variation in their formation, as depicted in figure 6.15, results in the need for different enzymes to process these two biopolymers. Upon close inspection of these two structures, the oxygen linkage between individual glucose molecules is discernibly different. In starch, this linkage is called an alpha linkage, while in cellulose the linkage is a beta linkage. This depiction of the chemical structure may make it seem like a minor structural difference, but in fact it makes a huge difference, as many organisms including humans are incapable of breaking the beta linkage, and are therefore incapable of digesting cellulose.

When a corn kernel is used to produce bioethanol, it is first broken open to allow the inside starchy portion to come in contact with the enzyme amylase. This enzyme can break the long chains of starch into smaller glucose components. The same occurs in the human digestive tract. When you eat a potato, which is made primarily of starch, amylase found within the digestive system can easily come in contact with the starch and begin to break it down into smaller glucose components. When eating corn, however, whole kernels can pass through one's digestive tract and be

discharged intact. This occurs because the outside of the corn kernel is made up primarily of cellulose. The enzymes in the human digestive tract including amylase are unable to break down the cellulose in the corn kernel wall to gain access to the starch inside. So, while bioethanol research in the past primarily focused on the alpha linkages found in starch to unlock glucose, recent studies have turned toward gaining a better knowledge of breaking down the beta linkage found in cellulose. In the end, once the alpha or beta linkage is released, the result is the same: a glucose molecule that can be used in fermentation to create bioethanol for use as a transportation fuel.

Biomass Sources for Cellulosic Ethanol

Because cellulose is one of the most abundant molecules on the planet, it should not come as a surprise that there are many sources of biomass having the potential to be developed into cellulosic ethanol. One source, discussed earlier but for a different function, is sugarcane bagasse, the leafy stalks that surround the actual sugarcane. Currently, the large amount of sugarcane bagasse left over once the sugarcane is extracted for ethanol fermentation is burned to provide the heat needed for distillation (Basso et al., 2011). This process results in a significant lowering of the energy inputs needed and an ethanol product that is both cheaper and environmentally friendlier as was shown in figure 6.13 in terms of emissions. Yet, bagasse is also a form of cellulosic biomass. The value of bagasse as a combustible fuel versus a feedstock for cellulosic ethanol will need to be carefully evaluated to determine which method results in the most economical and environmentally beneficial production of bioethanol.

TABLE 4
Approximate percent composition of structural biomass components for feedstocks
being considered in cellulosic ethanol production

Biomass Source	Cellulose (%)	Hemicellulose (%)	Lignin (%)
Softwood	35–40	25–30	27–30
Hardwood	45–50	20–25	20–25
Switchgrass	30–50	10–40	5–20
Corn Stover	38	26	23
Miscanthus	43	24	19

DATA FROM: Gonzalez (2011).

In the United States, the largest source of cellulosic biomass that is presently available is corn stover. Corn stover includes the stalks, leaves, and cobs that remain in the field after the harvest of corn. It is estimated that the United States produces nearly 120 million tons of corn stover every year (Koundinya, 2009). While some stover must be allowed to decay and return nutrients into the soil for the next crop, large amounts of this corn stover could be used in the production of cellulosic ethanol without impeding the use of the corn itself as a source of food.

Two other herbaceous grasses are also being studied for their potential in the production of cellulosic bioethanol, switchgrass, and giant miscanthus. Both of these grasses would be grown specifically for the production of bioethanol on marginal lands as opposed to fertile and arable land needed for agricultural food production. There are also some woody species of plants being considered in the development of cellulosic ethanol including poplar, willow, and eucalyptus. These woody tree species are known for being able to produce a large amount of biomass very quickly; for instance, poplar can produce up to 11.2 dry metric tons per hectare and willows can produce up to 10.5 dry metric tons per hectare. In comparison, corn grown in Iowa on average results in 9.4–11.3 metric tons per hectare with a large portion of this biomass being the corn kernel important in food production (Elbheri et al., 2008; Schnepf, 2010; Johanns, 2014).

Whether an herbaceous grass or a woody species is chosen for the production of bioethanol, the chosen feedstock should have certain qualities including a high rate of growth, an ability to grow in various soils and climatic regions, and the ability to be harvested using machinery. These qualities result in a cheaper biomass and translate into a cheaper cellulosic ethanol. However, the main consideration in choosing a plant species for the production of cellulosic bioethanol is its composition of cellulose relative to other structural components like hemicellulose and lignin. Table 4 compares the structural biomass composition of various plants being discussed in the production of cellulosic ethanol. As you can see, biomass materials such as hardwood, miscanthus, and switchgrass have the potential for higher percentages of cellulose. However, it is also very important to have a lower percentage of lignin, because high lignin content reduces the release of polysaccharides during the pretreatment process. In the end, a large amount of lignin often means significantly more work in obtaining the glucose from cellulose and thus a much higher price for the cellulosically derived ethanol.

Generating Cellulosic Ethanol

There are many technologies that are still being examined to identify the most economic method for the production of cellulosic ethanol, including concentrated and dilute acid hydrolysis, gasification, fermentation, catalytic conversion, and enzymatic hydrolysis (Dwivedi et al., 2009). In the end, all of these technologies can be grouped into one of two base technologies: **biochemical** and **thermochemical**. Thermochemical technologies generally use the feedstock or cellulosic biomass source in a gasification process to produce synthesis gas or syngas. Syngas is a mixture of gases, mainly carbon monoxide and hydrogen, and is often used as starting material in the production of transportation fuels and other products. In the case of cellulosic biofuel production, once syngas is produced from the cellulosic biomass source, it can be used to produce mixed alcohols by the Fischer–Tropsch process (Johnson et al., 2010).

In the biochemical approach, polysaccharides are broken down into individual sugar molecules enzymatically and then the sugar molecules are used in fermentation, similar to what has been discussed for corn or sugarcane previously in this chapter. There are six basic steps for the biochemical production of cellulosic ethanol (Johnson et al., 2010). The first step is the harvest. As discussed in the last chapter on industrial agriculture, harvesting biomass became much easier with the development of the plough, tractor, and other agricultural machinery. These technologies are important in the development of cellulosic ethanol; however, it

BREAKING DOWN CELLULOSE WITH THERMOPHILIC BACTERIA

Travis L. Johnson

One of the toughest challenges in producing cellulosic ethanol is breaking down the cellulose to fermentable sugars. Cellulose is found in plant cell walls and helps give plants their rigid structure. Bioethanol producers have found that raising the temperature during the breakdown phase can quicken the process and make it more efficient. So, researchers are bioprospecting for microbes that can decompose cellulose and withstand temperatures as high as boiling water: hyperthermophiles. Using these special bugs during the process will essentially decrease contamination because other organisms are not able to survive in these conditions. Some of the most promising of these organisms are *Caldicellulosiruptor bescii*, a bacterium found in a Russian volcanic spring that can break down raw unprocessed biomass; *Halanaerobium hydrogenoformans*, a salt-loving microbe found in Washington State's Soap Lake and capable of producing hydrogen from sugars in biomass; and *Pyrococcus furiosus*, a hyperthermophile found in a thermal vent off Volcano Island in Italy, whose cellulase (enzymes that break down cellulose) can tolerate 100 degrees Celsius.

SOURCE: Jaret (2015).

---

will be necessary to develop new mechanisms to efficiently harvest cellulosic biomass sources, be they trees or grasses, in a variety of land topographies different from conventional agriculture. In the end, harvestability may play a key role in choosing a biomass source for large-scale cellulosic ethanol production.

The next step in the production of cellulosic ethanol is size reduction. Many of the biomass sources being considered are rather large and have a more rigid structure than biomass sources considered in the past. As a result, many of these feedstocks will need to be mechanically cut in order to facilitate access by enzymes and reagents. The size reduction step is likely to figure in significantly to the total energy requirements needed in the production process, with different biomass sources requiring differing amounts of energy depending on how much an individual source needs to be cut. As an example, hardwoods such as cottonwoods and willows require about 130 kilowatt-hours per metric ton, while straw requires only 7.5 kilowatt-hours per metric ton to reduce their sizes to small accessible pieces (Carroll et al., 2009).

Once the feedstock has been reduced in size, it is ready to begin pretreatment, the third step in the biochemical production of cellulosic bioethanol. This step usually involves a treatment with steam or dilute acid (0.9% H_2SO_4) in order to soften the hard biomass and reduce the level of polymerization and structure within the feedstock. Softening is important because it will begin to break down the protective layers of lignin and hemicellulose in the plant cell walls to make the cellulose more accessible (Carroll et al., 2009).

Finally, the feedstock must undergo cellulose hydrolysis where the individual molecules of glucose locked in the cellulosic chains are released. Biochemical hydrolysis occurs through the use of three important types of cellulase enzymes: endocellulase, exocellulase, and beta-glucosidase. Endocellulase begins by disrupting the overall structure of the cellulose. As depicted in figure 6.14, cellulose is found as bundles of long chains of glucose. Separating the chains from one another is important in allowing the other enzymes access to their desired substrate. Once the overall structure is disrupted by endocellulase, exocellulase begins chopping the cellulose chains into shorter polysaccharide chains. These shorter chains are now a good substrate for the final enzyme beta-glucosidase, which cleaves away individual glucose molecules that can then be used for fermentation. Another enzyme known as hemicellulase can also be used to cleave hemicellulose into its various individual sugars including xylose, arabinose, galactose, glucose, and mannose. Using this enzyme in addition can sometimes increase the efficiency of fermentation.

Biochemical hydrolysis requires large amounts of enzymes, around 25 kilograms per ton of cellulose (Carroll et al., 2009). The extreme quantity of enzymes needed represents the single largest cost in the production of cellulosic fuels and one of the largest impediments to the economics of the production of cellulosic ethanol. Just as an example, the cost of the enzymes alone accounts for $0.10–0.40 per gallon of ethanol produced, and this is not considering the cost of the process to make sure the enzymes are working efficiently, like temperature or pH controls (Klein-Marcuschamer et al., 2012). Thus, the success of cellulosic ethanol is likely to hinge upon biochemical advances that decrease the cost of the enzymes involved in processing.

The last three steps in the processing of cellulosic feedstocks in the production of cellulosic ethanol are the same as the last steps used in the production of corn- or sugarcane-derived ethanol including microbial fermentation of the sugars, distillation to produce 95% pure ethanol, and finally dehydration of the ethanol to a concentration of 99.5% ethanol. While these steps are the same, there are some important considerations being studied in relation to improving the fermentation process specifically for the use of cellulosic feedstocks. One of the issues is the formation of toxic by-products during the acid pretreatment process of cellulosic feedstocks. This pretreatment process releases low molecular weight fatty acids, aromatic compounds, and dehydrated sugars that are toxic to yeast during fermentation (Carroll et al., 2009). In order to optimize bioethanol production from cellulosic biomass sources, it will be necessary to eliminate these toxic compounds perhaps through genetic modifications to the feedstock plants. Another concern with fermentation of cellulosic feedstocks is the variety of sugars present, specifically those that differ from the traditional glucose-derived 6-carbon sugars that yeast can accept. The percentage of total carbohydrate in cellulosic feedstocks range from about 61.5% to 69.6%, and cellulose represents on average only about 40%. This leaves 20–30% of the structural biomass wasted during the production of cellulosic ethanol (Gonzalez, 2011). To maximize the value of cellulosic biomass sources and keep the cost of cellulosic bioethanol as low as possible, it will be necessary to find specialized organisms capable of fermenting a variety of carbohydrate substrates, as well as identify a valuable use for the large amounts of lignin that are left over after cellulose degradation.

The Future of Cellulosic Ethanol as a Fuel Source

The technologies required for production of bioethanol from cellulosic biomass sources have not advanced to a level where ethanol production from cellulose can compete economically with ethanol production from corn. However, cellulosic sources offer a number of benefits over corn sources. For one, due to the inability of humans to digest cellulose, cellulosic materials represent inedible parts of a plant, sidestepping the food versus fuel tension. Cellulose is also one of the most abundant organic materials on Earth. Not only is it abundant, but also many regions of the world, including the United States, already have access to large amounts of cellulosic waste. Estimates suggest that 1.3 billion tons of cellulosic waste may be available yearly in the United States alone. If we predict a yield of about 50 gallons of ethanol per ton, then that is equivalent to 50 billion gallons of ethanol from cellulosic waste (Perlack et al., 2005). In 2014, the United States used 137 billion gallons of gasoline, so clearly there is a huge potential for ethanol derived from cellulosic waste to make an impact on renewable fuel use (RFA, 2014).

Cellulosic biomass sources also have a positive effect on the environment in a number of ways. Many of the dedicated cellulosic biomass sources being studied are known as C4 plants. C4 plants, which also include both corn and sugarcane, tend to grow efficiently and tolerate high temperatures. The plants being considered as cellulosic ethanol sources are generally deep-rooted plants capable of sequestering carbon in their roots as well as in their leaves, an important environmental benefit. **Sequestration** occurs due to extensive root systems that can access deeper water sources, and these deep roots also transfer carbon deep into the soil. A field of switchgrass can sequester 1,430 kilograms of carbon dioxide per hectare per year (Skinner and Adler, 2009). This sequestration of carbon means that cellulosic-derived fuels from sources like miscanthus and switchgrass have significantly reduced levels of net carbon dioxide emissions as compared to petroleum gasoline, as shown in figure 6.13. Although corn- and corn-stover-derived ethanol helps reduce greenhouse gas emissions, these new-dedicated cellulosic sources of biomass such as miscanthus and switchgrass can actually act as a net carbon sink, helping to decrease the amount of carbon dioxide already in the atmosphere.

Predictably, cellulosic ethanol also has some challenges that must be addressed before it will be competitive with fossil fuels or even with other biofuels. As mentioned, the largest challenges are finding ways to economically unlock cellulose from its entrapment by insoluble fibers such as lignin, and fermenting the variety of sugars found in hemicellulose. Finally, the cost of the enzymes required for cellulose hydrolysis must be lowered in order to create a cost competitive product.

Despite these challenges, cellulosic ethanol is still expected to continue to advance in the coming years. As can be seen in figure 6.16, the US Energy Independence and Security Act projects that the development of cellulosic ethanol will allow for a production expansion from 0.1 to 16 billion gallons of fuel per year within the next few years (Schnepf and Yacobucci, 2012). But there is presently no indication that this level of production can be achieved for cellulosic ethanol without some significant biological and technical breakthroughs. Reaching these production

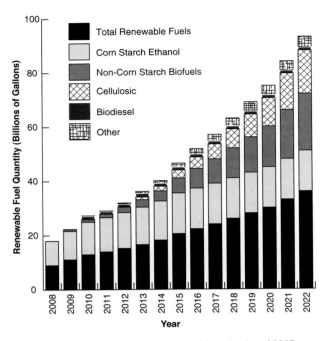

FIGURE 6.16 The Energy Independence and Security Act of 2007 adjusted renewable fuel standards (RFS2). These standards represent the goals for the production of renewable fuels between 2008 and 2022 culminating in the production of 36 billion gallons of renewable fuels annually (data from Schnepf and Yacobucci, 2012).

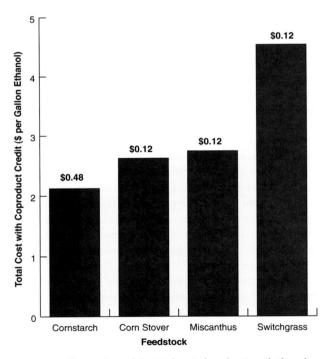

FIGURE 6.17 Comparison of the total cost of production of ethanol from cornstarch and other cellulosic feedstocks, 2008. Totals have been adjusted with coproduct credits and these credits are shown above columns. The price of corn assumes a $5 per bushel of corn market price (data from Khanna, 2008).

volumes will likely depend on the development of a diverse supply chain of cellulosic feedstocks and a decrease in the overall price of growing and preparing the feedstocks. By comparing the farm gate costs of high-level productivity for cellulosic crops in Illinois, Khanna (2008) found that switch grass ($96.88 per ton dry matter [t dm] per year) and miscanthus ($70.11 t dm per year) were slightly more expensive to produce than corn stover ($67 t dm per year). The main driving factors behind the cost differences are harvesting and storing the biomass material, both factors that could be improved with technological advances (Khanna, 2008).

The cost of ethanol production is also higher for cellulosic sources when compared to corn ethanol as shown in figure 6.17. Even with the Food, Conservation and Energy Act of 2008 resulting in $1 billion in incentives for cellulosic feedstocks and a $1.01 per gallon tax credit for cellulosic biofuel production and cost share payments, some cellulosic ethanol feedstocks, particularly switchgrass, are still not competitive with corn ethanol. As explained earlier, much of this difference in cost during ethanol production is tied to the cost of the necessary enzymes.

The third factor that will play a role in the success or failure of cellulosic ethanol will be biotechnological improvements in the plant breeds being used. These improvements will include better cell wall compositions, increasing growth rate, finding strains suitable for growth in different geographical regions, and identifying strains that more efficiently use resources like nutrients and water. Once these limitations for cellulosic ethanol production are addressed, cellulosic biomass could be an extremely valuable renewable resource to help lower the use of fossil fuels, particularly petroleum used for transportation fuels.

Future of Bioethanol

Bioethanol represents one of the major biofuels currently being produced and consumed both in the United States and around the world. Historically, this ethanol has largely been produced from either corn or sugarcane, but an ongoing debate about the impact that the use of a food resource to produce fuel will have on food availability and affordability has led to a desire to develop alternative bioethanol feedstocks, like cellulosic ethanol. Ultimately, while ethanol will always represent an excellent fuel additive that is likely to play a role in renewable transportation fuels for many years to come, it is an unlikely candidate to completely replace petroleum resources, due to its lower energy content, lack of sufficient scalable production, and its inability to be used at higher than 10% in our current petroleum-based infrastructure. In the next several chapters, we will discuss other biofuels that are a closer match to current fuel types including the other major biofuel produced around the world, biodiesel.

STUDY QUESTIONS

1. Briefly explain photosynthesis and why it is important to understand this concept when studying biofuels.
2. Why is corn a more highly used resource for bioethanol production in the United States compared to sugarcane?
3. Explain why ethanol and gasoline should not be directly compared as fuels.
4. Briefly describe the production of ethanol from corn. Why might fermentation of sugarcane be considered easier than that of corn?
5. How do governmental subsidies, taxes, and credits impact the corn and ethanol markets?
6. Give some examples of cellulosic ethanol feedstocks. Explain why the use of these feedstocks could be beneficial for the future of bioethanol.
7. Briefly explain how the biochemical and thermochemical treatment of cellulosic feedstocks differ for ethanol production.
8. What are the benefits and challenges of cellulosic ethanol production?

CHAPTER SEVEN

Biofuels from Fats and Oils: Biodiesel

The first use of bioethanol as a fuel for engines came at about the same time as the introduction of another first-generation biofuel, **biodiesel**. The use of biodiesel continues today apparent in the advertisements for biodiesel seen on some local buses sporting the phrase "powered by biodiesel." Unlike bioethanol, which is used as a low percentage additive to gasoline, biodiesel has a greater potential to replace petroleum-based diesel fuels because most diesel engines can utilize modern biodiesels directly without significant alterations to the engine design. In addition, biodiesel is fairly easy to obtain as it can be derived from vegetable oils and animal fats that are not always in direct competition as a human food or animal feed source. This chapter will introduce you to the properties of biodiesel and the potential of this fuel for the future.

Biodiesel is typically composed of long chains of hydrocarbons (12–22 carbons in length) that are derived from a simple chemical modification of animal fats or vegetable oils, and are chemically similar to the petroleum-based fuel molecules described in Chapter 2. In their natural state, these hydrocarbons are found as individual single-chain fatty acids or as groups of three fatty acids linked together, called **triacylglycerides.** Triacylglycerides will be discussed in more depth later in the chapter. As figure 7.1 illustrates, the structure of biodiesel is very similar to the structure of petrodiesel, and both molecules have significantly different chemical structures than ethanol. One of the main differences between diesels and ethanol is the ratio of carbon, hydrogen, and oxygen. It is easy to see that ethanol has a higher ratio of oxygen than the hydrocarbon-chain-based diesels because ethanol is a much smaller molecule and, as an alcohol, always contains a hydroxyl (–OH) chemical unit. This difference is one of the key factors in ethanol having a lower energy density than gasoline, biodiesel, and petrodiesel. A higher oxygen ratio lowers the energy density of a fuel molecule requiring a larger volume of that fuel

to travel an equivalent distance compared to fuels composed of molecules with a lower oxygen ratio. While the energy density of ethanol may be lower when compared to gasoline, making it an unlikely candidate as a total replacement for this petroleum-based fuel, the energy densities for biodiesel and petrodiesel are nearly equivalent. Because biodiesel has a similar chemical structure and energy density to petrodiesel, with proper preparation, biodiesel can be used as a direct replacement or blending component with petrodiesel. This is an important characteristic of biodiesel as a biofuel because biodiesel can immediately be available as a drop-in fuel to help replace or reduce the use of petrodiesel without the need for major technological advances or changes in the current fuel infrastructure.

If the characteristics of biodiesels make them easier to implement as a replacement for petrodiesel fuel, then why does bioethanol dominate the US biofuels market? The answer to this question comes in two parts. First, as discussed in the last chapter, the corn agricultural industry in the United States is enormous and politically and economically very strong, and the wide availability of corn and ease with which it can be turned into ethanol made the production of corn-based ethanol a logical step for biofuels development. Secondly, in the late 1800s, two different types of engines were developed and adopted by the transportation industry: the gasoline engine and the diesel engine. The gasoline combustion engine became the mainstay for most personal transportation and the diesel engine replaced large hauling and working vehicles like tractors. As the price of petroleum decreased in the first half of the twentieth century, there was an enormous expansion in the use of personal transportation vehicles, and these relied on the gasoline-powered combustion engine, resulting in an increase in the demand for gasoline. With gasoline leading the transportation market, ethanol, an additive that makes gasoline burn more efficiently, became an important

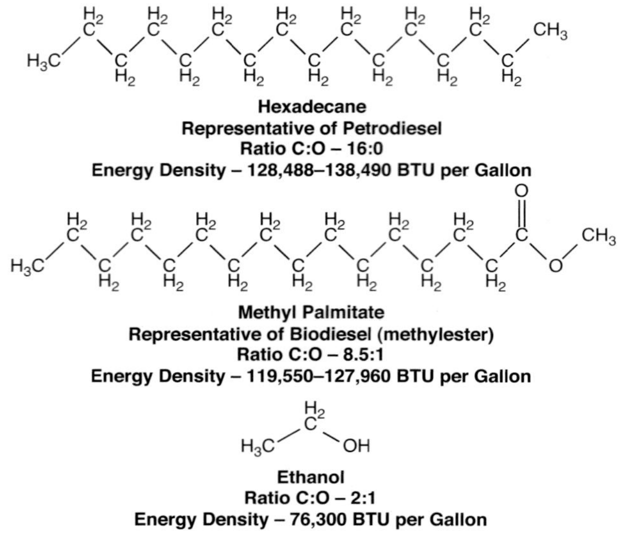

Hexadecane
Representative of Petrodiesel
Ratio C:O – 16:0
Energy Density – 128,488–138,490 BTU per Gallon

Methyl Palmitate
Representative of Biodiesel (methylester)
Ratio C:O – 8.5:1
Energy Density – 119,550–127,960 BTU per Gallon

Ethanol
Ratio C:O – 2:1
Energy Density – 76,300 BTU per Gallon

FIGURE 7.1 Comparison of the chemical structures of a hexadecane, a representative of petrodiesel fuel, methyl palmitate, a representative of biodiesel fuel, and ethanol (also called bioethanol). The energy content for diesel and biodiesel is much closer than that of ethanol largely due to the much higher carbon-to-oxygen ratio in ethanol (source: US Department of Energy Alternative Fuels Data Center).

commodity in the energy market and now dominates over other liquid biofuels (Melosi, 2010; Stafford, 2014). However, as the United States continues to seek a renewable replacement for transportation fuels, more and more research has focused on the development of biodiesel, due to its superior energy density and engine compatibility when compared to bioethanol.

History of Biodiesel

The history of biodiesel begins with the invention of the first diesel engine by Rudolf Diesel in 1896. The original diesel engine had an efficiency of nearly 75%, much higher than previous steam engines, and actually ran off a product familiar to most students: peanut oil (Stafford, 2014). Despite the initial use of this natural product as a biofuel, biodiesel was quickly overshadowed by petrodiesel due to the continued commercialization and falling prices of petroleum-based fuels in the early 1900s. Luckily, the development of cheap petrodiesel did not wholly stop research into the potential of biodiesel as a transportation fuel. The government and public once again became interested in the use of biodiesel due to the Oil Crisis of 1973. This crisis stemmed from an embargo placed on the United States by the oil cartel, OPEC, resulting in a lack of availability and climbing prices for petrodiesel and gasoline. Unfortunately, the early generations of biodiesels were simply too thick or viscous to be used in conventional diesel engines, thus reducing their large-

scale availability for replacing dwindling supplies of petrodiesel at the time. In the early 1980s, scientists in South Africa unlocked a clue to lowering the viscosity of these native biodiesels. These scientists developed a chemical transformation in sunflower oil that broke down the complex storage form of the hydrocarbon chains generally found in plant oils (triacylglycerides) into simple linear hydrocarbon chains referred to as methyl esters. These methyl esters had a lower viscosity than native biodiesel and thus functioned better in diesel engines (Biodiesel Energy Revolution, 2014; Pacific Biodiesel, 2014). The utilization of sunflower oil methyl esters by the South Africans set the stage for the development of **transesterification**, a process for creating these methyl esters from many different types of animal fats and plant oils. Today, these esterified biodiesels are still used successfully both as direct replacements and as blending agents with petrodiesels without a significant amount of altering to the diesel engine.

In 2013, the United States produced about 1,359 million gallons of 100% biodiesel used largely as a liquid fuel and representing an increase from previous years (EIA, 2014). Some countries such as Germany and Brazil have begun to require the blending of biodiesel with petrodiesel similar to the US requirement for blending ethanol with gasoline. This trend has spilled over into some states in the United States including Minnesota, Louisiana, and Washington with a mandate for a 2% blend and New Mexico mandating a 5% blend of biodiesel with all petrodiesel fuel (Gross, 2007).

Because first-generation biodiesel is largely produced from agricultural sources used for food such as soybeans and sunflowers, biodiesel also suffers from the same food versus fuel tension as corn-based ethanol. The remainder of this chapter will discuss the current production of biodiesel and introduce a number of biomass options for renewable, sustainable biodiesel production that should not impact the price and availability of food in the future.

Plant Reproduction Generates Biodiesel

Like the production of bioethanol from cornstarch, biodiesel is also derived largely from a storage form of energy found in plant seeds. During their reproductive cycles, almost all plants produce seeds. These seeds are typically designed for two functions: dispersal and germination. Seed dispersal occurs when birds, insects, and other animals are attracted to a plant due to its bright colors or sweet smell. These animals and insects can then pick up the seeds and carry them wherever they go or eat them and eliminate the seeds in their waste, either way dispersing the seeds. This role by animals and insects in dispersing seeds is crucial for maintaining plant life throughout the Earth. Once the seeds are dispersed, they are only valuable if they

BIODIESEL VERSUS RENEWABLE DIESEL

Travis L. Johnson

What's the difference between biodiesel and renewable diesel?

Biodiesel

Biodiesel is a fuel that is produced by transesterification, a reaction of fats and oils with an alcohol and a catalyst creating fatty acid methyl esters. The oils and fats are in the form of triacylglycerides, which as described in this chapter are three fatty acid molecules connected to a glycerol molecule. When the triacylglycerides react with alcohol, the fatty acids detach from the glycerol and combine with alcohol molecules to become fatty acid methyl esters. The process creates a glycerol by-product and the remaining biodiesel contains oxygen atoms.

Renewable Diesel

Renewable diesel is chemically similar to petrodiesel, but like biodiesel, it is derived from biological sources such as plant oils or animal fats. It differs from biodiesel in that it does not include the use of esters. Instead, renewable diesel generation processes include hydrotreatment, pyrolysis, and biomass to liquid with Fischer–Tropsch synthesis. These methods require high temperatures and pressure and in some cases catalysts to chemically convert biomass to renewable diesel. Renewable diesel has no oxygen atoms and is therefore more stable than biodiesel.

SOURCES: Pomeroy (2014); Yoon (2011); Kotrba (2008).

germinate and grow. Germination efficiency depends partly on the seeds' ability to survive in sometimes-harsh environmental conditions and their ability to provide nutrients and energy for a budding young plant. To improve germination efficiency, many seeds are protected from harsh environmental conditions like desiccation and temperature fluctuations by a hard outer shell. This shell also protects a reserve of energy biomolecules often made of lipids or oils needed to support the budding plant. These lipids are the same molecules used in the production of biodiesel fuel.

Introduction to Plant Biochemistry

Plant oils can be loosely classified into three categories: triacylglycerides, free fatty acids, and phytosterols. The basic chemical structure for each of these is shown in figure 7.2. Phytosterols may appear very different from both triacylglycerides and free fatty acids, but they are in fact very similar when you consider their elemental makeup. Despite having a nonlinear structure, a phytosterol is still composed primarily of carbon and hydrogen, thereby making it a good candidate to become a fuel. Although triacylglycerides and fatty acids are also different, there is an important structural similarity. Can you see it? The similarity comes as individual fatty acids make up the three tails found in the triacylglyceride. It is these carbon and hydrogen "tails" that are so important when it comes to energy production. These three fatty acid tails are held together by an oxygen-containing glycerol-based backbone that lends stability to the individual fatty acids. Because triacylglycerides are more stable molecules when compared to the individual fatty acids, they are the preferred energy storage units for plant cells and almost all seeds contain some form of these triacylglyceride molecules. The glyceride backbone also significantly increases the size and viscosity of these molecules (this is the reason for the viscosity problems in the initial biodiesel fuels in the early 1900s) (Singh and Singh, 2010). In order to be used either as an energy source for the plant or as a biodiesel feedstock, the glyceride backbone of the triacylglycerides must be removed to release the free fatty acids.

Another important question to consider when develop-

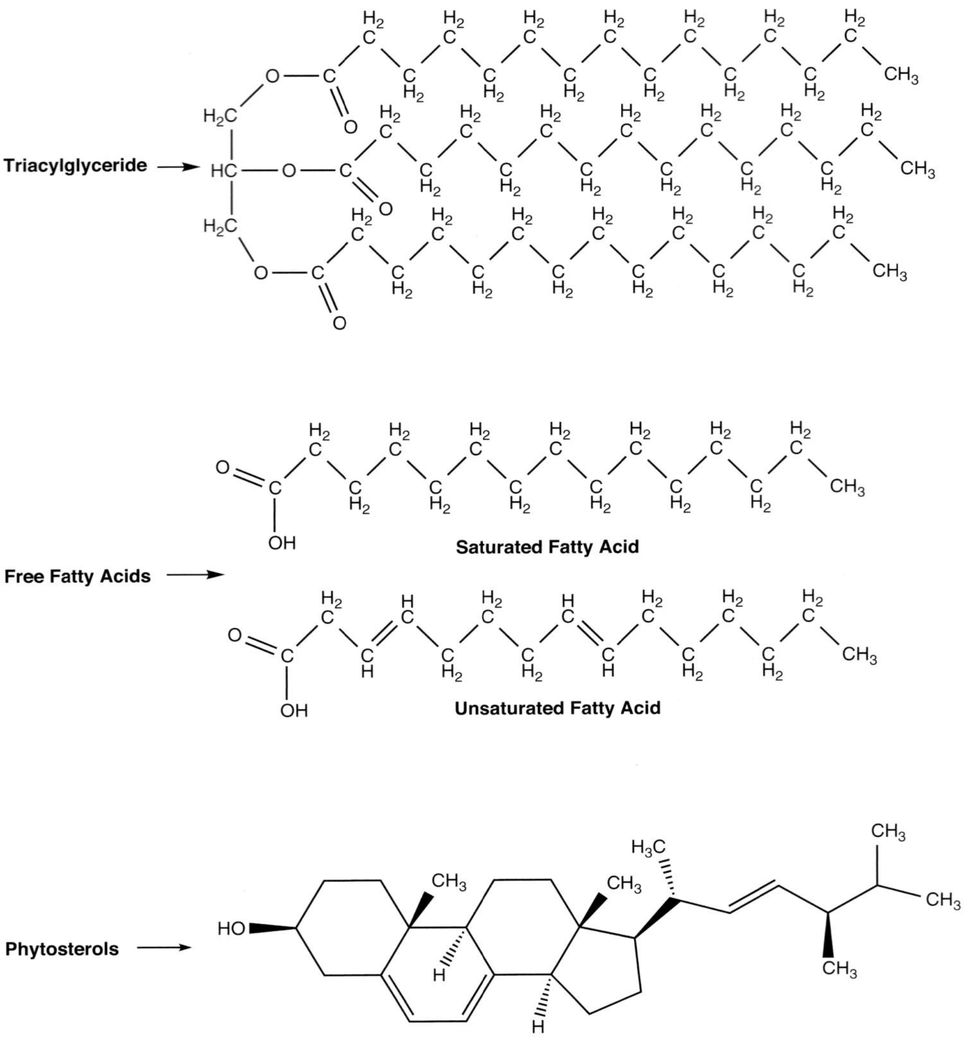

FIGURE 7.2 Chemical structures for a triacylglyceride, saturated and unsaturated free fatty acids, and a phytosterol. Notice that all of these molecules have a basic hydrocarbon skeleton despite the clear structural differences between the phytosterol and the more obvious fatty acid based triacylglycerides and free fatty acids. (*Remember that lines represent carbon–carbon bonds with hydrogen bonds being assumed at each intersection to fulfill the carbon octet*).

ing new biodiesel technologies is, "How do these plants actually make these triacylglycerides?" Understanding the process of triacylglyceride biosynthesis within seeds and plants is important because it can provide insights into mechanisms that can be used to engineer specific types of fatty acids or produce larger quantities of fatty acids within seeds, thus producing larger volumes and higher-quality biodiesel.

As diagramed in figure 7.3, the fatty acid tails found in tri-acylglycerides are synthesized in the chloroplast (where photosynthesis occurs) and then assembled into triacylglyc-erides within the endoplasmic reticulum. The fatty acids are built from the precursor acetyl-CoA, a by-product of gly-colysis. Aceytl-CoA is an important molecule not only

for the biosynthesis of fatty acids but also for general cellular metabolism. Individual acetyl-CoA units are assembled together to ultimately form a fatty acid chain through a four-step process driven by enzymes within the plant, with fatty acid synthase (FAS) being among the most important of these enzymes. This four-step assembly process combin-ing individual acetyl-CoA units is repeated over and over again until the fatty acid reaches an appropriate length. However, the addition of an acetyl-CoA unit during each round of assembly results in oxygen atoms within the grow-ing fatty acid chain. As you may recall from the previous chemical structures, fatty acid tails do not contain oxygen. This is due to a second set of reactions specifically designed to remove oxygen by reduction and dehydration of the oxy-

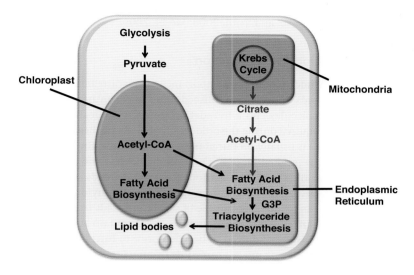

Part of Plant Cell

FIGURE 7.3 Diagram showing the flow of biomolecules within a plant cell used in the formation of fatty acids and triacylglycerides. The individual fatty acids are generated within the chloroplast and then shuttled to the endoplasmic reticulum where they are combined with glyceraldehyde 3-phosphate (G3P) to create the more stable triacylglyceride.

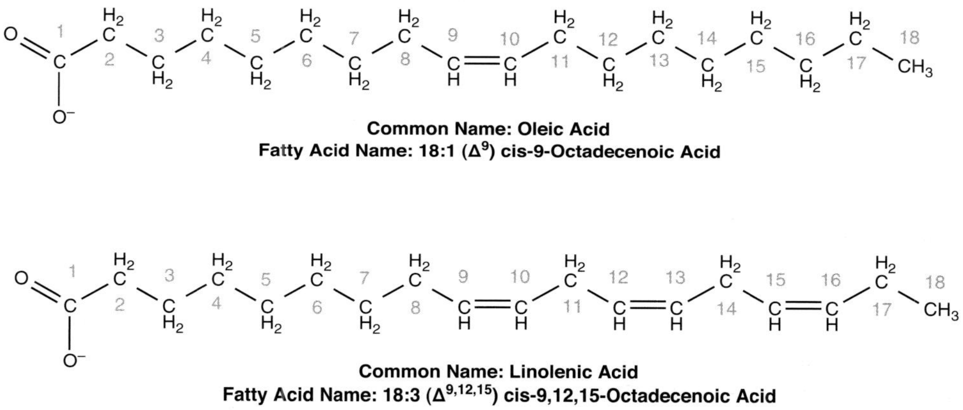

Common Name: Oleic Acid
Fatty Acid Name: 18:1 (Δ^9) cis-9-Octadecenoic Acid

Common Name: Linolenic Acid
Fatty Acid Name: 18:3 ($\Delta^{9,12,15}$) cis-9,12,15-Octadecenoic Acid

FIGURE 7.4 Naming fatty acids. Notice that the name is based on the number of carbon atoms as well as the number and locations of the unsaturations or double bonds.

genated fatty acid (Fishbach and Walsh, 2006). Although fatty acid chains do not contain oxygen, they can have double bonds between carbons, and these double bonds impact the physical properties of biodiesel, especially chemical stability.

Once the fatty acid is assembled within the chloroplast, it is shuttled to the endoplasmic reticulum where individual fatty acids are combined with glyceraldehyde 3-phosphate, generated from metabolic by-products from the chloroplast and mitochondria, to create triacylglycerides. Here, double bonds or unsaturations are also added in cases where polyunsaturated fatty acids are needed. These types of fatty acids have become important when considering

healthy food options because they are not as dense as the saturated fatty acids.

Fatty acids can vary significantly in their length (number of carbons) and the quantity of unsaturation points within the molecule; thus, in order for scientists and engineers to study the many variations of fatty acids created through this biosynthetic process, a naming system is needed in addition to the common names often used for the most prevalent forms of fatty acids. The number of carbons and the number and location of double bonds or unsaturations are the two main factors in learning how to name fatty acids. We can get a better grasp of how to name these fatty acids by looking at the examples in figure 7.4. The first fatty acid molecule is commonly known as oleic acid. This

TABLE 5
Names and abbreviated chemical structures of common
fatty acids found in oilseeds

Fatty Acid	Chemical Structure
Myristic acid (14:0)	$CH_3(CH_2)_{12}COOH$
Palmitic acid (16:0)	$CH_2(CH_2)_{14}COOH$
Palmitoleic acid (16:1)	$CH_3(CH_2)_5CH=(CH_2)_7COOH$
Stearic acid (18:0)	$CH_3(CH_2)_{16}COOH$
Oleic acid (18:1)	$CH_3(CH_2)_7CH=CH(CH_2)_7COOH$
Linoleic acid (18:2)	$CH_3(CH_2)_4CH=CHCH_2CH=$ $CH(CH_2)_7COOH$
Linolenic acid (18:3)	$CH_3CH_2CH=CHCH_2CH=CHC$ $H_2CH=CH(CH_2)_7COOH$

molecule has 18 carbons and one point of unsaturation beginning at carbon 9. The base name, Octadecenoic acid, results from the total number of carbons in the chain (18) including the first carbon known as the alpha (α) carbon. Once the base of the name is established, the points of unsaturation or double bonds can be located. Oleic acid has one point of unsaturation across the 18 carbons denoted by 18:1 in the name and the location of the single unsaturation begins at carbon 9 (Δ^9). This unsaturation is in the cis configuration; thus, the cis-9 designation in the name. Ultimately, this results in the name 18:1 (Δ^9) cis-9-Octadecenoic acid. You can then look at linolenic acid and see that the molecules are very similar, with the only differences being the number and locations of the unsaturations (3 unsaturations beginning at carbons 9, 12, and 15). Because both of these molecules are true fatty acids and not attached to the glyceride backbone, they are designated with the term "acid" in their names. The common names and chemical structures for the most common fatty acids found in oilseeds are shown in Table 5.

Each species of plant has a different ratio of fatty acid types and this ratio can play a role in the quality of biodiesel produced from a particular plant or seed oil. For instance, the length of the carbon chain and the level of saturation of a fatty acid impact the stability and energy density of that molecule when used as a fuel. Plants that have a higher percentage of fatty acids that are more fully saturated and have a longer chain of carbons typically contain a higher energy density and can be used to create a more efficient fuel (Demirel, 2012). For instance, most common feedstocks used in biodiesel production like soybeans, canola, and sunflowers have a fatty acid composition of 85–95% 18-carbon-length fatty acids with a saturation level of 9–15%. As will be discussed later in the chapter, grease, tallow, and lard are also important potential biodiesel sources that contain a much higher saturation level ranging from 45% to 76%. These are especially good feedstocks in

the instances where these resources can be recycled from wasted cooking materials and agricultural by-products (Canakci and Sanli, 2008; Moser, 2009).

Biodiesel Production

Understanding the importance of the distribution of fatty acids in different oilseeds and the synthesis of these molecules lays a foundation for the chemical techniques required during the production of biodiesel. Because biodiesel is usually produced from a plant source, the first step in the production of biodiesel is the harvest and preparation of the seeds. The method of harvest depends on the plant species but ranges from handpicking individual seeds to using a thresher capable of separating seeds from unusable plant stalks. In most cases, the seeds contain about 13–15% moisture and need to be dried prior to further processing. The drying process can either occur naturally using the sun where the seeds are laid out to dry for one to two weeks or it can be done artificially where the seeds are exposed to heated air, similar to how a blow-dryer is used on hair (Herbek and Bitzer, 1997). Once dried, the seeds must be separated from any impurities or detritus present within the harvest. This cleaning process usually involves suctioning off lighter impurities and then sifting the seeds to remove small detritus and contaminants like sand and dirt. Once harvested, dried, and cleaned, the seeds are ready for oil extraction.

The method of extraction is largely based on the type of seed and the amount of effort needed to break the outer shell. The easiest form of extraction is chemical extraction where a seed can simply be soaked in a solvent (chemical) that will dissolve the lipids. In chemical extraction, it is important to remember one simple concept in the choice of solvents, "like dissolves like." This concept is based on the polarity of solvents and solutes: if a solute is **nonpolar**, then it will dissolve in a nonpolar solvent, and if a solute is **polar**, then it will dissolve in a polar solvent. In the case of lipids, lipids are very nonpolar; therefore, most chemical extraction processes for lipids involve a nonpolar chemical solvent, like hexane. While the chemical extraction method is simplest, the hard outer shell of many seeds can prevent the lipids inside these seeds from ever being exposed to the solvent. In these cases, the seeds must first be physically broken or pressed to squeeze out the lipids. These lipids can then be chemically extracted in a manner similar to that described above, resulting in a lipid-dominated extract (Nowatzki et al., 2007).

As you learned earlier, one of the main problems with using lipids from oilseeds directly as a biodiesel is that these lipids are almost all found as part of a triacylglyceride rather than as free fatty acids. Triacylglycerides are typically much more viscous than normal petrodiesel and do not function well in conventional diesel engines. To overcome the viscosity of triacylglycerides in biodiesel, transesterification was developed to detach the glyceride backbone from the

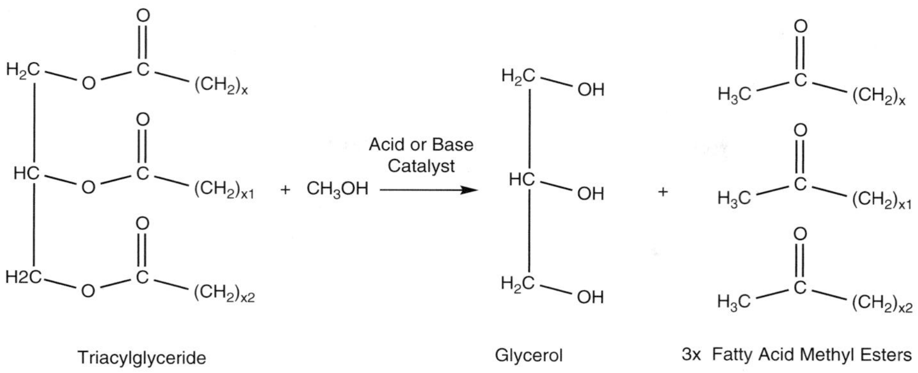

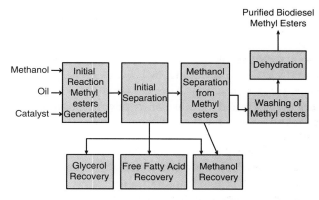

FIGURE 7.5 Basic diagram of the transesterification reaction showing the triacylglyceride reacting with the alcohol in this case methanol (CH_3OH) to produce glycerol and three fatty acid methyl esters as denoted by $(CH2)_x$.

fatty acid chains and thus lower the viscosity of the fatty acids, creating a better replacement for petrodiesel. Transesterification, as shown in figure 7.5, can occur in either a base catalyzed or acid catalyzed manner in the presence of an alcohol. Through either mechanism, the alcohol (ROH) attacks the alpha carbon of the fatty acid chain separating the fatty acid from the glycerol using a methyl group (CH_3) resulting in the formation of a fatty acid methyl ester, and leaving the glycerol in solution with no attached fatty acids. Both of the acid and base catalyzed reactions can also occur directly on free fatty acids to help stabilize them by adding the methyl group (CH_3) to the acidic end of the hydrocarbon chain. In the end, whether the transesterification is acid catalyzed or base catalyzed, the product is the same: three methyl ester fatty acids and glycerol (Schuchardt et al., 1998).

The process of transesterification looks pretty simple when considering these chemical reactions on a single triacylglyceride; however, for the commercial production of biodiesel, this reaction must be scaled up to allow large quantities of triacylglycerides to undergo the reaction simultaneously. In addition, the separation and recycling of individual components used during the processing, like the glycerol product and alcohol substrate, are important in adding increased value to the overall biodiesel production process. In an example of the commercial-scale process, diagramed in figure 7.6, methanol, the extracted oil, and a catalyst (either acid or base) are combined and reacted for one hour at high temperatures. In the next step, the methyl esters are separated from the glycerol. Once the methyl esters are separated, the methanol can be removed, any remaining free fatty acids neutralized into methyl esters, and the water removed to create the final biodiesel product. The glycerol can also be recycled and developed as a value-added product (Gerpen, 2005). Together, this process can efficiently produce biodiesel fuel while also lowering the cost of the fuel due to the use and recycling of the non-fatty acid products.

FIGURE 7.6 Basic flow diagram for the production of biodiesel at a commercial level. Commercial-scale biodiesel production will likely require the recycling of methanol and glycerol in order to make the process more cost competitive (source: Gerpen, 2005).

Fuel Properties of Biodiesel

Diesel and gasoline engines are fairly similar in the sense that they both use pressure to ignite a fuel that drives a shaft up and down powering the engine. The main difference between a diesel and gasoline engine is when the fuel is introduced into the system. In a gasoline engine, the gasoline fuel is introduced with the air and then compressed. Following compression, a spark ignites the fuel, causing an explosion that drives the shaft back down. In a diesel engine, the air is first compressed then the diesel fuel is introduced into the already compressed air. Since compressed air is heated, it results in an explosion, without the need for a spark, that once again drives the shaft down and powers the engine. The mechanism of a diesel engine results in about a 15% increase in efficiency compared to a gasoline engine for converting the chemical energy of fuel into mechanical energy. Since diesel fuels contain slightly longer carbon chains than gasoline molecules, usually 12–20 carbons for diesel compared to 4–10 carbons for gasoline, these fuels have a slightly higher

TABLE 6
Important fuel properties of diesel fuels, their effects on engine performance, and lipid conditions that change these properties

Fuel Property	Effect on Engine Performance	Lipid Effects on Property
Viscosity	Effects fuel injection and atomization	Increases with greater carbon chain length; decreases with larger number of unsaturations (double bonds)
Density	Effects fuel injection, atomization, heating content, and exhaust emissions	Increases with greater carbon chain length and with lower number of unsaturations
Cetane number	Effects ignition delay	Increases with greater carbon chain length; decreases with greater carbon branching and larger number of unsaturations
Flash point	Effects volatility of burning	Decreased upon tranesterification and with higher level of saturation
Cold flow	May cause blockage of fuel lines and filters	Increases with greater carbon chain length; decreases with larger number of unsaturations
Oxidation stability	May cause gum formation in the combustion chamber	Increases with larger number of unsaturations

NOTE: Fuel properties based on American Standard Specifications (ASTM).

energy density and thus are more reactive in the pressurized environment without the addition of a spark (Motor Trend, 2005).

A diesel fuel is generally considered any fuel used in a diesel engine and usually derives from petroleum. A biodiesel is a renewable fuel that can be used directly in a diesel engine without modification to the engine. This is different than some converted diesel engines that can run off of vegetable and waste oils. A true biodiesel can be used as a direct replacement for petrodiesel. However, to be able to use biodiesel as a direct replacement or drop-in fuel, the biodiesel must meet certain fuel standards. These fuel standards are outlined in Table 6 and described below (Gerpen et al., 2004).

CETANE NUMBER   The cetane number measures the quality of the fuel for ignition. In general, the higher the cetane number, the more easily the fuel will ignite. If the cetane number is too low, then there will be a delay in the ignition. The cetane number increases with increased carbon chain length and decreases with branching and unsaturation within the hydrocarbon chains. Biodiesels tend to have higher cetane values than typical petrodiesels.

ENERGY CONTENT   The efficiency of a fuel for producing power is a direct result of its energy content. In general, the energy content of biodiesel is slightly lower than that of petrodiesel fuels.

FLASH POINT   The flash point characterizes the volatility of a fuel for burning. This fuel property is measured as the lowest temperature at which a fuel can vaporize to form an ignitable mixture in air. A flash point that is too low can spontaneously ignite, while a flashpoint that is too high will not ignite under the pressurized conditions within the

diesel engine. The flash point is reduced through transesterification or by having unsaturations within the fuel molecules.

VISCOSITY AND DENSITY   Both viscosity and density play an important role in fuel because they affect the atomization or droplet size of the fuel and thus the ease of injection of the fuel into the engine. A denser and more viscous fuel will not flow as easily and will not vaporize in a manner sufficient for maximum ignition efficiency. Both viscosity and density increase with increased carbon chain length and saturation.

OXIDATIVE STABILITY   A methyl ester is susceptible to oxidation when exposed to atmospheric oxygen. Oxidation of the fuel leads to formation of gums that can damage an engine and potentially block the filters within the engine. Thus, creating a stable fuel that has a low likelihood of being oxidized during storage is important in increasing the shelf life of the biodiesel, allowing it to be stored.

CLOUD POINT AND COLD FLOW   Cloud point is the temperature at which a diesel fuel begins to precipitate. Small wax crystals will begin to form and make the fuel appear cloudy. These crystals can result in blockage of the fuel filter. At the same time, these crystals will begin to prevent the flow of the fuel; thus, cold flow is the temperature at which the flow of fuel begins to slow. Both of these criteria are important when considering biodiesels because if a fuel is clouded or does not flow due to low temperature, the engine will not run. Cold flow typically increases with the number of carbon atoms in the chain and decreases with the number of double bonds.

Aside from these important fuel properties, a diesel fuel is also tested for the percent of fatty acid methyl esters, glycerol, methanol, water, and other salts such as sodium,

TABLE 7
Comparison of fuel properties for traditional diesel fuel and common biodiesel feedstocks as pure oils and following tranesterification

Feedstock	Viscosity (mm²/s)	Density (g/cm³)	Cetane Number	Flash Point (°C)	Cloud Point (°C)	Pour Point (°C)
Diesel	2–4.5	0.820–0.860	51.0	55	–18	–25
Soybean	33.1	0.914	38.1	254	–3.9	–12.2
Soybean methylester	4.08	0.884	50.9	131	–0.5	–4
Rapeseed	37.3	0.912	37.5	246	–3.9	–31.7
Rapeseed methylester	4.83	0.882	52.9	155	–4	–10.8
Sunflower	34.4	0.916	36.7	274	7.2	–15.0
Sunflower methylester	4.60	0.880	49.0	183	1	–7

DATA FROM: Canakci and Sanli (2008).

potassium, calcium, magnesium, sulfur, and phosphate that are in the final product. There are limits on all of these component parts in order for a fuel to meet the required technical specification (Gerpen et al., 2004). For instance, a diesel fuel cannot have a pour point below –10 degrees Celsius and expect to be used in colder climates because the fuel will simply solidify in these climates and prevent the engine from starting. However, the perfect combination of properties is actually a moving target, particularly with biodiesel. In most cases, one biodiesel product is not perfectly ideal in each fuel property category, but the variations are within a range acceptable for that fuel (Canakci and Sanli, 2008). A comparison of fuel properties for some of the pure oils and transesterified products of common biodiesel feedstocks is shown in Table 7. In general, a vegetable oil processed into its methyl ester typically has a lower viscosity, which allows it to flow easier. It also tends to have a higher cetane value, lower flash point, and worse cold flow properties than the vegetable oil itself. In fact, the cold flow properties of biodiesels can be a significant problem. When compared to petrodiesel, the pour point temperature for biodiesel is much higher. Therefore, in regions where temperatures regularly drop below –10 degrees Celsius, most biodiesel fuels would actually be solid and could not be used as a liquid fuel.

Ideally in the future, petrodiesel could be replaced completely with biodiesel, but currently most biodiesel comes blended into petrodiesel. In fact, a blend of just 1–2% biodiesel can improve petroleum-based diesel by increasing the lubricating properties of the final fuel. Generally, biodiesel has a slightly lower energy value than petrodiesel, but these values are largely dependent on feedstock and it may be possible to develop new feedstocks with higher energy values. Blending biodiesel with petrodiesel at a range of 6–20% is also beneficial because it maintains fuel properties including cetane number and viscosity much closer to pure petrodiesel as compared to 100% biodiesel fuels (Gerpen, 2005).

Biodiesel is a valuable, renewable form of transportation energy that can help lower dependence on petroleum-based diesel fuels. Further, biodiesel fuels tend to have better lubricity and reduced emissions, and are considered biodegradable. However, there are also some concerns with biodiesel that will need to be addressed before these fuels gain wide-scale use as a replacement for petrodiesel. Biodiesel tends to have increased water content that can result in microbial contamination and degradation during long-term storage. As mentioned, these fuels generally have poorer cold climate performance, lower oxidative stability, slightly lower energy content, and most importantly a higher cost compared to petrodiesel. The success of biodiesel as a fuel replacement for petrodiesel will likely depend on the discovery of novel oil feedstocks or methods to improve the properties and storage of biodiesel as well as the cost. In the next sections, we will discuss some of the traditional biodiesel feedstocks as well as some of the new-generation feedstocks being studied for biodiesel production.

First-Generation Biodiesel Feedstocks

Biodiesel feedstocks can be classified into two loosely organized oilseed crop categories: traditional and alternative. In general, the traditional crops are considered first-generation biodiesel feedstocks because these are the oilseed crops that have been used for over a century. Figure 7.7 compares the global production levels of major

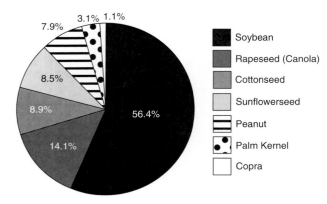

FIGURE 7.7 Global major oilseed production in 2013/2014. Major oilseed commodities are shown as a percentage of the total global oilseed production of 505.21 million metric tons. Soybeans are the most important global source of oilseeds (data from USDA, 2014a).

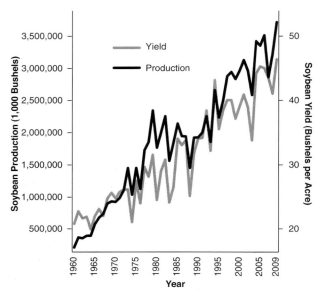

FIGURE 7.8 Comparison of US soybean production and oilseed yield between 1960 and 2009. Both steadily increased during this time (data from USDA, 2014c).

oilseeds during the 2013/2014 year. Clearly, soybeans are the dominant source of oilseeds around the world.

Soybeans make up the largest percentage of the global oilseed crop as well as 90% of the oilseed production in the United States. Soybeans are the second highest valued crop in the United States exceeded only by corn. The farm value from soybean crops was $29.6 billion in 2008/2009 with 77.5 million acres under cultivation, and approximately 54% of the crop exported (Ash, 2012). The most valuable product of soybean is soybean meal. This meal accounts for 50–75% of the value of soybean crops and provides two-thirds of the total protein feed in the United States. Soybeans are also the second largest source of vegetable oil, accounting for 55–65% of all vegetable oils and animal fats consumed in the United States (Hay, 2012). Soybeans are a species of leg-

umes that often grow in a crop rotation with corn due to their nitrogen fixation properties that help in replenishing soil nutrients. This property of nitrogen fixation reduces the levels of fertilizers needed on these crops; in fact, soybeans require about 40% less fertilizer than many other crops. The utilization of less fertilizer plays an important role in lowering the ecological footprint of soybeans (Ash, 2012).

Soybean production and the yield of oilseed from the harvest have steadily increased over the past 50 years as shown in figure 7.8. This increase has resulted from changes in planting flexibility, identification of new seed varieties, and improved crop management practices such as fertilizer and pesticide applications. The productivity of soybean crops has also improved due to bioengineering of these crops. Biotechnological advances have produced crops that are nearly herbicide resistant, resulting in higher yields and lower production costs (Ash, 2012). Today, biotechnology is also focused on creating specific oil profiles within the seeds that will help lower trans-fat content and create healthier oil for human consumption. Without further biotechnological modifications, the use of soybean as a food oil is thought to have peaked and is declining. At the same time, the utilization of soybean oil as biodiesel is gaining momentum. In 2009, 77.5 million acres of soybean resulted in 3.4 billion bushels of seeds in the United States. Each bushel can produce about 1.5 gallons of biodiesel; therefore, if all soybean crops were used in the production of biodiesel, about 5.1 billion gallons of biodiesel could be produced each year, equivalent to almost 15% of the yearly total diesel fuel consumption in the United States (Hay, 2012). While replacing 15% of petroleum-based diesel with biodiesel would be helpful environmentally, it would also clearly leave no soybean oil resources available for human food consumption and animal feed production. Thus, other resources will also be needed for large-scale biodiesel production in the future.

The second most abundant oilseed crop is rapeseed or canola, accounting for 10–15% of the total oil crop. Rapeseed is usually grown in temperate latitudes with dry weather and short growing seasons. Therefore, a large majority of the production of rapeseed is centered in the European Union, China, Canada, and India. The uses of rapeseeds are similar to those of soybeans, with rapeseed representing the second largest feed meal and 13–16% of the world vegetable oil production. In the United States, canola has become the term synonymous with the edible form of rapeseed. This crop is typically grown in the northern regions of the central United States without competing for land space with corn or soybean. The value of this oil is climbing due to its use as a food source in areas such as China and India and as a major crop for biodiesel production in the European Union. In 2008/2009, the use of rapeseed oil for industrial uses was 5.791 million metric tons, representing 69% of the total production. Most of this was used in the production of biodiesel due to mandates and policies requiring the blending of biodiesels with petrodiesels. While the United States only produces a small amount of rapeseed, a large amount of rapeseed oil and meal utilized in the United States is imported from Canada. A rapeseed typically contains 38–45% oil compared to 18–19% oil in the soybean, but the high price of rapeseed oil in the United States makes it unlikely that it will be a leading player in the biodiesel market in the future (Ash, 2012; USDA, 2012).

Sunflower seeds are used primarily as birdseed, snacks, and for baking. Once processed, the main product of these seeds is oil, accounting for about 11% of the world vegetable oil trade. The growth of sunflower seed crops rose in the 1970s, but, due to competition by other oilseed crops like soybean, has since declined. There are two types of sunflower seeds—an oil type and a non-oil type. In all, 90% of the US production is the oil type. The oil type sunflower seed contains about 35–55% oil and about 20% protein by weight (Ash, 2012).

Cottonseed and peanuts are the other two crops that in addition to rapeseed, soybeans, and sunflower seed make up a majority of the traditional global oilseed crop considered as potential biodiesel sources. Both of these commodities are used primarily for food and feed but still represent potential sources of biodiesel. The price of oilseeds tend to fluctuate in a similar pattern with one another, but sunflower seed and cotton seed are usually the highest priced oilseed commodity, making it unlikely that these will be highly utilized as future biodiesel resources (Ash, 2012).

Alternative Biodiesel Feedstocks

Most of the traditional oilseed crops are food crops used either as a direct source of nutrition or as cooking oil. Soybeans, rapeseed, sunflower seed, cotton seed, and peanut oil together make up 95.8% of the global oilseed crop (USDA, 2014a). Table 8 shows a comparison of the oilseed yields for each of these crops. Despite the fact that many of these crops are prolific oil producers and can be used to produce high-value biodiesel methyl esters, many do not produce enough oil to supply for both food and fuel. Thus, researchers are studying other alternative oilseed crops, mainly *Jatropha* and oil palm, that can be grown in addition to the traditional food crops as biodiesel crops that have the potential to produce even higher yields of biodiesel as shown in the Table 8.

Price will play a major role in the success of these alternative feedstocks for biodiesel production. Over the past decade, there has been a steady rise in the cost of first-generation traditional oilseed. For example, soybean oil increased in price from 14.1 cents per pound in 2000 to 35 cents per pound in 2009. Much of the price increase for these commodities is likely due to the demand for the oilseed feedstock itself (Pimentel and Patzek, 2005). Unfortunately, as the cost of the feedstock and oilseed rises, so does the cost of biodiesel. In order to be competitive with petrodiesel, biodiesel must be equivalently priced to petroleum-based fuels. The demand for these first-generation feedstocks as a food and feed source in the future due to the rising popula-

TABLE 8

Comparison of potential oil yields from biodiesel feedstocks including first-generation traditional feedstocks as well as alternative feedstocks

Oil Feedstock	Oil Yield (gallons/acre)
Traditional Feedstocks	
Corn	18
Cotton	35
Soybean	48
Mustard seed	61
Sunflower	102
Canola/rapeseed	127
Alternative Feedstocks	
Jatropha	202
Oil palm	635

DATA FROM: Pienkos (2007).

tion may hold their price above the price point competitive for biodiesel production. To combat this issue of price, feedstocks that are not competitive as food products and/or that have higher yields of oil are being explored.

One feedstock that is readily available in the United States but possibly underutilized as a source of biodiesel is waste oil, particularly wasted cooking oil, tallow, and lard. The average American consumes 29 pounds of French fries every year (*The Week*, 2012). This means Americans in total consume about 4.5 million tons of French fries in a year. Interestingly, the used fryer oil has the same molecular backbone as biodiesel. Imagine the production potential when taking into account the many other common fried foods in the American diet.

There are specifically two types of waste vegetable oils: yellow grease and brown grease. Yellow grease is grease associated with foods like French fries that are cooked in a deep fryer. This type of grease typically has less than 15% free fatty acids and a slightly higher saturation level at 38%. Brown grease, in contrast, is the waste recovered from grease traps, typically having greater than 15% free fatty acids and a slightly lower saturation level at 37% (Canakci and Gerpen, 2001). Both types of grease require a dedicated refining process to deal with variations in water content and free fatty acid content. The United States easily surpasses other countries in the quantity of waste cooking oil produced at 10 million tonnes per year compared to 4.5 tonnes per year for China and 0.85 tonnes per year for the European Union. In a 2006 analysis, the price of biodiesel from waste grease was estimated to be approximately $1.54 per gallon of diesel equivalent. This was lower than the estimated costs of petrodiesel

($2.17), soybean biodiesel ($2.44), and rapeseed biodiesel ($2.54) at that time (Lam et al., 2010).

Another source of waste oil produced in the United States comes from meat. According to the US Census Bureau, in 2010 the United States produced over 12 million metric tons of beef, 10 million metric tons of pork, and 16 million metric tons of broiler meats such as poultry (US Census Bureau, 2012). In most cases, these meat sources are not used in their entirety. Instead, the fat is removed to produce a leaner cut of meat. In total, about 11 billion pounds of animal fat are produced annually as tallow rendered from beef or lamb and as lard rendered from pig fat (Groschen, 2002). All of this animal fat could be a valuable source of biodiesel, producing nearly 1.5 billion gallons of biodiesel per year (Greene, 2010). However, in order to use either waste vegetable oils or animal fats, refining techniques will need to be developed to help enhance some of the fuel properties including oxidative stability, cetane value, and an improvement in cold flow properties due to high saturation levels.

While waste vegetable oils and animal fats represent an excellent opportunity to recycle a waste product into a fuel, all of this waste oil and animal fat collectively could only produce a small percentage of the current US demand for on-highway diesel fuel; therefore, scientist are also considering other oilseed crops as potential dedicated biodiesel producers to supplement the use of both waste and food crops (Groschen, 2002). Three crops being studied for their potential are *Jatropha*, oil palm, and algae. While algae will be treated separately in Chapter 9, *Jatropha curcas* and oil palm crops will be discussed here.

Oil Palm for Biodiesel Production

One of the highest yielding oilseed crops is oil palm. Compared to other first-generation oilseed crops shown in Table 8, palm is a mega oil producer. Palm oil is primarily used as food oil but is being considered as a biodiesel crop due to its high oil yields. With a seed yield of about 6,251 pounds per acre, it is estimated that oil palm could produce about 635 gallons of biodiesel per acre per year, a volume significantly higher than other oil-producing crops (Collins et al., 2012). Even with these already high yields, biotechnological advances are also expected to continue to increase. However, the use of oil palm as a global biodiesel crop is limited by its very restricted growth range. Oil palm grows in a very narrow range around the equator, with most palm oil production in equatorial countries including Indonesia and Malaysia which together accounted for 87% of the palm oil produced in 2006 (USDA, 2007).

Despite its very limited range, the potential for the development of palm oil as a biodiesel feedstock is considerable. Figure 7.9 shows a comparison of the cost of production for biodiesel as well as an energy input and output comparison for oil palm, rapeseed, and soybeans. In this comparison, palm oil biodiesel is significantly better in terms of hav-

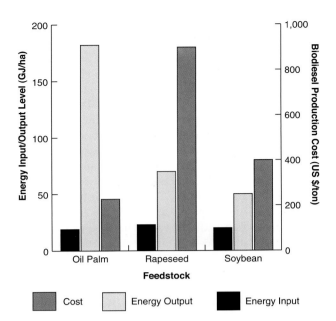

FIGURE 7.9 Comparison of energy input, energy output, and cost for the production of biodiesel from three feedstocks including from traditional soybean and rapeseed feedstocks and the alternative oil palm. Oil palm has a much higher energy output over input and its cost of production is lower than biodiesel produced from rapeseed or soybean (data from Lam et al., 2009).

ing both a much higher energy output compared to energy input and a lower cost of production when compared to traditional rapeseed and soybean. Clearly, price could be a huge factor separating these biodiesel resources, as palm oil costs about $228 per ton, while rapeseed and soybeans cost between $400 and $900 per ton (Lam et al., 2009).

Palm oil may represent a very prolific resource for biodiesel production, but a few issues need to be considered. First, limited range of growth means that oil palm would not be available as a more global fuel feedstock and would require significant shipping around the world, negatively impacting its ecological footprint. Also, when considering the fuel properties of biodiesel derived from palm oil, a high percent of saturation means palm oil has extremely poor cold flow properties. With a pour point of 15 degrees Celsius (59 degrees Fahrenheit), most places in the world would be unable to utilize palm oil as a biodiesel feedstock; therefore, either varieties will need to be developed that produce an oil composition with a more acceptable temperature range or the oil will need additional chemical processing to impart these characteristic to the fuel. Finally, palm oil is a food source especially in those countries capable of producing a significant amount of the oil. Therefore, it would be necessary to strike a balance once again between food and fuel.

Jatropha for Biodiesel Production

Jatropha curcas is an undomesticated small tree or shrub that can grow on nonarable land and still produce a higher yield of biodiesel than traditional crops as shown in Table 8

FIGURE 7.10 Pictures of *Jatropha curcas*. (A) Picture of the *J. curcas* tree on a plantation in Paraguay. (B) Picture of the seeds produced by the *J. curcas* tree (images by "Jatropha in Paraguay Chaco" by Helmut von Brandenstein, von Brandenstein Plantations Agroconsulting. Licensed under Attribution via Wikimedia Commons, http://commons.wikimedia.org/wiki/File:Jatropha_in_ Paraguay_Chaco.jpg#/media/File:Jatropha_in_Paraguay_Chaco.jpg; "J curcas seed ies" by Frank Vincentz, Own work. Licensed under CC BY-SA 3.0 via Wikimedia Commons, http://commons.wikimedia.org/wiki/File:J_curcas_seed_ies.jpg#/media/File:J_ curcas_seed_ies.jpg).

TABLE 9

Select biodiesel properties for biodiesel produced from various first-generation and alternative feedstocks including soybean, rapeseed, palm, and jatropha compared to the properties of petrodiesel

Feedstock	Viscosity (mm²/s)	Pour Point (°C)	Flash Point (°C)
Petrodiesel	2.6	−20	68
Soybean	4.08	−3	69
Rapeseed	4.5	−12	170
Palm	4.42	15	182
Jatropha	4.8	2	135

DATA FROM: Gui and Lee (2008).

due to its large seeds as shown in figure 7.10. The two main characteristics making *Jatropha* a potentially valuable biodiesel crop are that it is not a food source and that it can be grown on marginal lands rather than competing for arable lands with food crops. This perennial tree grows to about 5–10 meters and can survive for long periods without water. Honeybees pollinate this species and the seeds can mature after about three to four months. Jatropha plants can produce oilseeds for 40–50 years once established and the cost of producing oil from these plants is cheaper than soybeans; however, its cost is highly dependent on labor costs as its seeds are hand-harvested. Jatropha grows from about 30°N to about 35°S, a range significantly larger than oil palm (Kumar and Sharma, 2008; Jingura et al., 2010). Thus, biodiesel produced from *Jatropha* could be produced within a much wider global range.

The oil composition of *Jatropha* is a balance of saturated and unsaturated hydrocarbons similar to that found in soybeans. Table 9 shows some of the biodiesel properties of *Jatropha* as compared with biodiesel produced from other feedstocks and petrodiesel. Here it is evident that the biodiesel from *Jatropha* has similar properties to other feedstocks that separate it from petrodiesel including its higher viscosity and higher pour point (Gui and Lee, 2008). A pour point of 2 degrees Celsius would need to be lowered to be able to use this biodiesel fuel in temperate and colder climates.

While *Jatropha* is being studied as a dedicated biodiesel crop, it also has a number of other uses. It has been used for centuries as a medicinal agent and it is also commonly used as a hedge to assist in erosion control. One benefit of *Jatropha* as a biodiesel feedstock is that it will not compete with the other oilseed crops as a food source. In fact, *Jatropha* contains two major toxins, curcin and phorbol esters, that would actually prevent it from being used as a source of food (Achten et al., 2008).

HISTORICAL USE OF *JATROPHA*

Travis L. Johnson

Although *Jatropha* is now being researched and used as a feedstock for biodiesel production, it has been used as a medicine around the world for hundreds of years. *Jatropha* likely originated from Central America, and was transported to South America and Africa by Portuguese explorers and subsequently to India and Southeast Asia. The name *Jatropha* comes from "jatrós" (doctor/physician) and trophé (food/nutrition) and was used for purging due to its toxicity. It has also been used to treat arthritis, gout, jaundice, toothaches, allergies, scabies, small pox, HIV, and tumors. The oils produced in *Jatropha* seeds are an excellent source for biodiesel, and the *Jatropha* plant can be grown on marginal land, making it a potentially good source of sustainable biofuel production.

SOURCE: Schmidt (2014).

Because *Jatropha* has never been domesticated, superior seed stock and best agricultural practices remain largely unestablished. Identifying ideal irrigation methods, fertilizer combinations, and plant cultivars through selection and breeding all have the potential to greatly increase *Jatropha*'s agricultural productivity. For instance, domestication of the rubber tree over the course of 100 years increased the yield of rubber significantly (Schultes, 1993). While 100 years may seem way too long, it is expected that the time needed for the domestication of *Jatropha* would be vastly decreased due to advances in biotechnology and genetic screening methodologies.

Environmental and Economic Impacts of Biodiesel

While there are many potential sources of biodiesel being developed, biodiesel production will still depend on the most common feedstocks including rapeseed, soybean, and even oil palm, for the foreseeable future. Thus, it is important to review the environmental impact of biodiesel production in order to understand the ramifications of increased production of biodiesel. One way to analyze this environmental impact is through a life cycle analysis for soybean-derived biodiesel. The three most energy-consuming processes during soybean biodiesel production are the agricultural production of the feedstock itself, the crushing of the soybeans to release the oil, and the conversion of the oil into biodiesel. The fossil energy used for producing the feedstock is largely bound up in irrigation, applied fertilizers, planting, and harvest, but since soybeans are capable of producing their own nitrogen, the level of fertilizer needed is much lower than many other crops. In both crushing and oil conversion, the fossil energy used is largely for powering machinery and heating. When comparing total energy input with total output, soybean biodiesel production does ultimately generate energy with an energy returned on energy invested of 4.56. This means that the output of energy exceeds the input of fossil energy by 4.56 times (Pradhan et al., 2009). This comparison can be taken one step further when you compare the energy return of biodiesel from soybean to that of bioethanol from corn. The production of biodiesel from soybean oil is much more energy efficient than the production of bioethanol from corn, the latter of which has an energy return of only 1.67 (Taylor et al., 2010).

As discussed in this chapter, the cost of biodiesel is directly linked to the cost of the feedstock. The feedstock makes up the largest portion of the cost of oilseed and this cost carries over to the production of biodiesel. Many governments, including the United States, offer tax incentives to help make biodiesel more economically competitive. In the United States, the tax incentives include $1 per gallon for agricultural biodiesel blending sources and $.50 per gallon for blending a recycled source like waste vegetable oil or animal fat (Yacobucci, 2012). These tax incentives are important for the competitiveness of biodiesel on the open market. In fact, when the tax credits expired in 2009, production of biodiesel diminished below what was needed to meet the RFS mandate (Schnepf and Yacobucci, 2012). These incentives have since been reinstated. Even with these incentives, biodiesel is still more expensive than pet-rodiesel and is only competitive with petrodiesel if the price of the feedstock and production is kept very low.

The Future of Biodiesel

In order for biodiesel to become truly competitive in the United States or world market as a replacement fuel for petro-diesel, biodiesel production capabilities will need to increase significantly. In 2010, only 10 million barrels of biodiesel were actually produced. This dwarfs in comparison to the 3.87 million barrels per day (1.4 billion barrels per year) of diesel used for highway transportation alone (Habiby, 2011).

The ability to match the high demand of petrodiesel and replace it with a renewable biodiesel will depend on the optimization and biotechnological advances of new feed-stocks as well as the recycling of waste oils and fats already available. Developing an infrastructure for recycling waste oils and fats would be one step toward lowering the con-sumption of petrodiesels while building a larger market for biodiesels. However, the future success of biodiesel will undoubtedly largely depend on the production of a dedi-cated and inexpensive crop for its production. Both oil palm and *Jatropha* could represent this type of crop in the United States. As will be discussed in Chapter 9, algae also represent a potential high-yielding source of oil that could be used to transform the future of biodiesel.

STUDY QUESTIONS

1. Explain how the chemical structure of ethanol, diesel, and gasoline molecules impact their energy density.
2. Briefly explain why understanding the biochemistry involved in the biosynthesis of fatty acids may be important for the future development of biodiesels.
3. Outline the process of biodiesel production from a plant source. Why is transesterification such an important part of this process? What needs to happen during the transesterification process to make biodiesel more cost-competi-tive at a commercial scale?
4. Briefly describe the major fuel properties that should be considered when producing biodiesel fuels. Which of these properties are likely to represent the biggest obstacle for the development of biodiesels?
5. What biomass feedstocks are currently being used to produce biodiesel? Why are these unsustainable? What are some of the alterna-tive feedstocks that may be valuable in biodiesel production and why might they be better than currently utilized feedstocks?

CHAPTER EIGHT

Gaseous Biofuels: Biogas and Biohydrogen

The term "biofuel" is most commonly associated with ethanol and biodiesel because these renewable liquid fuels are often thought of as replacements for the fossil-derived gasoline and diesel that we use every day. You have probably gone to a gas station and seen the "10% ethanol blend" sticker displayed on the pump. However, the term biofuel actually refers to a plethora of materials used as fuel sources including biologically derived combustible materials such as dead plant matter, cow dung, or wood, as well as the gaseous by-products from biological metabolism including the digestion of plants by cows and the breakdown of organic materials in landfills. You may have personally experienced the production of these gaseous by-products after eating certain intestinally challenging foods like beans. The production of gas in the human intestines generally consists of oxygen, nitrogen, and carbon dioxide, but flammable methane and hydrogen gases can also be produced. Methane (biogas) and hydrogen (biohydrogen) are generated within the human intestines by anaerobic bacteria, and similar reactions occur in a number of natural environmental processes, some of which can be scaled to relatively large volumes and thus hold the potential to be a source of renewable gaseous biofuels.

While both biogas and biohydrogen represent renewable and environmentally friendly alternative fuels, most gaseous fuels today are derived from nonrenewable fossil fuels, particularly natural gas. The United States withdrew over 30 trillion cubic feet of natural gas in 2013 from 487,286 wells across the country. During this same period, the United States consumed 26 trillion cubic feet, largely for electricity generation, industrial uses, and residential uses such as heating (EIA, 2014a). Of the three main fossil fuels, natural gas is considered the cleanest burning of these fuels, emitting only 117,000 pounds CO_2 per billion BTU of energy compared to 208,000 pounds from coal and 164,000 pounds from petroleum (King, 1999). However, natural gas,

like the other fossil fuel resources, takes millions of years to develop deep in the Earth, and thus is not a sustainable fuel for the future due to the high consumption levels today. While considered a cleaner burning fuel, obtaining and using natural gas are associated with environmental damage. You may recall that methane, the main component of natural gas, is 21 times more potent as a greenhouse gas than carbon dioxide. While methane that is burned ultimately generates carbon dioxide, there is still the threat of raw methane leaking from the natural gas wells and pipelines. In addition, recent technological advances have made horizontal drilling and hydraulic fracturing (fracking) economically competitive, especially in the United States, for the production of natural gas.

Hydraulic fracking combined with horizontal drilling allows fossil energy companies to make one vertical well and then branch out from this well horizontally deep within the Earth. Once the horizontal wells are drilled, high-pressure water combined with chemicals can be shot into the wells to crack open the surrounding rock. This fracking process ultimately allows the natural gas trapped within the rock to escape and travel to the surface through the well. Hydraulic fracking is a growing industry that now allows access to natural gas reserves previously unavailable due to a lack of economic return from these reserves. However, hydraulic fracking is a controversial process due to the potential to contaminate groundwater with the chemicals used in the fracking process and the potential to stimulate earthquakes due to the disruption of the rock surrounding the wells.

Currently, the global reserve-to-production ratio predicts that there is a little more than 55 years worth of natural gas remaining (British Petroleum, 2014). As discussed above, the recent technological advances in hydraulic fracking have allowed previously unattainable natural gas associated with shale deposits in the United States to become a viable resource. The continued development and extraction of these new sources of natural gas could have a dramatic impact on the proved reserve calculation in the United States (currently only 13.6 years) and extend the number of years of viable natural gas in the United States considerably. However, rising consumer demand and concern for the environment indicate that the development of renewable biogas will be extremely valuable in the future.

Another gaseous fuel is hydrogen. Hydrogen is an energy carrier currently generated mostly from natural gas. Hydrogen is not a common fuel source in the United States but can be utilized for electricity generation and as an automobile fuel. Unfortunately, since hydrogen is currently derived mostly from natural gas, it depends on the availability of this fossil fuel for production (DOE, 2013). Thus, as natural gas supplies become limiting or too expensive, the availability of fossil-derived hydrogen will also suffer. For hydrogen to be a viable fuel source for the future, there is a need to develop alternative biohydrogen production methods, including the production of this gas from microorganisms. The remainder of this chapter will focus on understanding both biogas and biohydrogen as energy resources for the future.

Production and Use of Biogas

While carbon dioxide attracts the most attention as the anthropogenic greenhouse gas with the highest concentration in the atmosphere, there are other more potent greenhouse gases that also contribute to radiative forcing in the atmosphere. In 2007, the IPCC reported that about 28% of global radiative forcing since the pre-industrial era is caused by the emissions of greenhouse gases other than carbon dioxide including methane and nitrous oxide. Of this 28%, methane makes up the largest portion at 20.7%. The atmospheric concentration of methane has risen steadily although at lower concentrations. But don't let the lower concentrations fool you; every molecule of methane is more powerful than any single molecule of carbon dioxide. Therefore, a small change in atmospheric methane concentrations can have a relatively bigger impact on global warming. Thus, limiting emissions of methane will be crucial to reducing climate change in the future (Alsalam and Ragnauth, 2011).

We most often associate the emission of methane with the use of natural gas related to energy consumption, but the production of anthropogenic methane gas comes from a variety of other sources as well. Figure 8.1 shows that 40.5% of methane emissions result from energy-related activities; however, both agriculture sources (35.5%) like enteric fermentation and waste disposal (20.7%) including landfills and waste treatment represent large sources of emissions as well (Hockstad et al., 2014). Interestingly, one of the largest sources of anthropogenic methane in the

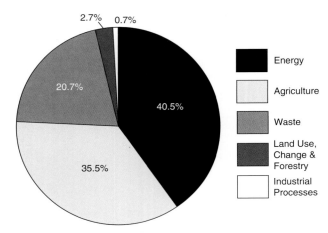

Total Methane Emissions – 567.29 Million Metric Tons of Carbon Dioxide Equivalents

FIGURE 8.1 Comparison of US anthropogenic methane emission sources as a percentage of the total US methane emissions of 567.29 million metric ton CO_2 equivalent. Anthropogenic methane is largely a result of three sectors: energy, agriculture, and waste management (data from Hockstad and Weitz, 2014).

United States comes from **enteric fermentation**. Enteric fermentation is the digestion of carbohydrates in animals by microorganisms resulting in the release of methane gas. The amount of methane produced by an individual animal depends on the digestive tract of that animal. Ruminant animals like cattle, buffalo, sheep, and goats are the largest emitters of methane due to the presence of a rumen or forestomach. In 2012, global livestock were responsible for emitting an equivalent amount of methane to 141.0 teragrams or 141.0 million metric tons of carbon dioxide with 71% of this coming from beef cattle (Hockstad et al., 2014). In 2013, the cattle inventory in the United States alone was 89.3 million head, allowing for the export of 10.0% of this inventory, and providing the 25.5 billion pounds of beef consumed in the United States (Mathews, 2014). Clearly, the desire to consume large quantities of beef in the United States has an impact on anthropogenic methane emissions and the subsequent potential for these emissions to play a role in climate change.

Whether derived from livestock enteric fermentation, natural gas wells, or poor waste management, the anthropogenic methane gas entering our atmosphere represents a wasted energy source. Just as fossil-fuel-derived natural gas is used for a variety of energy-consuming activities including power generation and heating, all of the biological sources of methane gas can also be used as fuel, or more precisely as biogas. Biogas results from the conversion of organic matter directly into a gas, usually a mixture of methane and carbon dioxide. This gas can be used in the same manner as natural gas; yet, it is generated from sustainable activities and waste that already exist in society as opposed to being extracted from the ground. Thus, biogas represents a renewable resource available to replace the finite and environmentally challenging sources of fossil-fuel-derived natural gas.

Anaerobic Digestion for Methane Production

As you can imagine, it is not practical for all methane to be trapped and used as biogas. For instance, collecting enterically produced gas directly from individual cows is probably not a viable option (or very comfortable for the cow). However, many methane-producing activities can be adapted to trap the biogas and utilize it as a source of energy. The production of biogas relies on a process known as **anaerobic digestion**. Anaerobic digestion is a bacterial fermentation process that occurs in the absence of oxygen. This process is different than **aerobic conversion** typically used during commercial composting and the breakdown of activated sludge wastewater because aerobic conversion utilizes oxygen and often produces non-useful gases in terms of fuel use, such as carbon dioxide and water. Anaerobic digestion, in contrast, occurs in oxygen-free (anaerobic) environments and produces biogas, a mixture of methane and carbon dioxide. Anaerobic digestion occurs naturally in swamps, water-saturated soils, rice fields, and, as mentioned previously, in the digestive tract of some animals. Yet, anaerobic digestion can also be accomplished in specialized facilities that use a variety of substrates including wastewater, food wastes, and landfill materials to produce biogas, making this natural process a powerful option for the production of a biofuel.

The anaerobic digestion process has four major steps as outlined in figure 8.2 including **hydrolysis, acidogenesis, acetogenesis**, and **methanogenesis** (Weiland, 2010). Each of these steps represents a different set of chemical reactions that takes complex polymers (mixtures of molecules) found in organic materials and breaks them down to the simplest carbon-containing gases: methane and carbon dioxide. Hydrolysis is the first step where these complex mixtures of organic molecules such as proteins, carbohydrates, and lipids found in plant biomass and municipal type wastes are broken apart by enzymes to produce simpler units termed monomeric or oligomeric units. These simple units may include individual sugar molecules, amino acids, alcohols, and long-chain fatty acids. Hydrolysis can be the slowest (rate-limiting) step of the anaerobic process because of the complexity in recognizing and breaking down so many different types of substrates. Imagine disassembling a car and a bike; the car is going to take many more tools and much longer to disassemble because it is more complex. This is also true for organic molecules. An organic material that contains a large amount of lignin (an extremely complex and tough natural molecule discussed in Chapter 6) will be significantly harder and take much longer to break down than a simpler material made of starch (repeating sugar units). Just as was the case for cellulosic ethanol development, the balance of biomolecules

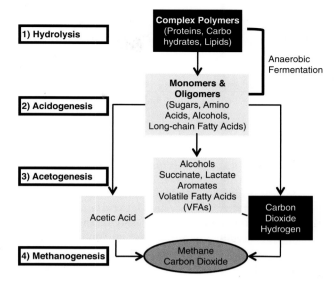

FIGURE 8.2 Schematic diagram of the anaerobic conversion process showing the use of four steps including hydrolysis, acidogenesis, acetogenesis, and methanogenesis for the conversion of complex polymers into methane and carbon dioxide gas (source: Weiland, 2010).

to be digested must be considered in the design of the anaerobic digesting facilities, and in the selection of the bacteria involved in the process to maximize digestion efficiency.

Once simpler units are generated, they are exposed to acid-forming bacteria in the step termed acidogenesis. The goal of acidogenesis is to produce organic acids including acetic acid and volatile fatty acids (VFAs). This process also generates some nonfuel gases such as carbon dioxide and hydrogen. In many cases, the first two steps—hydrolysis and acidogenesis—are combined and known as the anaerobic fermentation process.

The acetic acid produced during anaerobic fermentation is basically a food for the microorganisms in later steps, providing them the energy they need to continue breaking down substrates and producing biogas. While the acetic acid is consumed, the VFAs are not. In order to create a more efficient production process for biogas, a third step, acetogenesis, is added. This step involves the addition of different microorganisms that can digest the VFAs to produce acetic acid, carbon dioxide, hydrogen, and sometimes water.

Once these VFAs are converted to acetic acid, the acetic acid, carbon dioxide, and hydrogen can all be fed into the fourth and final reaction known as methanogenesis. Methanogenesis is powered by methanogens, or methane-forming microorganisms belonging to the very simple, single-celled organisms called Archaea. During this reaction, these methanogens convert acetic acid and water into methane, carbon dioxide, and water while also converting carbon dioxide and hydrogen into methane and water (Demirbas, 2009). In the end, this four-step anaerobic conversion process can

METHANOGENIC ARCHAEA

Travis L. Johnson

There is a group of bugs that are used in the renewable biogas production process called methanogenic archaea. These archaea take carbon dioxide and hydrogen and convert them into methane and water. Archaea are classified as different from both eukaryotic organisms (plants, insects, animals) and bacteria, but they are more similar to bacteria in their size and structure. Methanogenic archaea are considered extremophiles, meaning they are found in extreme temperature environments that have little oxygen like hot springs, deep-sea thermal vents, and the guts of animals and humans. Methanogenic archaea are currently the only known organisms that can be used to create methane. These organisms are also very efficient at using energy to produce methane, using very little of the carbon dioxide substrate for their own metabolism, resulting in a better yield of renewable biogas.

SOURCE: Hein (2014).

take almost any type of biological material and break it down into carbon dioxide and methane that can be used as a replacement for natural gas.

Commercial-Scale Biogas Production

The production of methane through anaerobic digestion on a small scale is complicated, and the commercial-scale deployment of such biological processes presents even bigger and more complex problems. To put this into perspective, think of the difference between cooking a meal for yourself and cooking a meal for a large party. It is a lot easier to cook for one or two people than for 100 or 200 people. An example of the complexity could be envisioned when considering the quantity and complexity of agricultural products used in a small-scale process compared to those used at the commercial level. A commercial facility is going to want to accept as many types of agricultural waste as possible and may range from livestock manure to plant detritus to corn silage. All of these products may require slightly different modifications to the overall process to produce the maximum amount of methane. At the same time a much smaller livestock ranch could utilize manure alone in an anaerobic digester to provide enough biogas to power their home, irrigation systems, and other operations. While this might be a much simpler process, the amount of biogas produced is also much lower than what a commercial facility can produce (Braun et al., 2010).

Regardless of the method chosen for biogas production, the rate of digestion and the level of biogas production will depend on a few key factors. The first and most important of these parameters is temperature. Ideal temperatures will result in the proper conditions for efficient bacterial metabolism. The anaerobic digestion process can occur at two different temperatures known as **mesophilic** and **thermophilic**. In the mesophilic process, digestion occurs at 95 degrees Fahrenheit for 15–30 days, whereas in the thermophilic process, digestion occurs at 130 degrees Fahrenheit for 12–14 days. One may ask why a biogas generation facility would prefer the longer period of time; it all comes down to energy utilization. It takes more energy to heat a digester to 130 degrees Fahrenheit than to 95 degrees Fahrenheit; therefore, some facilities would be better off maintaining a lower temperature for a longer period of time (Hamilton, 2009). When considering energy generation from any renewable resource, it is extremely important to obtain significantly more energy from the process than is required to power the process.

Another important operational parameter for anaerobic digestion is pH or acidity level. The pH for this process must be slightly acidic and maintained between 5.5 and 8.5. The challenge here comes from the fact that acetic acid and VFAs are generated and have acidic properties that will work to lower the pH within the digester, creating an even more acidic environment. If a pH is generated that is outside the optimal range, then the microorganisms involved in the digestion process will not work as efficiently and the production of biogas will be reduced. The same issue can occur with the amount of nutrients within each of the digestion chambers, known as the **organic load**. If levels of individual organic components are not maintained within a specified range, then microorganisms within a digester could be overwhelmed or underwhelmed, both resulting in a decrease in biogas production (Hamilton, 2009).

Another important parameter for the production of biogas through anaerobic digestion is the carbon-to-nitrogen ratio. This ratio should be maintained at around 20–30 carbon molecules to 1 nitrogen molecule. The reason for this is that methanogens use the carbon-based organic molecules' acetic acid and carbon dioxide to produce methane. Since these reactions do not utilize nitrogen, a high nitrogen level can actually reduce the production of biogas (Hamilton, 2009).

Finally, it is also important to consider the sustainability of this biogas production cycle. A potentially sustainable cycle for biogas production through anaerobic digestion begins with agricultural waste materials being transported to a centralized biogas plant. Here, anaerobic digestion is used to produce biogas, but there is still leftover digested biomass. This additional biomass contains valuable nutrients like nitrogen, sulfur, and other microelements, which can be utilized as fertilizers for fields including those fields from which the original agricultural waste came, thus providing a full-circle sustainable cycle for biogas production and adding further value to the biogas (Weiland, 2010).

Landfills and Biogas Production

In 2012, Americans generated 251 million tons of trash. This is equivalent to about 4.4 pounds of trash per person per day and represents one of the highest levels of per capita trash generation in the world. In fact, the global average is only 1.42 pounds per person per day (Hoornweg and Bhada-Tata, 2011; EPA, 2014). Most of this trash is classified as **municipal solid waste** (MSW), and the amount of MSW has been steadily rising for the past 50 years from 88.1 million tons in 1960 to 251 million tons in 2012 (EPA, 2014). In the United States, MSW comes in a wide variety of forms including paper, glass, metals, plastics, rubber, leather, wood, yard trimmings, food waste, and other sources. Most of the waste (62.2%) comes from organic materials like paper, yard trimmings, wood, and food scraps. Since these materials are organic and composed of biomolecules, they all represent potential substrates for anaerobic digestion. Some of these materials like paper can be recycled, but most MSW is simply transported to landfills where it decomposes over time without being used to generate energy (EPA, 2014).

Landfills are large outdoor sites designed specifically for the disposal of waste. These sites are designed to bury gar-

TABLE 10

Typical components found in landfill gas

Component	Content (% by volume)
Methane	40–60
Carbon dioxide	20–40
Nitrogen	2–20
Oxygen	<1
Heavier hydrocarbons	<1
Hydrogen sulfide	<1

DATA FROM: Demirbas (2009).

bage and avoid environmental contamination rather than allowing for optimal decomposition of the materials. For this reason, it can often take many types of materials much longer to decompose within a landfill than in other places. Because they are usually covered over with dirt, removing the supply of oxygen, and because they contain large amounts of organic materials, landfills often naturally result in the production of high levels of methane through anaerobic decomposition, sometimes even posing a risk of explosion. Table 10 presents an analysis of raw landfill gas. Methane composes 40–60% by volume of this gas followed by carbon dioxide and nitrogen. Despite constituting a high percentage by volume of landfill gas, the concentrations of methane and carbon dioxide actually peak at very different times. A study by Demirbas in 2009 showed that the percentage of carbon dioxide in landfill gas peaks fairly quickly at about 20 months, while the percentage of methane in the gas did not begin to increase until about 20 months and continued to increase until leveling out at about 120 months. This makes sense if you think back to the four-step process of anaerobic digestion—methanogenesis occurs last. From this study, it was concluded that the exploitation of landfill gas would be most economically feasible after the MSW had been in the landfill for two years (Demirbas, 2009).

In order to utilize the MSW-sourced biogas, the landfill would have to be specifically designed for this process. Figure 8.3 diagrams what a modern landfill with a methane gas recovery system might look like. In this case, the landfill contains a liner at the bottom that prevents liquids from the MSW from leaking into the groundwater supply; any nutrients or contaminants that build up are collected above this liner and pumped out of the landfill. The MSW is placed within receptacles built into the landfill space; these receptacles are filled and then covered over with dirt and clay. Between these receptacles is piping designed to capture the biogas generated during the natural anaerobic decomposition that occurs within the landfill. The biogas can then be pumped out of the landfill and used for varying combustion processes. Regardless of method, the important feature is

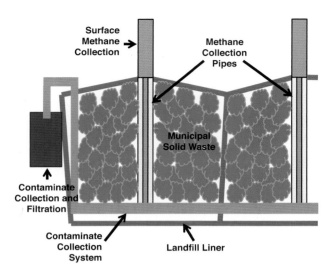

FIGURE 8.3 Example of a modern landfill with a methane collection system. The landfill should consist of a liner to protect the environment as well as a contaminant collection system to prevent leakage after rain into the groundwater. The landfill should also include methane collection pipes that collect the biogas as it is produced.

that the methane biogas is recovered from the landfill and an important renewable energy resource is used rather than wasted.

Biogas in the Future

As discussed, many forms of waste can be used in biogas generation; in fact, pretty much any organic substrate can be anaerobically digested into biogas. Not all substrates, however, are equal in their biogas yields. Weiland (2010) found that cow and pig manure had low average biogas yields (25 and 30 cubic meters per ton fresh material [FM], respectively), while food residues (240 cubic meters per ton FM), waste fats (400 cubic meters per ton FM), wheat or corn (630 cubic meters per ton FM), and used grease (800 cubic meters per ton FM) had the highest average biogas yields (Weiland, 2010). These high yields are likely due to the latter types of waste sources having high levels of fatty acids and sugars that can easily be converted into VFAs and acetic acid. Many of the agricultural wastes could simply be collected and used directly for biogas production rather than being placed in landfills, which requires a longer timeframe for biogas production.

The utilization of biogas as a source of energy has many positive impacts on the environment. Biogas use is particularly important in that it has a dual greenhouse gas emissions benefit. First, this gas is being produced regardless of whether it is collected for energy generation or not; therefore, by deliberately recycling waste for biogas production, the emission of atmospheric methane will be reduced. The second benefit comes in the reduction in CO_2 emis-

sions because rather than burning fossil fuels, particularly coal, to produce heat and electricity, biogas is being used. In a study by Börjesson and Mattiason in 2007, biogas production from liquid manure in Sweden resulted in a 180% average reduction of greenhouse gases when fossil fuels were replaced by this biogas in vehicles. This reduction in greenhouse gas emissions is significantly more than the biodiesel produced from rapeseed (reduction of 60%) or ethanol from wheat (reduction of 70%) (Börjesson and Mattiasson, 2008).

Sweden uses "boiled gut biogas" in a number of transportation applications. In a first for the world, a train named Amanda runs solely off biogas on her 166 kilometer journey. To produce the biogas, the intestines, udders, blood, kidneys, and livers of cattle used for meat production are heated and then submitted to anaerobic digestion for one month. The entrails of one cow can move the train 4 kilometers (CNN, 2009). Is this sustainable for the future of biogas in the world? Probably not, but it is a step in the right direction where biogas production is concerned. It proves that recycling organic waste is both feasible and profitable in the production of biogas.

Hydrogen Production and Use

Hydrogen is one of the most abundant elements on Earth, and an essential component for both industrial and food processing. But hydrogen gas is not found in large quantities on Earth because hydrogen gas is lighter than air and rises very quickly through the atmosphere before being released into space. Hydrogen gas (H_2), made by coupling two atoms of hydrogen together, will be referred to as hydrogen for the remainder of the chapter. The United States produces about 9 million metric tons of hydrogen annually. Almost all of this hydrogen is used by industry for oil refining, treating metals, and processing foods. Recently, hydrogen has generated publicity for its potential use as a transportation fuel in hydrogen fuel cell cars that emit no carbon dioxide. These fuel cells are fairly expensive, and the United States lacks the infrastructure to distribute hydrogen to a large number of consumers; thus, the number of hydrogen fuel cell vehicles today is limited to about 500 in total in the United States (EIA, 2014b). However, research continues on methods to generate and distribute hydrogen in an efficient and cost effective manner.

Currently, hydrogen is largely produced in one of two methods: steam reforming of methane or electrolysis of water. Most hydrogen production (95%) comes from steam reforming. It is the least expensive method in which hydrogen atoms are separated from carbon atoms in natural gas. Not only does this method rely on finite fossil fuel resources, but it also results in the emission of the greenhouse gas carbon dioxide. Hydrogen can also be generated by thermochemical methods discussed in Chapter 11; however, most of these methods also rely on the use of fossil fuels like coal (DOE, 2010; EIA, 2014b).

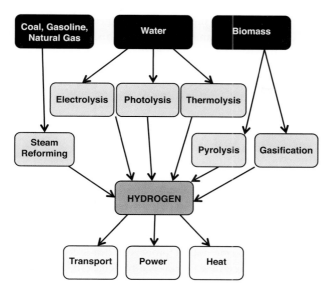

FIGURE 8.4 Flow diagram of methods used in hydrogen production. Hydrogen gas is generally produced from one of three sources, fossil fuels, water, or biomass, through a number of different methods that include steam reforming, electrolysis, photolysis, thermolysis, pyrolysis, and gasification (source: Manish and Banerjee, 2008).

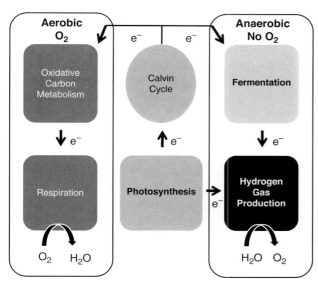

FIGURE 8.5 Diagram of aerobic and anaerobic metabolic pathways in relation to hydrogen production. When taken together, these pathways coordinate to provide electrons and a source of hydrogen (water) for the anaerobic production of hydrogen gas (source: Carrieri et al., 2008).

Due to its high cost and infrastructure requirements, hydrogen has had limited use as a transportation fuel; however, research continues in an effort to develop a cost-effective, renewable method for producing and distributing hydrogen. While fossil fuels represent the most used source for the production of hydrogen, both water and biomass can also be used to produce hydrogen. Figure 8.4 provides

an overview of the methods that can be used to produce hydrogen from fossil fuels, water, and biomass for use in transportation, power, and heat generation.

Water (H_2O) can be used for the production of hydrogen in a process called electrolysis. During electrolysis, water is split into its component parts, oxygen and hydrogen gas, by passing an electrical current through water. However, the high cost of this process, due to the cost of electricity and the expensive required catalysts, has limited the use of electrolysis as a large-scale process to make renewable hydrogen (EIA, 2014b). Nevertheless, this remains a potential source of renewable hydrogen, as both wind- and photovoltaic-generated electricity can be used in this process, once a cheaper catalyst is developed.

Recently, research has focused on using biomass as a source for hydrogen production. There are three methods in which biomass can be used for the production of hydrogen: photolysis, fermentation, and thermochemical methods. The thermochemical methods use biomass as a substrate for pyrolysis or gasification, which will be discussed in Chapter 11; however, both photolysis and fermentation use metabolic properties in photosynthetic and non-photosynthetic living microorganisms to produce biohydrogen in a completely renewable manner. These processes will be discussed in the next sections.

Fermentative Biohydrogen Production

Whether single-celled or multi-celled all living organisms are powered by metabolic pathways that provide the energy and organic molecules needed for their function and growth. These same metabolic pathways when coupled in the proper manner in select microorganisms can also work to produce hydrogen. Figure 8.5 summarizes these metabolic pathways and how they can interact to produce hydrogen. It is important to note that hydrogen gas is generally produced only in an anaerobic environment, despite using substrates that are found in other aerobic metabolic pathways. The main reason for this is that the final enzyme needed to generate hydrogen gas, hydrogenase, can be poisoned by oxygen. As shown in the figure, both photosynthesis and fermentation can directly generate the substrates used by proton-reducing enzymes to produce hydrogen. The direct relationship between these two metabolic pathways and hydrogen production leads to two types of microorganisms being capable of hydrogen production: photosynthetic and anaerobic microorganisms.

There are four main pathways for the production of hydrogen from biomass, which include **dark fermentation**, **photo fermentation**, **indirect photolysis**, and **direct photolysis** as shown in figure 8.6. Fermentative bacteria that employ the dark fermentation method of hydrogen production such as *Enterobacter* and *Clostridium* use a process similar to that described for the production of methane in biogas. The dark fermentation process generally results in hydrogen production from anaerobic metabolism.

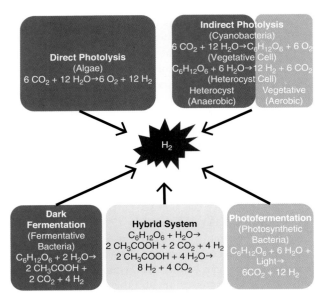

FIGURE 8.6 Microbial metabolic pathways for the production of hydrogen gas. Four major pathways utilized by different types of bacteria are shown including direct photolysis, indirect photolysis, dark fermentation, and photofermentation. In addition, a fifth potential pathway that uses a hybrid system combining dark fermentation and photofermentation is shown. This hybrid system may prove to be more efficient than the individual pathways for hydrogen gas production (source: Kotay and Das, 2008; Manish and Banerjee, 2008).

While the first three steps, hydrolysis, acidogenesis, and acetogenesis, are the same between the biohydrogen fermentation process and that of methane production, for biohydrogen fermentation, any type of methanogenic reaction must be inhibited to avoid the hydrogen gas being combined with carbon dioxide to form methane rather than remaining as hydrogen gas. The basic equation for hydrogen production by dark fermentation is shown below.:

$$C_6H_{12}O_6 + 2H_2O \rightarrow 2CH_3COOH + 2CO_2 + 4H_2$$

$$1 \text{ Glucose} + 2 \text{ Water} \rightarrow 2 \text{ Acetic acid} + 2 \text{ Carbon Dioxide} + 4 \text{ Hydrogen}$$

In general, dark fermentation is a fairly efficient method for producing biohydrogen from a variety of carbon sources without the presence of light. *Enterobacter cloacae* has been shown to produce biohydrogen at a rate of 75.6 millimole hydrogen per liter per hour (Kotay and Das, 2008; Manish and Banerjee, 2008). However, the quantity of biohydrogen produced by any organism can depend largely on the substrate provided to the organism.

Photo fermentation is a similar anaerobic metabolic process to produce biohydrogen. The main difference between it and dark fermentation is that photosynthetic bacteria are responsible for photo fermentation. These bacteria, including *Rhodopseudomonas* and other purple non-sulfur bacteria, use a wide spectrum of light energy as well as a variety

of waste materials to supply the energy needed to produce biohydrogen (Kotay and Das, 2008; Manish and Banerjee, 2008). The basic equation for photo fermentation using glucose is shown below where glucose ($C_6H_{12}O_6$) combines with water (H_2O) and light to produce six molecules of carbon dioxide (CO_2) and 12 molecules of hydrogen gas (H_2):

$$C_6H_{12}O_6 + 6H_2O + \text{Light} \rightarrow 6CO_2 + 12H_2$$

This process should not be confused with photosynthesis. During the photosynthetic process, oxygen is generated through the splitting of water during the light reactions; however, in photo fermentation light energy is solely used to drive the conversion of sugar molecules like glucose and water into carbon dioxide and hydrogen gas.

The photo fermentation process is dependent on the enzyme **nitrogenase**. Usually nitrogenase is associated with the conversion of nitrogen into ammonia and other organic forms of nitrogen that are particularly important for plant growth. But this enzyme is strongly inhibited by oxygen and stimulated by light; thus in most cases, normal photosynthesis would inhibit nitrogenase activity (Kotay and Das, 2008; Manish and Banerjee, 2008). However, in an oxygen-free environment where there is also no nitrogen for the nitrogenase enzyme to convert into organic nitrogen, this enzyme is capable of a biohydrogen-producing reaction. When nitrogen gas (N_2) is present, the enzymatic reaction results in the formation of ammonia (NH_3); however, if nitrogen gas is not present, the enzymatic reaction will shift to converting protons (H^+) into hydrogen gas (H_2) while using cellular energy in the form of ATP as shown below:

$$\text{Nitrogen Fixation} = N_2 + 8H^+ + 8e^- \rightarrow 2NH_3 + H_2$$

$$H_2 \text{ Production} = 2H^+ + 2e^- + 4ATP \rightarrow H_2 + 4ADP + 4Pi$$

Photo fermentation is not as prolific as dark fermentation in biohydrogen production. For instance, *Rhodobacter sphaeroides* only produces 0.16 millimole hydrogen per liter per hour. Yet, ideally dark fermentation and photo fermentation could be combined in a hybrid system. This process could use the acetic acid, produced as a by-product of dark fermentation, as a substrate for photo fermentation by photosynthetic bacteria to produce more hydrogen and carbon dioxide, thus increasing the efficiency of these hydrogen-generating processes (Kotay and Das, 2008; Manish and Banerjee, 2008).

In a study by Manish and Banerjee, the fermentative methods for biohydrogen production are compared to the commonly used steam methane reforming technology to better understand the energy requirements and greenhouse gas emissions of these different processes. While steam methane reforming produces 12.8 kilogram of CO_2 per kilogram of hydrogen produced and uses 188 megajoules of nonrenewable energy, the fermentative process produces on average 4.1 kilogram of CO_2 per kilogram of hydrogen produced and uses 47 megajoules of nonrenewable energy. When comparing dark fermentation, photo fermentation and the hybrid process to each other, dark fermentation produces the most carbon dioxide and uses the most non-renewable energy, indicating that continued research into fermentative hydrogen production should use the photo fermentation process for the best results in consideration of nonrenewable energy use and potential for greenhouse gas reduction (Manish and Banerjee, 2008).

Photolysis and Biohydrogen Production

While some bacteria can use photo fermentation in an anaerobic process to produce biohydrogen, true photosynthetic microorganisms such as algae and cyanobacteria can also produce biohydrogen using a process known as **photolysis**. Photolysis is a chemical process that uses light to catabolize or break apart larger biomolecules into smaller biomolecules. In the case of photolysis in photosynthetic microorganisms, water can be catabolyzed to produce hydrogen gas.

This process was first observed in cyanobacteria and later shown to occur in eukaryotic algae. Photolysis, whether direct or indirect, utilizes the enzyme hydrogenase. Hydrogenases function to produce hydrogen gas by breaking down water to produce oxygen and hydrogen. As described earlier, other species of bacteria can use the enzyme nitrogenase in nitrogen-free conditions to produce biohydrogen; however, a dedicated hydrogen-catalyzing enzyme like hydrogenase is more efficient. Hydrogenase enzymes have been shown to have species diversity, thus also giving them some basic diversity in functionality (Melis and Happe, 2001). However, generally hydrogenase enzymes function best in anaerobic conditions where no oxygen is present.

If you recall from your reading about photosynthesis, one of the major products of photosynthesis, particularly the light reactions, is oxygen. So how do photosynthetic organisms that are producing oxygen use hydrogenase enzymes that function best in minimal oxygen conditions? One way is known as indirect photolysis. Indirect photolysis typically occurs in filamentous cyanobacteria such as *Anabaena variabilis* that have specialized cells known as heterocysts. These heterocysts are separate individual cells that have an anaerobic environment allowing for both nitrogen fixation and the production of hydrogen at the same time that oxygen is being generated by photosynthesis in different vegetative cells within the filament. Generally, hydrogen production uses a metabolic process where the first photosynthetic step of converting carbon dioxide and water into glucose and oxygen occurs in the normal vegetative cells. Then the glucose and water migrate into the heterocyst where a second round of reactions catalyzed by the hydrogenase enzyme will produce hydrogen gas and carbon dioxide (Kotay and Das, 2008; Manish and Banerjee, 2008). The chemical reaction formulas for this process are shown below where the vegetative cell hosts the fixation of carbon dioxide (CO_2) and water (H_2O) to

form glucose ($C_6H_{12}O_6$) and oxygen (O_2), while the heterocyst reaction uses the glucose ($C_6H_{12}O_6$) and water (H_2O) to produce hydrogen gas (H_2) and carbon dioxide (CO_2):

Step 1: $6CO_2 + 12H_2O \rightarrow C_6H_{12}O_6 + 6O_2$ (vegetative cell /aerobic conditions)

Step 2: $C_6H_{12}O_6 + 6H_2O \rightarrow 12H_2 + 6CO_2$ (heterocyst /anaerobic conditions)

A second type of photolysis, direct photolysis, does not use separate cells but rather photosynthesis and hydrogen production occur in the same cell. This is the type of photolysis used by many microalgae such as *Chlamydomonas reinhardtii*. During direct photolysis, hydrogen production occurs when the production of oxygen from water splitting by photosystem II is inhibited. In direct photolysis, the electrons and protons formed during the light reactions are used to provide the substrate for the production of biohydrogen. Direct photolysis is generally less efficient than indirect photolysis, and photosynthetic growth cannot proceed when photolysis is occurring (Kotay and Das, 2008; Manish and Banerjee, 2008). Thus, continued development of photolysis for biohydrogen production likely will require a balance of photosynthetic growth and hydrogen production with a focus on the strains and processes developed to allow for both. While biohydrogen production offers an opportunity to potentially lower the use of fossil-fuel-derived natural gas, it has some barriers that must be overcome before it will be competitive at a commercial scale. These barriers include a need for a better understanding of the basic biology of the hydrogen-producing organisms and their enzymes, better fermentative capabilities with better fermentative strain development, and a need for advancements in engineering including the development of reactors and capture methods to maximize hydrogen production.

Biogas and Biohydrogen in the Future

In a period of time when fossil fuels are likely to become less available and more expensive, it is important that we develop alternative and renewable methods for the production of all different types of fuels. Natural gas makes up about 23% of energy consumption and thus represents a significant percentage of fossil fuel use. In order to replace natural gas as a feedstock for electricity generation and heating, it will be necessary to produce renewable gaseous fuels. Two options for these fuels have been described in this chapter, biogas and biohydrogen. Both of these renewable gaseous fuels rely on feedstocks that are either generated as biomass or result from the recycling of large amounts of digestible waste. Biogas offers a more direct replacement for fossil-fuel-derived natural gas without the requirement of major biotechnological and technological advances. Through continued development, both biogas and biohydrogen could prove to be a valuable piece of the puzzle in meeting the energy needs for the future.

STUDY QUESTIONS

1. Briefly explain the advantages and disadvantages of using fossil-fuel-derived natural gas as an energy resource. How is using biogas different?
2. Why could emissions from the use of natural gas play a bigger role in the level of atmospheric greenhouse gases and the greenhouse effect in the future? Where are the emissions of methane likely to come from?
3. Explain the differences between anaerobic digestion and aerobic digestion. Briefly describe the steps in anaerobic digestion and their roles in producing biogas.
4. How can landfills be used for biogas production?
5. Explain how hydrogen gas can be used as an energy resource. Why hasn't hydrogen become a more prevalent resource in the United States?
6. Briefly describe the four main biological pathways for biohydrogen production.

CHAPTER NINE

Aquatic Versatility for Biofuels: Cyanobacteria, Diatoms, and Algae

When asked about photosynthesis, most people envision green leaves and root systems of common plants like trees or shrubs. We might even picture these plants photosynthesizing to produce sweet fruit, colorful flowers, or the oxygen needed for us to survive. But despite our initial notions of the importance of photosynthesis in these common plants, land plants only contribute about half of the oxygen found on Earth (Bosch et al., 2010). The other half of this life-sustaining gas is generated from a group of much smaller photosynthetic organisms known as **phytoplankton**. Phytoplankton is a general term for microscopic autotrophs (meaning they produce the organic materials needed for growth such as is the case with photosynthesis) that inhabit any aquatic systems including lakes, ponds, rivers, and oceans. Due to their autotrophic nature, these tiny creatures are generally at the bottom of the aquatic food chain, and thus provide nutrients to higher-level organisms including some of the largest, whales. Due to their photosynthetic metabolism, phytoplankton play an important role in sequestering carbon in aquatic environments. Yet, despite their natural abundance in these environments, their immense role in the planet's ecosystems, and their vital importance in the balance of atmospheric gases allowing life to thrive on Earth, these versatile microscopic organisms are often overlooked for their potential in the development of biofuels. This chapter will introduce three types of phytoplankton and their potential to produce a wide variety of biofuels including biodiesel, bioethanol, biogas, biohydrogen, and other renewable hydrocarbons.

Introduction to Aquatic Biomass

Aquatic biomass includes all organisms and their by-products that use photosynthesis within an aquatic environment such as a lake, river, pond, or ocean. Since the organisms that compose aquatic biomass are autotrophic,

they play an important role in the food chain within each of these ecosystems. Aquatic biomass includes the microscopic phytoplankton mentioned earlier as well as the large macroscopic aquatic plants often referred to as seaweed. In general, the most important forms of aquatic biomass being studied in biofuel research fall under two general categories: **microalgae** and **macroalgae**.

Microalgae is a term that has been adopted to include all photosynthetic aquatic microscopic (thus the Latin prefix micro) organisms including "true" algae, diatoms, and cyanobacteria. **Cyanobacteria** are photosynthetic prokaryotic bacteria thought to be responsible for much of the original oxygenation of the planet. These microscopic bacteria are found not only in aquatic environments but also in almost every other environment on Earth including many extreme environments like the Antarctic. To survive in so many different ecological niches, cyanobacteria have evolved a large array of unique metabolic functions that, combined with their photosynthetic capabilities, make them valuable targets for the production of biofuels and the identification of genetic traits that could benefit other biofuel-producing organisms (Sorrels, 2009). The second two examples of microalgae—true algae and diatoms—are usually microscopic in nature, yet are significantly more complex than the cyanobacteria. True algae and diatoms are true eukaryotes that possess membrane-bound organelles separating many of their metabolic functions. For instance, both true algae and diatoms possess a chloroplast functioning in photosynthesis with a cellular design similar to that seen in higher plants (see figure 6.1).

The second major group of aquatic biomass important in biofuels research is macroalgae or seaweeds. Macroalgae are generally macroscopic (thus the Latin prefix *macro*), multicellular eukaryotic algae that closely resemble a traditional terrestrial plant and inhabit aquatic environments. While the relatively large size of macroalgae and the conditions

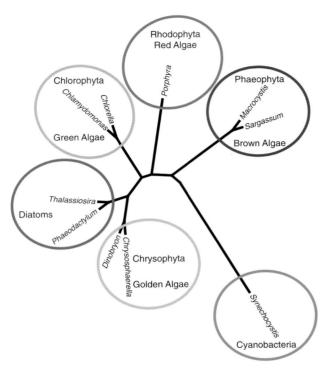

FIGURE 9.1 Basic evolutionary comparison of the types of algae. Although often referred to as algae, cyanobacteria are prokaryotic organisms and are evolutionarily separated from the other true algae and diatoms. Diatoms and true algae are eukaryotic and have a chloroplast responsible for photosynthesis. True algae are sometimes classified based on their dominant color—brown, red, green, or golden (tree based on 18S and 16S rRNA sequences obtained from the National Center for Biotechnology Information).

required for growth make their use in biofuels more difficult, they still represent a productive biomass source with the potential for biofuels development, particularly in fermentative, digestible, and combustible processes.

While all of these aquatic organisms are related due to their photosynthetic nature, an evolutionary comparison shown in figure 9.1 takes a closer look at this relationship. In this comparison, the microalgae and macroalgae that are often described based on their coloring (green, red, golden, and brown) are more closely related to diatoms than to cyanobacteria. This is likely a result of the ancient evolutionary history associated with cyanobacteria and their potential symbiotic relationship with other organisms resulting in the incorporation of the chloroplast seen in plants and algae today. This evolutionary history lasting hundreds of millions of years further suggests their potential for unique adaptive traits required during the transition between different planetary environments over this time period (Sorrels, 2009). Throughout the remainder of this book, we will refer to cyanobacteria, diatoms, true microscopic algae, and macroalgae collectively as algae; however, it is important to realize that each of these types of algae represents very distinct organisms that offer different traits and characteristics useful in the production of a wide variety of different types of biofuels including gases, liquids, and solids. Moreover, these algae have a wide array of unique genetic properties that can be more easily and quickly studied for biotechnological applications than many similar properties found in more complex land plants.

While corn, spinach, or strawberries might seem more familiar when it comes to plants important for fueling human metabolism, algae are also not uncommon within our everyday lives. For instance, sushi is a fairly common food consumed by people around the world. In many cases, a sushi roll is wrapped in seaweed, a type of macroalgae. Another example of an algal species used in food products can be found in the Naked Brand smoothie known as the "Green Machine." While this smoothie contains conventional fruits and vegetables like apples, bananas, spinach, and ginger, it also includes both cyanobacteria and true microalgae (Naked Juice Company, 2012). The use of algae is not limited to food products; some hair and cosmetic products use algae within their formulas due to the natural concentration of antioxidants. While algae provide a host of beneficial nutrients, they also offer a suite of metabolic processes that can be leveraged in medicinal and biotechnological applications. For instance, many cyanobacteria are being studied for their production of secondary metabolites, or "natural products." Unlike primary metabolites that are used by organisms simply to survive, **secondary metabolites** are metabolic products produced by an organism for a specialized function like adaptation to a unique environment or protection from a potential predator. These metabolites are not only valuable to the organism but also in many cases have the potential to be medicinal drug candidates important in advancing human health (Sorrels, 2009). Moreover, many algal species, particularly microalgae, have evolved other unique genetic characteristics, allowing them to live in a range of different environments. Because these organisms are generally simpler than higher plants and animals, these characteristics can often be exploited by biotechnological methods and used for the production or the enhancement of other products. For instance, a cyanobacterium isolated from a hypersaline lake may possess genes allowing for the maintenance of osmotic balance within the cell even under high salt conditions. These genes, when applied to other organisms, may allow them to survive in water at higher saline levels as well. Because of their versatile and wide-ranging applications, many algal species are already being produced at large scale around the world today. The production levels for some common algal species around the world are given in Table 11. The commercial applications and large-scale growth conditions already used for algae today combined with their versatile ability to produce many different biofuels and potential to provide novel genetic and biotechnological information make these organisms extremely valuable for the future of biofuels.

TABLE II

Commercial scale production, use and price of algae species in 2010

Microalgae	Annual Production	Producer Country	Application/Product	Price (US$)
Spirulina	3,000 tonnes dry weight	China, India, Myanmar, Japan	Cosmetics Phycobiliproteins	53 per kilogram 16 per milligram
Chlorella	2,000 tonnes dry weight	Taiwan, Germany, Japan	Cosmetics Aquaculture	53 per kilogram 74 per liter
Dunaliella salina	1,200 tonnes dry weight	Australia, Israel, United States, Japan	Cosmetics β-carotene	318–3,186 per kilogram
Haematococcus pluvialis	300 tonnes dry weight	United States, India, Israel	Aquaculture Astaxanthin	74 per liter 10,596 per kilogram
Crypthecodinium cohnii	240 tonnes DHA oil	United States	DHA oil	64 per gram
Schizochytrium	10 tonnes DHA oil	United States	DHA oil	64 per gram

DATA FROM: Brennan and Owende (2010).

Why Biofuels from Aquatic Biomass?

For those who are used to hearing about biofuels in terms of corn-based ethanol and soybean-derived biodiesel, the first question that may arise is: why algae? To begin answering this question, it is important to look back on the commercial potential and sustainability of the production of biofuels in the United States today. In 2013, the United States consumed about 289.8 billion gallons of petroleum, mostly in the form of gasoline (134.4 billion gallons) and diesel (58.8 billion gallons) (EIA, 2014). In 2013, the main biofuel produced in the United States was corn ethanol with about 13.3 billion gallons produced from 35 million acres of corn (Capehart and Vasavada, 2014; EIA, 2014). If we consider that ethanol is at the moment our most viable alternative fuel for gasoline, replacing 140 billion gallons of gasoline per year with ethanol would require the production of about 500 million acres of corn, an area about the size of Texas, California, Montana, New Mexico, and Arizona combined. Currently, the United States harvests a total of about 87.7 million acres of corn; therefore, a significant increase in land area would be needed to produce all of this fuel from corn (Capehart, 2014). Moreover, this only represents the replacement of gasoline in the United States, without taking into consideration diesel or jet fuel.

Now, let us consider the production of oil from algae. Currently, the focus of most research and production of algae-based fuel is in the form of diesel type fuel molecules, due in part to the structure of the Renewable Fuel Standards, although research and production of gasoline and jet fuel type molecules is growing. For simplicity's sake, let us assume all fuel consumption is gasoline; the United States needs, then, about 180 billion gallons of fuel (combining current levels of gasoline and diesel consumption). Scientists predict that algae could on average result in oil yields of nearly 5,000 gallons per acre, a yield much higher than other oil crops (Pienkos, 2007). Therefore, at this productivity level algae would only need an area of about 36 million acres to produce the 180 billion gallons of gasoline, an area about half the size of the state of New Mexico. Further, the actual total amount of acreage needed would likely be much less because algae generally grow faster than land plants; therefore, a single acre could be continuously used and harvested for a number of algae crops over the course of a year. This is perhaps the most compelling reason to consider algae as a biofuels platform.

Algae production is scalable and outcompetes other biofuel feedstocks in terms of yield per acre, saving valuable land space. Figure 9.2 compares the estimated productivity between different biofuels producers. This graph demonstrates how algae are predicted to be significantly more productive than corn grains, switchgrass, and rapeseed. Algae are believed to have a 6- to 12-fold higher solar energy yield than terrestrial plants due largely to three factors. First, algae generally have higher solar energy conversion efficiency at 3–9% compared to 2.4% for crops with a C3 photosynthetic pathway such as wheat and 3.7% for crops with a C4 photosynthetic pathway such as corn or switchgrass. Second, algae can typically grow over a greater range of light environments. This increased solar range compared to other plants allows these organisms to take advantage of a larger quantity of photons or a larger amount of energy available from the sun. Finally, algae are also efficient at capturing light. Algae can grow year-round and they are considered full canopy absorbers with a growth cycle allowing for one to three doublings per day (Dismukes et al.,

AQUATIC VERSATILITY FOR BIOFUELS   117

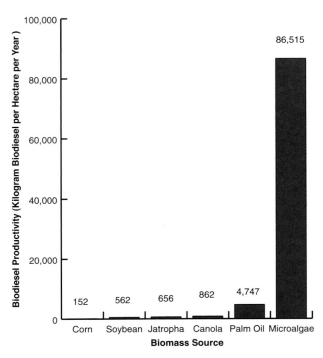

FIGURE 9.2 Comparison of biodiesel productivity between microalgae and other common feedstocks. On average, microalgae are thought to be capable of producing 86,515 kilograms biodiesel per hectare per year. As is shown in the graph, this is significantly higher than any of the other biodiesel producers, making microalgae the most productive source of biodiesel known from common feedstocks. Values shown above columns represent the kilogram biodiesel per hectare per year originally reported. Microalgae data comes from a medium oil content species (data from Mata et al., 2010).

2008). Indeed, algae are among the most photosynthetically efficient plants on Earth.

Another reason algae hold great potential to impact the biofuels industry is the wide range of biofuel applications algae are suitable to be used to address including gasoline, diesel, and jet fuel. Many types of renewable energy technologies already exist as outlined in Chapter 4; however, most of these technologies are replacements for electricity and heat generation from coal and natural gas. While algae biomass can do this as well, these organisms uniquely have the potential to also produce liquid transportation fuels. With transportation making up more than a quarter of the energy consumption in the United States and the market price of oil subject to volatility and high prices, the need for a viable replacement liquid fuel is imminent. Ethanol and biodiesel are both liquid fuels with potential; however, neither can easily replace petroleum-derived gasoline or diesel due to issues with infrastructure compatibility, commercial scalability, and environmental sustainability. In contrast, algae are capable of producing **fungible fuels**. Fungible fuels are also known as "drop-in" fuels, meaning that the oil they produce, once extracted and purified, can directly replace gasoline and diesel without significant

modification to the global transportation infrastructure. Renewable, fungible hydrocarbons are referred to as renewable gasoline, diesel, and jet fuel. Since our multitrillion dollar transportation infrastructure could not easily be replaced, the fungibility of biofuels is a critical component of their being integrated into the American and global energy economy.

As discussed with other biofuels, we must also consider the environmental impact and sustainability of algae-derived fuel. Two of the main environmental concerns with many agricultural products including biofuel feedstocks are the amounts of water and nutrients needed and the ensuing runoff and environmental pollution. With increasing concerns over water supply worldwide, it is unlikely that any biofuel source could take precedence over feeding an ever-increasing global population. Therefore, it is important that a biofuel source is chosen that requires a limited amount of freshwater. Algae generally need less water than corn or rapeseed crops but more water than sugarcane or switchgrass crops (Dismukes et al., 2008). However, many algal species can grow in reclaimed and non-freshwater sources, potentially even saline water, creating an entirely new set of production opportunities. In fact, algae-based biofuels could be commercially produced using municipal wastewater, making it not only competitive with biofuels from sugarcane and switchgrass, but also turning a liability—nutrient rich wastewater—into an asset (Kong et al., 2010). That said, many species of algae require a much lower nutrient load or level of fertilizing than terrestrial plants during the production of biomass for biofuels. For some species of algae, total fatty acids and triacylglycerides produced can actually increase under nitrogen-limiting conditions (Cakmak et al., 2012).

Finally, since algae are versatile in the types of biofuels they can produce, an integrated biorefinery system could be developed to take advantage of different component parts of algae to convert into a range of fuel products. For instance, one strain of algae could be used to produce a gaseous fuel that would generate the power needed to later extract the lipids from another strain, or perhaps ethanol can be fermented from algal sugars before the extraction and processing of its lipids (Jones and Mayfield, 2012). The high level of growth productivity and the ability to produce versatile and fungible biofuels all under environmentally friendly conditions together constitute the incredible potential for the development of algae as a platform for biofuels.

Aquatic Biomass Sources and Biofuels

Just like the large number of terrestrial plants that are used or being considered for the production of biofuels, there are also many different species of algae being researched and developed. While algae may have different colors and morphologies, they all represent biochemical factories that can efficiently photosynthesize in aquatic environments to cap-

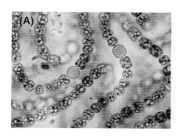 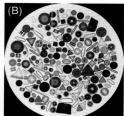  

FIGURE 9.3 Examples of aquatic photosynthetic organisms being studied for biofuels development including cyanobacteria, diatoms, microalgae, and macroalgae.

A Picture of filamentous cyanobacteria showing typical photosynthetic cells and larger heterocyst cells found in some species (image by California Environmental Protection Agency, State Water Resources Control Board, www.waterboards.ca.gov).

B Picture of the diverse structural formations of diatoms. Each individual shape within the large white circle represents a single diatom species (image by "Diatom2" by Wipeter, Own work. Licensed under CC BY-SA 3.0 via Wikimedia Commons, http://commons.wikimedia .org/wiki/File:Diatom2.jpg#/media/File:Diatom2.jpg).

C Picture of Dunaliella, a green microalgae (image by "CSIRO ScienceImage 7595 Dunaliella" by CSIRO. Licensed under CC BY 3.0 via Wikimedia Commons, http://commons.wikimedia.org/wiki/File:CSIRO_ScienceImage_7595_Dunaliella.jpg#/media/File:CSIRO_ ScienceImage_7595_Dunaliella.jpg).

D Picture of Sargassum, an example of a brown macroalgae (image by "Sargassum weeds closeup" by Unknown, Ocean Explorer/NOAA. Licensed under Public Domain via Wikimedia Commons, http://commons.wikimedia.org/wiki/File:Sargassum_weeds_closeup.jpg#/media /File:Sargassum_weeds_closeup.jpg).

ture solar energy, fix carbon dioxide, and produce a wide array of metabolites including fuel molecules. Many of these organisms also offer the potential to provide insight into unique genetic adaptations that can be used in biotechnological and biomedical applications. Examples of aquatic photosynthetic organisms being studied for biofuels production are shown in figure 9.3.

Cyanobacteria

Of all algal species, cyanobacteria are some of the simplest, yet they also hold some of the greatest potential for biotechnological and biomedical advances. Cyanobacteria are true prokaryotic organisms in that they do not contain membrane-bound organelles. Cyanobacteria are gram-negative photosynthetic bacteria thought to be among the most ancient organisms on the planet and largely responsible for the oxygenation of Earth's atmosphere. Cyanobacteria are globally distributed in a wide variety of environments where they have a number of ecological roles and a wide variety of morphologies. In many cases, cyanobacteria can even be found in the most extreme of environments ranging from arid deserts to the frigid arctic. The ecological versatility of these organisms stems from the evolution of diverse and unique genetic characteristics. Many cyanobacteria are also able to fix nitrogen from the air, produce protective pigments against harsh levels of radiation, and regulate flotation. They are also valuable in biotechnological applications including secondary metabolite or drug candidate production, desiccation tolerance, and the use of an expanded range of solar radiation. In addition, some species of cyanobacteria are naturally transformable, allowing for the deliberate enhancement of various traits useful in

the production of valuable products like medicines and biofuels (Sorrels, 2009; Ducat et al., 2011).

Specific to biofuels research, the metabolic diversity of cyanobacteria, their ability to be genetically manipulated, and the availability of many already sequenced genomes allow researchers to alter these organisms to produce specific fuel products or adjust the ratio of carbohydrates to lipids generated in the cell body (Ducat et al., 2011). Since many species are capable of nitrogen fixation, cyanobacteria can actually provide a source of nitrogen to their surrounding environment and could potentially be employed commercially as an alternate source of nutrient supplementation. Some cyanobacteria already do this in nature, by growing alongside plants, providing these plants with their nitrogen allotment and reducing the amount of nutrients needed from the soil. In agriculture, this commensal relationship can cut down on the need for nitrogen-based fertilizers. It is no stretch of the imagination, then, to envision them being used to provide a source of nitrogen for other aquatic environments or simply for themselves in ponds designed specifically for biofuel production (Chisti, 2007). Some forms of cyanobacteria also grow as filaments. While these species often grow significantly slower than the aforementioned unicellular types, they offer the potential to be easily harvested through skimming rather than filtration. As will be discussed later in the chapter, reducing the cost and energy consumption involved in harvesting algae and the extraction of lipids or sugars will be essential to the future competitiveness of algae biofuels. Finally, since cyanobacteria grow in such a wide variety of environments, species are available with unique tolerances to certain extreme growth conditions, known generally as extremophiles. Some species can grow in high pH levels, while

CYANOBACTERIA: ARCHITECTS OF OUR ATMOSPHERE

Travis L. Johnson

You can thank cyanobacteria for most of the breathable air we now have in Earth's atmosphere. Long before humans, more than 3.5 billion years ago, the Earth's atmosphere was most likely made up of carbon dioxide, carbon monoxide, water, nitrogen, and hydrogen. The atmosphere had little oxygen, and anaerobic organisms inhabited the planet. Then between 3.5 and 2.8 billion years ago, cyanobacteria appeared on earth and began producing oxygen and fixing carbon dioxide via photosynthesis. The produced oxygen was absorbed by iron in the oceans and then subsequently by organic matter on land surfaces. After about 200 million years of this, these oxygen-absorbing sinks became saturated, and the oxygen gas started accumulating in the atmosphere, creating what is known as the Great Oxygenation Event. This caused one of the biggest extinctions in Earth's history, wiping out most of the anaerobic organisms that could not tolerate oxygen. But because of the oxygenated atmosphere, aerobic organisms began to appear, consuming the oxygen and starting the evolutionary process that led to all of the oxygen-breathing organisms on Earth—including humans. Even today, cyanobacteria all around the world produce around 30% of the oxygen we breathe.

SOURCES: Golden (2014); British Broadcasting Company; Rice University; Zimmer (2013); Holland (2006).

others are more tolerant to salt. These types of characteristics allow such organisms to grow in places that other organisms cannot, thus reducing competition and predation, key factors for enhanced agricultural production.

While these properties are important, the most important property to consider in cyanobacteria for purposes here is their ability to produce biofuels. Although cyanobacteria have been largely overshadowed by biofuel production from terrestrial plants and other algal sources, cyanobacteria are uniquely versatile in their biofuels production. Cyanobacteria have been studied mostly for their production of hydrogen and ethanol. Many of these bacterial species have specialized nitrogen fixation cells called heterocysts and are capable of fixing nitrogen using the enzyme nitrogenase. When presented with nitrogen gas (N_2) limited conditions, nitrogenase will actually switch to producing hydrogen as discussed in Chapter 8 (Kotay and Das, 2008; Manish and Banerjee, 2008). Therefore, cyanobacterial species containing heterocysts presented with nitrogen-limiting conditions can be fairly efficient producers of hydrogen gas. While cyanobacteria could also be used in cellulosic ethanol production (a process detailed in Chapter 6) due to their deposition of nearly 25% of their dry weight as cellulose, cyanobacteria contain the more novel capability to ferment ethanol naturally. In other words, many species of cyanobacteria can actually produce ethanol directly without the need for the traditional yeast fermentation. This capability could prove beneficial in lowering the cost of bioethanol production in the future. Research on cyanobacteria as a production platform for biodiesel is more limited than many other feedstocks including other microalgae species. While cyanobacteria can produce biodiesel, it is thought finding a suitable strain for biodiesel production may be more challenging than in other types of algae. Finally, cyanobacterial biomass can be processed through anaerobic digestion to produce methane gas. However, this biomass source tends to be less efficient for gas production than other types of biomass including cattle waste; therefore, research on the use of cyanobacteria for biogas is also limited (Quintana et al., 2011).

While this profusion of possibility may be exciting, there are a number of challenges that must be addressed to allow cyanobacteria to be successful as a biofuels platform. One of the main challenges comes in terms of how they store their excess carbon. As covered in Chapter 7, oil-producing biofuels feedstocks should ideally store their lipids as triacylglycerides or free fatty acids. Cyanobacteria, in contrast, have a large number of polar membrane lipids and minimal triacylglycerides. Put simply, cyanobacteria lock up their lipids within their membranes, making them less accessible to traditional biofuels extraction, purification, and conversion methodologies. Another challenge is that most cyanobacteria do not store their excess carbon as lipids; instead, they store it mainly as glycogen (Hu et al., 2008; Quintana et al., 2011). Finding a way to transition more of the carbohydrates into gasoline- and diesel-like lipids will be important in increasing the yields of fungible biofuels from cyanobacteria in the future.

Diatoms

Diatoms are another important algal species being studied for biofuel production. They are classified as eukaryotic microalgae and are responsible for an estimated 20% of global carbon fixation (Hildebrand, 2008). Diatoms are highly abundant and considered one of the dominant primary producers in the ocean. They are a significant carbon exporter and play an important role in the natural sequestration of carbon in the ocean. Like cyanobacteria, diatoms are also very diverse in morphology, genetics, and environmental distribution. Diatoms can be found in both terrestrial and aquatic environments and in a number of extreme environments including areas with variable salinity, temperature, and pH. Their wide-ranging distribution is likely derived from their metabolic diversity, which stems from a rapid and widespread evolutionary divergence. In fact, diatom genomes can differ from one another to the same degree that the genomes of fish and mammals differ. This genetic diversity could provide a wealth of novel genetic and metabolic targets for biotechnological applications, directly benefitting biofuels research. One of the most distinctive morphological differences between species of diatoms comes from their use of silica. Diatoms have cell walls made of silica, giving them a wide variety of interesting and unique structural formations; many are shown in figure 9.3B (Hildebrand et al., 2012).

Diatoms have much to offer in the production of biofuels. Thirty of the top 50 microalgae strains identified in a survey of over 3,000 aquatic candidates for biofuels production potential were diatoms. There are a number of reasons why diatoms hold so much promise as a biofuels production platform. First, diatoms have very high growth rates as a result of their ability to productively fix carbon and sequester nutrients. In fact, diatoms bloom to very high densities under nutrient-replete conditions because they are able to sequester and store nutrients better than most of their competitors. Clearly, their ability to grow to high densities and their carbon-fixing ability is attractive for the high-yield production of biofuels, but being able to grow in nutrient-replete conditions is also important. Since large-scale algal cultures are likely to be grown in outdoor ponds for biofuels production, the ability of diatoms to suck up all the nutrients quickly leaves few nutrients remaining to help feed contaminating organisms.

Diatoms also promise to help one of the most expensive and difficult aspects of algal biofuels production: harvesting and preparation of the algal biomass. An algal species that either flocculates at the surface or sinks to the bottom would allow the concentration of algae to increase in a smaller amount of water, thereby making harvesting a simpler and cheaper prospect. Since diatoms are known to be a major sink for carbon in the ocean by fixing carbon dioxide and then sinking through the water column, these increased settling rates in nature relative to other algal species would likely carry over to production ponds. In addition, some species of diatoms are known to naturally flocculate through direct adhesion or the production of cell surface exopolymers. **Flocculation** of these organisms would allow much of the biomass to be skimmed from the water within a pond rather than the use of more expensive and energy-intensive filtration methods (Hildebrand et al., 2012).

Another significant attribute of diatoms relevant to biofuels production is their trophic flexibility. Diatoms are known to be autotrophic, mixotrophic, and heterotrophic. Trophic level is determined by how an organism obtains its metabolic energy. Autotrophic organisms can solely use inorganic substances to produce their metabolic energy, as is the case with photosynthesizing plants using carbon dioxide and sunlight. Heterotrophic organisms require their metabolic energy to come from other organic sources such as when humans eat food to gain energy. Mixotrophic organisms use a combination of both trophic traits. During heterotrophic growth conditions, diatoms are capable of using a variety of complex carbon sources, which may be applicable to fermentation for ethanol production in the future. Diatoms are also efficient accumulators of neutral lipids that could be used for the production of lipid-based biofuels like biodiesel. Their ability to produce lipids is thought to be a characteristic stemming from their need for extra buoyancy due to the relative heaviness of their silicon cell wall. The most efficient accumulation of lipids, particularly triacylglycerides, tends to occur when the cell cycle is stopped either biochemically or through nutrient limitation, with silicon limitation resulting in higher lipid yields than nitrogen limitation. Although it may seem counterintuitive because of their hard outer shell, the extraction of lipids from these organisms is actually quite easy without the requirement of additional steps to break the cell wall as can be seen in other oilseed biodiesel feedstocks. Aside from producing lipids with chemical structures important for biofuels, these organisms also produce other important hydrocarbons such as eicosapentaenoic acid (EPA), docosahexaenoic acid (DHA), and other isoprenoids that are important to human health. These additional hydrocarbons may represent value-added coproducts that can be extracted from diatoms to lower the overall cost of biofuels production (Hildebrand et al., 2012).

Like cyanobacteria, diatoms also have some limitations. One of the major perceived limitations in using diatoms at large commercial scales for biofuels production is their need for silicon. Silicon is an essential part of their morphology and is an essential nutrient for growth. The requirement for the addition of silicon into large ponds would result in an increase in cost for the production system. However, some of this cost may be recovered after biofuels extraction if the remaining biomass is sold as diatomaceous earth or a method is devised to recycle the silicon and thus lower the overall cost of nutrient additions to the culture (Hildebrand et al., 2012).

True Algae

The final group of algae to be discussed is the true algae. True algae can be broken into two groups: macroalgae and microalgae. As their names imply, the differences between these types of algae have largely to do with their complexity and size. Macroalgae such as seaweed and kelp usually have morphologies much closer to our general description of a terrestrial plant except they grow in the water. These organisms are usually attached to a solid structure such as the sea floor and then grow upward toward the light at the surface. In general, many macroalgae also have small air-filled sacs that help their stems and leaves float upright in the water column. Although many species of macroalgae can grow very quickly, they are usually outpaced by their much smaller cousins the microalgae. True microalgae are microscopic photosynthetic organisms that have a plant-like cellular structure. These organisms have membrane-bound organelles including a chloroplast and metabolically function similar to that of the macroalgae and terrestrial plants but without stems and leaves. Like cyanobacteria and diatoms, these true microalgae are thought to play a significant role in the production of atmospheric oxygen and the uptake of carbon dioxide on the Earth (Wageningen UR, 2011). Both macroalgae and microalgae have the potential to be used in biofuels development and grow in a variety of aquatic environments including freshwater and saltwater.

Much of the development of macroalgae has been for the production of non-biofuel-related products such as agar, alginate, and food; however, these organisms are now being studied for their potential as a biofuels platform. The larger and multicellular-like structure of macroalgae could be advantageous for biofuels production simply because they can quickly grow large amounts of biomass that can easily and cheaply be harvested. Macroalgae have not been found to be particularly proficient lipid accumulators; thus, their potential for biodiesel production seems limited. However, macroalgae do contain a significant quantity of sugar molecules that can be used for the cellulosic fermentation of ethanol. Additionally, despite having an upright morphology like terrestrial plants, macroalgae actually do not contain the same lignin cross-linking found in other cellulosic ethanol feedstocks, likely a result of growing in water without the need to grow against gravity. The lack of lignin cross-linking means that macroalgal cellulosic sugars would be more easily accessible to cellulases during the production of ethanol, resulting in increased ethanol yields and potentially lower costs. While macroalgae could prove to be a valuable biomass resource for cellulosic ethanol, their biomass could also be useful simply for combustion (John et al., 2011; Jones and Mayfield, 2012).

Even though microalgae are much smaller than macroalgae, their diversity leads to an abundance of opportunity for the production of various biofuels. Their small size does not translate into less overall biomass because these microscopic organisms can multiply quickly. Once harvested, a new batch can be grown in a matter of days. As mentioned earlier, the process of harvesting is not easy or cheap; therefore, microalgal species with flocculation and/or sinking characteristics similar to diatoms may be important to allow for maximum biomass production. Microalgae can also tolerate a range of water qualities from saline to brackish to freshwater. This tolerance allows these algae to be grown using wastewater as a source of nutrients, a form of nutrient recycling likely crucial going forward. Unlike many of the terrestrial plant biomass sources, microalgae can be grown and harvested year-round from nonarable lands (Li and Wan, 2011; Singh et al., 2011). This characteristic is important for both minimizing the food versus fuel tension in terms of land use and producing a constant supply of biofuels.

True microalgae can produce a wide array of biofuels including ethanol, biodiesel, hydrogen gas, biogas, or simply biomass for combustion. Like cyanobacteria, some species of microalgae can directly produce hydrogen gas anaerobically through photofermentation; however, the efficiency of biohydrogen production depends on the efficiency of the hydrogenase enzyme that is highly variable between species. Biogas production and combustible biomass material can both be produced from microalgae even after extraction of other biofuel molecules (Jones and Mayfield, 2012). While large amounts of biomass can be grown in semi-controlled environments such as open ponds, it may also be possible to utilize massive algal blooms that occur in natural aquatic environments around the world. In many cases, these blooms are toxic. By using the algae that cause these blooms as biomass, the toxicity could be reduced due to a lower quantity of algal cells producing toxins.

While microalgae are composed of carbohydrates, proteins, and lipids just like all other organisms, the focus of microalgal research has largely been on their lipid compositions. Microalgae are known to be capable of producing significant quantities of lipids as storage products ranging as high as 50–60% of dry weight (Griffiths and Harrison, 2009). These lipids have a similar chemical structure to other lipids derived from oilseed crops and can be used for biodiesel production. However, while high-yielding algal species may represent a more prolific source of biodiesel, their fuel is still more expensive than other forms of biodiesel or petrodiesel due in large part to this industry being relatively undeveloped.

RAINBOW ALGAE

Travis L. Johnson

Scientists have created the first rainbow-colored algae. And beyond looking brilliant, these new genetically altered algae show that we are now ready to domesticate algae similar to other crops such as corn, wheat, and rice. The modern corn crop originates from a small grain called teosinte, and over the course of 7,000 years, humans have bred and selected this crop to become a high-calorie staple food, feeding billions of people all around the world. Unfortunately, we do not have thousands of years to domesticate algae to provide meaningful products like fuels and medicines, but the latest advances in biotechnology have sped up this process. The new rainbow algae are created through genetic engineering by inserting different recombinant proteins that fluoresce different colors when exposed to light. Although rainbow-colored algae are not necessarily valuable on their own, the fluorescent proteins are used to show that inserted genes are actually being expressed. When proteins are targeted to certain parts of the cell, and the modified algae are selectively bred to produce rainbow algae, it gives us a proof of the concept that we now have an advance genetic toolbox to create many different valuable products from algae.

SOURCE: Mayfield (2014). Image by Dr. Beth Rasala, Triton Health and Nutrition.

Algae Biofuel Production

Being a relatively new field, the production of biofuels from algae is focused on improving overall efficiency and reducing cost. The production chain for algal biofuels can be broken into five basic steps: strain development, algal growth/ production, harvest, extraction, and processing/refining. Unlike many of the terrestrial plant sources that can rely on thousands of years of agricultural evolution—like corn, wheat, or rice—the optimal production process of algal sources of biomass remains relatively unknown, thus requiring a significant amount of research at each step of the production chain and a massive burden on start-up enterprises. Still, there is reason to believe that recent developments in biotechnology, such as genetic engineering, can expedite the domestication of algae for agricultural purposes.

The first aspect of the production chain is strain development. The previous section overviewed the most common types of algae and their potential for biofuels production, but research continues to find and develop new strains of algae that can produce high yields and molecularly specific biofuels. In a process called **bioprospecting**, biologists can collect algae directly from the environment in order to find novel species that may have better characteristics for growth and survival than those currently known. Bioprospectors will often seek out environments with conditions similar to those anticipated for agricultural algae production, in which the native algae may have evolved an environmental advantage. For instance, if an algae pond will be placed in the southwestern portion of the United States, scientists would want to identify algal strains with a higher salt tolerance due to rapid evaporation or a higher tolerance to heat due to the warm desert-like conditions.

While bioprospecting can offer a wide array of unique organisms with specific adaptations that benefit biofuels production, it is widely thought that the success of algae biofuels in the near term will depend greatly on biotechnological advancements in genetics and breeding to create algal strains with specific characteristics that can most precisely maximize their productivity in fuel molecule generation (Hannon et al., 2010). Some of these biotechnological applications will be discussed in the next chapter. Whether through strains already used commercially, bioprospecting or biotechnological applications, once an algal strain has been selected, the next step requires optimization of agricultural production.

FIGURE 9.4 (Left) Photo of algae raceways showing how a paddlewheel can circulate the water to keep the algae well mixed and evenly exposed to the sunlight (image by California Center for Algae Biotechnology, University of California, San Diego).

FIGURE 9.5 (Right) Photo of an algae photobioreactor. Here the algae are grown inside closed tubes supplied with bubbling air (including carbon dioxide) while being exposed to light (image by Dr. Brenda Parker, University College London).

The second key aspect of algae biofuel production is productivity, which means producing the maximum quantity of fuel per gallon of culture. This factor translates into the price per gallon of finished fuel and has a number of standard metrics: grams per square meter per day, tons per acre per year, or gallons per acre per year. No matter how it is measured, it all comes down to producing the largest quantity of algal biomass as quickly as possible. The photosynthetic nature of algal cultures makes their growth to high densities at small scales fairly easy, but in order to supplant any significant quantity of petroleum, the growth of algae would need to occur at gigantic commercial scales, significantly complicating growth conditions. There are currently two main methods being studied for large-scale growth of algae: open ponds and photobioreactors.

Open ponds are just that: large outdoor ponds usually designed in the shape of a raceway as shown in figure 9.4. Raceways are typically in the shape of closed oval loops and relatively shallow at 0.2–0.5 meters in depth. The water and algae growing in these ponds are slowly mixed and circulated through the raceway often by a paddle wheel. Open ponds are already used in places such as Hawaii for the growth of algae and the production of food and health-related products. By contrast, a photobioreactor system is typically closed as a set of flat panels, horizontal tubes, or columns as shown in figure 9.5. These closed systems are not exposed to outdoor conditions to the same extent as open ponds and can often result in a better control of growth, mixing conditions, and predatory species. The design of the photobioreactor depends largely on the location and the ability to take advantage of natural sunlight as well as the cost of the system (Scott et al., 2010).

Whether using an open pond or photobioreactor, a few production characteristics have been shown to greatly impact the growth of algae in these systems. In general,

commercial algae production should mimic natural growth conditions as much as possible. This usually means a high level of exposure to light and variations in temperature and mixing. These conditions are most easily controlled in the closed environment of the photobioreactor, but due to capital costs, producing large volumes of algal biomass in photobioreactors can be prohibitively expensive. Therefore, open ponds remain highly attractive as likely being the least capital intensive in start-up. Yet, in order to reach maximum efficiency in open ponds or photobioreactors, growing algae commercially will largely depend on the control and efficiency of four parameters: temperature, light, mixing, and crop protection.

The ideal locations for algae production—nonarable land with abundant non-freshwater sources—expose the organisms to large fluctuations in temperature and therefore evaporation. While temperature and evaporation can be completely controlled within a photobioreactor, these conditions are not as easily controlled in an open pond. Both evaporation and temperature of the ponds will depend entirely on diurnal cycles and seasonal variations. For instance, levels of evaporation will change based on changes in wind speed, humidity, and temperature. Maintaining ideal salinities and nutrient availability for the growing algae in open ponds requires precisely matching water input to that of the evaporation rate on a daily basis. Because air temperatures can fluctuate greatly between different seasons and even between day and night in places like the arid southwestern region of the United States, the algae will be exposed to fluctuations in water temperatures (Brennan and Owende, 2010; Singh and Gu, 2010). Therefore, production strains will need to be suitable to a wide range of water temperatures. In addition, to be able to assure year-round productivity, open ponds will need to use different strains most suitable to each of the different seasons. Invariably, temperature variation will still cause fluctuations in biomass yields, with warmer months yielding higher levels of biomass and thus larger volumes of biofuels.

Since algae use light to sequester carbon to produce lipids, the highest levels of productivity in either photobioreactors or open ponds will depend on the availability of light. Light can be understood in two dimensions: the quantity of radiation and the quality of radiation. In a photobioreactor under artificial light, the quantity of radiation can be carefully measured and maintained; however, quality can be a bit harder. In order to mimic natural growth conditions, algal cultures will need to be exposed to a wide array of radiation wavelengths similar to the spectrum emitted by the sun. Many algal species specifically contain a wide range of pigments designed to harvest this array of radiation for a number of metabolic and protective functions (Sorrels, 2009). In a photobioreactor under natural or artificial light conditions, the glass is likely to absorb some of these wavelengths, preventing the organism from exposure to the full spectrum of light. Continued research will be needed to understand how variations in radiation quality will impact productivity and efficiency. With respect to natural light conditions using only the sun as a source, there is a limited ability to control the quantity of light radiation, but in open ponds, algae are likely exposed to all necessary wavelengths. Diurnal cycles, weather patterns, and seasonal light variations will all impact the quantity and quality of light radiation available in different locations. For instance, days are shorter in the winter than in the summer and an algal pond would only be productively photosynthesizing during daylight hours. Also, certain regions, like the southwestern United States, will offer more sunny days that foster better algae growth than locations like the Pacific Northwest where clouds block the sun for much of the year.

Maintaining optimum temperature and light conditions within a culture depends heavily upon mixing. In the case of photobioreactors, effective mixing is easily accomplished, but with large open ponds, thorough mixing presents a more significant challenge. Two methods of mixing in open ponds include the low-speed paddle wheel and the propeller aspirator pump. The paddle wheel shown in figure 9.4 is usually placed at one end of a raceway with the algae seeded into the pond directly in front of the wheel. The wheel will then slowly rotate at about one to three rotations per minute, pushing large volumes of water and the algae through the raceway. This method of mixing allows for moving large volumes of water without requiring extensive energy inputs. The second method is the propeller aspirator pump. This pump creates a current along the bottom of the pond with a return flow along the top of the pond in an opposite manner to that of the paddle wheel. These types of pumps may be more valuable for non-raceway type ponds but will likely require more energy without moving as much water, thus reducing overall efficiency of the mixing (Hargreaves, 2003).

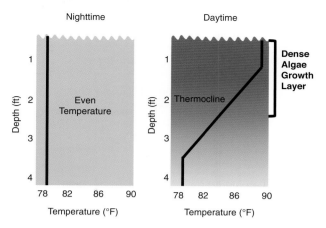

FIGURE 9.6 Diagram of potential pond stratification due to higher daytime temperatures compared to nighttime temperatures. Similar to the ocean and other large bodies of water, warm daytime temperatures can cause stratification in water temperature called a thermocline. The thermocline can result in an uneven distribution of algae growth where the algae grow into a thick layer on the surface in the warmer water, blocking the light from reaching the deeper layers. The relatively stable nighttime temperatures result in a more even distribution of temperature, but there is no sunlight to promote growth, likely limiting the density of algae growth throughout (source: Hargreaves, 2003).

The importance of mixing ponds should not be understated due to its significant impact on biomass productivity. A stagnant pond can develop stratifications where the highest temperature of water is located within the top foot of the water column due to the sun warming this upper layer, and the temperature will gradually be reduced over about 2 feet until it drops off in the last 1–2 feet. The potential for daytime stratification in ponds is diagramed in figure 9.6. Stratification in the pond can expose biomass at the bottom of the pond to much cooler temperatures and less light, thus reducing productivity. In these conditions, a little over a third of the biomass would grow at suboptimal productivity levels. Without mixing, a pond will also likely develop a thick layer of algae in the top layer due to the higher photon flux of light. This layer will also block the light from penetrating into the lower depths. By maintaining a well-mixed pond, these stratifications and layers are avoided and the light exposure is increased. Mixing will also allow for better gas exchange with the surface of the water increasing biomass productivity due to the introduction of higher levels of carbon dioxide (Hargreaves, 2003).

Finally, the last major consideration with respect to algal production in photobioreactors or open ponds is contamination. Just as with terrestrial agricultural crops where pesticides are the second largest expense, algal crops need to be protected from pests that consume the biomass and reduce productivity. The issue of contamination is predictably less significant for photobioreactors than for open ponds. Contamination in open ponds from pathogens such as *Chytrid* fungi or predators such as rotifers readily occurs from exposure to the natural environment and results in a

TABLE 12
Comparison of the advantages and disadvantages of different production systems for algae cultures

Production System	Advantages	Limitations
Raceway Pond	Relatively cheap	Requires large land space
	Utilizes non-agricultural land	Poor biomass productivity
		Limited in algae strains
	Low energy inputs	Cultures easily contaminated
		Poor mixing, light and CO_2 utilization
Tubular Photobioreactor	Relatively cheap	Fouling
	Suitable for outdoor cultures	Requires large land space
	Good biomass productivities	pH, O_2 and CO_2 gradients along tubes
Flat Plate Photobioreactor	Suitable for outdoor cultures	Difficult scale up and temperature control
	High biomass productivity	Small amount of hydrodynamic stress
	Easy to sterilize	
Column Photobioreactor	Compact	Small illumination area
	Low energy consumption	Expensive
	Easy to sterilize	Shear stress

NOTE: Included are examples of algae species used in these production systems and their productivity levels.

SOURCE: Brennan and Owende (2010).

significant reduction in biomass. Crop protection in algae ponds will likely require the application of herbicides and pesticides, adding to production costs. Ideally, the first line of defense will be to find strains naturally resistant to pests specific to the region. This is where bioprospecting can be particularly valuable. These traits can be selected for and bred into production strains or potentially genetically engineered into these strains to provide them with increased resistance to natural pests. Ultimately, open pond algae production will require establishing a set of "best practices" to create the healthiest pond environment including crop rotations and the use of algal consortia to combat the onslaught of pest attacks (Hannon et al., 2010).

Because of the divergent advantages and disadvantages to photobioreactors and open ponds, it is likely that a combination approach will prove most effective, as each complements different stages of the cultivation process. Table 12 summarizes the advantages and limitations of the different production systems.

The third step in biofuel production after algal growth focuses on the harvest. Harvesting an algae crop has additional complications beyond terrestrial crops because algae are submersed in large volumes of liquid. Separating the algae biomass from such large volumes of water can be logistically difficult, cost prohibitive, and energetically demanding. In general, the technique used for harvesting will largely depend on the characteristics of the production strain including the size and concentration of the organisms. However, it will generally consist of two steps: bulk harvesting and thickening. First, the biomass is separated from the bulk of the water. Second, the biomass is thickened to further reduce the water concentration in preparation for extraction. One method for this process already discussed is flocculation. As with diatoms, flocculation results from algal cells interacting and aggregating with each other or another substance to create a much heavier molecule that then sinks to the bottom of a pond. The process of flocculation is outlined in figure 9.7. Generally, most microalgae do not flocculate naturally because they carry a negative charge on their cells that prevent them from aggregating when in suspension. However, if a flocculant such as a multivalent cation or cationic polymer (a substrate with a positive charge) is added to the culture, the flocculant interacts with and reduces the negative charge on the cells and enhances aggregation. The heavier algae aggregates will fall to the bottom of the pond, and the water level can be lowered and the algae harvested from much smaller volumes (Brennan and Owende, 2010).

Once the amount of water in which the algae are submerged is reduced, the culture can be thickened. Thickening is likely to occur using one of two methods: centrifugal sedimentation or biomass filtration. Centrifugal sedimentation uses an industrial-scale centrifuge that relies on the fact that solid particles under high levels of centrifugal force settle into the sides and bottom of the centrifuge. Thus, during centrifugal sedimentation, the algal cultures are separated along the sides of the centrifuge, allowing for the additional water to be drained away. The remaining thickened algal biomass can then be scraped out and is ready for extraction. While centrifugal sedimentation is

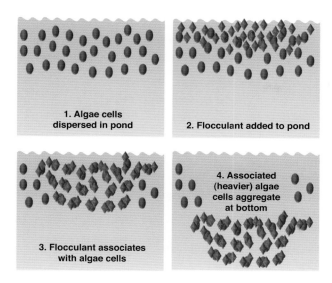

1. Algae cells dispersed in pond

2. Flocculant added to pond

3. Flocculant associates with algae cells

4. Associated (heavier) algae cells aggregate at bottom

FIGURE 9.7 Diagram of flocculation where dispersed algae cells are exposed to a flocculant (example, positively charged substrate) that will associate with these cells. This flocculant creates a heavier cell that will then sink, aggregating at the bottom of the water column where it can be harvested in a more efficient way by allowing for the removal of the bulk of the water.

very rapid especially for large volumes of water and dense algal cultures, it is an extremely energy-intensive process, and its efficiency depends on the density and radius of algal cells (Brennan and Owende, 2010).

The alternative to centrifugal sedimentation is biomass filtration. This method works in the normal manner of a filtration system such as that found in a pool: water passes through a filter membrane trapping the particles and releasing the clean water back into the pool. In the same way, biomass filtration operates under pressure or suction with the biomass becoming trapped in the filter, while the water flows through. Filtration as a potential method for harvesting of algae depends on the size of the algal cell. This method would only be effective for relatively large microalgae species because those approaching dimensions of a bacterium could not easily be separated at large scales in this manner (Brennan and Owende, 2010).

Once the algae have been grown and harvested, the fourth step is to extract and purify the fuel molecules, particularly lipids, from the algal biomass. Just as with soybeans and other feedstocks of biodiesel, algae can be exposed to nonpolar solvents like hexane. Since water and most other biomolecules such as proteins and carbohydrates are polar in nature, the use of a nonpolar solvent will cause a separation where the lipids that will not mix with water dissolve into the nonpolar solvent, leaving the remaining parts of the biomass in the water portion of the extract. In addition, nonpolar solvents tend to be less dense than water; thus, the lipids and the solvent will float to the surface and can be skimmed from the water, purifying it from the other biomolecules. One challenge to the extraction process is that many of the lipids in algae are part of

cell walls and are bifunctional, with polar head groups containing sugars, sulfur, phosphate, or amines and nonpolar lipid-like tails. To maximize the extraction of lipids from this biomass, a chemical process will be needed to break the cells open and separate the lipid tails from these polar heads (Brennan and Owende, 2010).

Lastly, after the lipids have been extracted from the algal biomass, they are ready to be refined into fuel using the same distillation methods as discussed in previous chapters. Once the lipids are extracted, the other biomolecules such as sugars and proteins are still available for the production of other types of biofuels using either biochemical conversion technologies such as fermentation or anaerobic digestion or thermochemical conversion technologies that will be discussed in Chapter 11. To reiterate, the use of this additional biomass for the production of a value-added product is likely to be important in making algal biofuel production profitable relative to petroleum fuel products in the future.

Benefits and Challenges to Algae Fuel

Algae are promising candidates as biofuels platforms for a variety of different biofuels including bioethanol, biodiesel, biohydrogen, biogas, and renewable hydrocarbons. However, most of the research and development of algae-based biofuels has focused on the production of biodiesel specifically from microalgae, diatoms, and cyanobacteria. As shown in figure 9.2, algae are prolific producers of lipids and could be very efficient biodiesel production hosts. Table 13 shows a comparison of biodiesel production in algae to other feedstocks in terms of gallons per acre. In terms of oil yield, it is evident algae can produce significantly more oil per acre than all other biodiesel producers including palm. At an average of 5,000 gallons per acre, algae could produce 150 billion gallons of fuel over just 30 million acres, a value that nearly matches the current consumption of both gasoline and diesel in the United States in a geographic area about the size of Pennsylvania (Pienkos, 2007).

The biodiesel extracted from algae is also at a quality similar to that of other first-generation biodiesel feedstocks. Table 14 shows a comparison of fuel properties between traditional biodiesels, algal biodiesels, and petrodiesels. This chart indicates that while algae-derived biodiesel is comparable in quality to conventional biodiesel, both products are still below the quality of petrodiesel in terms of viscosity and pour point. This fact will continue to drive research and development into conversion technologies valuable in converting biodiesel methyl esters into drop-in renewable hydrocarbons.

To summarize from earlier in the chapter, using algae as a feedstock for biofuels production also has a number of positive environmental impacts. First, many species of algae can be grown in much lower-quality waters than the freshwater needed for traditional agricultural crops. Some algae can even grow in saline water, raising the prospect of an entirely new form of agribusiness: saltwater agriculture. Growing algae in

TABLE 13

Comparison of potential oil yields from biodiesel feedstocks including first-generation traditional feedstocks as
well as alternative feedstocks and algae

Oil Feedstock	Oil Yield (gallons/acre)
Traditional Feedstocks	
Corn	18
Cotton	35
Soybean	48
Mustard seed	61
Sunflower	102
Canola/rapeseed	127
Alternative Feedstocks	
Jatropha	202
Oil palm	635
Algae	5,600

DATA FROM: Pienkos (2007).

TABLE 14

Comparison of fuel properties between algal-derived
biodiesel, first-generation biodiesel, and petrodiesel

Fuel Property	First-Generation Biodiesel	Algae Biodiesel	Petrodiesel
Viscosity (square millimeter per second)	3.6–9.48	5.2	1.2–3.5
Density (kilograms per liter)	0.86–0.895	0.864	0.83–0.84
Flash Point (degrees Celsius)	100–170	115	60–80
Pour Point (degrees Celsius)	–15 to –10	–12	–35 to –15

DATA FROM: Brennan and Owende (2010).

non-pure water sources eliminates the tension between the availability of potable water and fuel production. Uniquely, the growth of algae in wastewater can supply required nutrients in an alternative manner to conventional fertilizers. Not only does this cut down on the cost and use of fossil fuel energy, but it also recycles and cleans wastewater, helping with bioremediation and pollution. While algae will not compete for freshwater, they will also likely not compete for arable land space. With a worldwide population that has already passed 7 billion people, arable crop land is only becoming more precious. When algae are grown in desert regions, for instance, not only do they protect arable land for food cultivation, but the conditions also provide the algae maximal exposure to sunlight (Hannon et al., 2010).

Algae-based biofuels do come with significant challenges, particularly in energy use and cost. Currently, the harvesting, extraction, and refining of algae into fuel use a significant amount of fossil energy. Developing efficient harvesting methods such as flocculation and the use of extracted biomass for the production of biogas could help alleviate the high energetic costs associated with algae biofuel production. Today, algae fuels are not cost competitive with fossil fuels. Lowering the cost will likely result from a combination of bioprospecting, genetic engineering, nutrient recycling, and breakthroughs in harvesting and extraction methods. For any biofuel to truly supplant fossil fuels long term, it will need to be able to compete on the open market. That said, petroleum markets remain volatile and prices, despite the recent shale boom, remain stubbornly high.

Every indicator points to a world of increasing demand outstripped by our ability to globally meet that supply. These factors, taken together, will likely continue incentivizing research, development, and commercialization of biofuels resources like algae.

STUDY QUESTIONS

1. Briefly explain some of the roles of algae today and why they may help demonstrate the feasibility of algae as a biofuels source.
2. Explain why algae represent a good biofuels platform compared to some other potential sources based on scalability, fuel production, productivity, and environmental impacts.
3. Briefly describe the advantages and disadvantages of using each of the algae biofuels sources including cyanobacteria, diatoms, macroalgae, and microalgae.
4. Explain how and why light, temperature, mixing, and crop protection will all play an important role in the commercialization of algae biofuels.
5. What are the main differences between the production processes for biofuels from algae compared to other terrestrial plants? How do these differences influence the energy efficiency of production and the price of fuel?

CHAPTER TEN

Biochemistry and Biotechnology for Biofuels Development

The products of fossil fuels have become the mainstay of energy consumption around the world due to their high energy densities and cheap prices. This is particularly evident in the United States where the generally high standard of living to which people are accustomed stems directly from the availability and consumption of cheap fossil energy. However, sustained concern for a stable supply of fossil fuels, particularly petroleum, and the rising awareness of environmental damage from the combustion of these fuels are incentivizing the development and use of alternative and renewable forms of energy. While solar and wind power are being pursued for the replacement of heating and electricity from coal and natural gas, one of the major challenges still facing science and engineering is the replacement of liquid transportation fuels including gasoline, diesel, and jet fuels. The last few chapters introduced a number of options being considered in pursuit of an efficient and cost-competitive biofuel including biodiesel, cellulosic ethanol, biohydrogen, or algae-derived biofuels. Despite great potential, their ultimate success will likely depend not only on the technical feasibility of these fuels but also on the ability for scientists and engineers to enhance the production of these fuels, the organisms that produce these fuels, and the production of other potential coproducts through biotechnological applications. In this chapter, we will focus on how biotechnology, metabolic engineering, and synthetic biology can contribute to the future success of biofuels.

Biotechnology and Biofuels

As discussed in Chapter 5, traditional agricultural methods have evolved for thousands of years to allow agricultural products, particularly food crops, to feed an ever-growing population. Early agricultural methods including irrigation, fertilization, and crop rotation revolutionized society and likely led to the clustering of populations and the emergence of cities. Yet, in the twentieth century, it became increasingly clear that traditional agriculture would not suffice to feed the exploding population. Thus, a second revolution took place, the Green Revolution, driven largely by genetic screening and the availability of fossil fuels. Part of this Green Revolution included building a better understanding of living organisms and the metabolic bioprocesses that allow these organisms to survive and produce bioproducts. This field of science is now referred to as **biotechnology** and focuses on the human and environmental applications of bioproducts—any product produced naturally such as chemical molecules, proteins, DNA, lipids, and other products produced within the cells of various organisms. Biotechnological applications, such as the introduction of genetic material making a crop resistant to a pest and the discovery of novel growth traits in microorganisms and plants to help improve crop plant productivity, have greatly impacted the industrial agricultural sector. Unfortunately, even these remarkable advancements in agriculture may prove inadequate in sustaining both the food and energy requirements of the even larger, more prosperous global population on the horizon.

Striking a balance between the use of agricultural products for food and fuel will be tricky. However, the demand for the use of these products may diminish due to the development of alternative nonfood source organisms. These alternative organisms, including bacteria, algae, and certain trees and shrubs, are being sought as both direct production platforms for generating biofuels and a source for novel and unique genes that can enhance the creation of biofuels from traditional organisms. While traditional agricultural methods will play an important role in providing a base of understanding for biofuel technologies, recent biotechnological breakthroughs are likely to have the largest and most rapid impact on the success of these platforms. It will probably be these biotechnological advances that help scientists take biofuel production to commercial-level scales in the next 50 years compared to the nearly 6,000 years of agricultural advancements needed to produce the level of productivity seen in crops such as corn today. There are many aspects of biotechnology that will lead to this advancement and examples such as bioprospecting and metabolic engineering will be discussed in this chapter to give you a better idea of the role biotechnology could play in the future of biofuel development.

Taking What Nature Offers: Bioprospecting

The inspiration for many human products and applications important in chemistry and medicine comes from nature itself. Nature provides an array of organisms living in diverse environments with unique ecological relationships and adaptations. Researching organisms inhabiting specific niches or environments including deserts, hot springs, and the arctic can lead to the discovery of novel applications, processes, and products from nature that can benefit both humans and the environment. The search and discovery of these organisms and their products is known as **bioprospecting**. Bioprospecting, as originally discussed in the previous chapter, can lead to the identification of organisms that have a natural advantage in the production of an important biomolecule like a biofuel and can also offer a vast amount of genetic information related to their ability to survive and thrive in certain environments. This genetic information can be used to help other organisms survive in those same environments using biotechnological methods discussed later in this chapter. These adaptations and genetic information could prove important when considering the growth of biofuel-producing organisms in alternative environments such as growing algae in the desert and as the impacts of climate change on agricultural productivity become known in the future.

Biotechnology can exploit genetic technology from a variety of environments ranging from deep-sea hydrothermal vents to cow dung to acid mine drainage and even landfills. Organisms live in all of these environments and each has special adaptations allowing them to thrive in sometimes very harsh and specific conditions. With respect to biofuels, scientists are focusing on a number of environmental parameters to identify organisms containing unique adaptations, such as temperature extremes, desiccation, and varying salinity and pH levels. Each environmental parameter is likely to play an important role in developing ideal biofuel producers on nonarable land (land that cannot be used for the production of food) or may help expand the growth range of an individual crop plant. For instance, an organism that prefers higher temperatures (a **thermophile**) may offer unique genetic traits or metabolic processes that can be used to grow a crop plant in a warmer climate. Or perhaps, an organism that can grow in a hypersaline (salty) environment known as a **halophile** has the genetic potential to allow an algal species to grow in water

with a higher salinity level as an alternative to scarce freshwater sources. Across the entire planet, there is a nearly limitless opportunity to discover these unique traits. However, as technology increases and transportation to isolated environments becomes easier, it will be critical that environmentally beneficial procedures are used when bioprospecting to limit the impact of organism collection on ecosystems. It will also be increasingly important for agreements to be in place when bioprospecting that prevent an individual or company from removing biological property from one location (such as a developing country) and not returning proper economic benefits should the biological property result in financial gain. Such agreements will need to specify the importance of maintaining a healthy ecosystem and provide incentives and specific protocols to prevent bioprospectors from ruining an ecosystem and walking away without taking responsibility for the loss of services from the ecosystem and the impact of this loss on the indigenous population.

When done in an environmentally responsible way, bioprospecting combined with biotechnology can identify traits that could enhance the production of biofuels within a variety of different organisms. For instance, scientists are looking for organisms with traits that allow for rapid growth to high densities with minimal amounts of added nutrients. This is particularly relevant to algal biomass sources. The faster a particular algal species can grow, the faster it can be harvested, extracted, and a new culture started. Critically, quick growth must be matched with the ability to grow to a high density. A thicker algal culture makes the process of culturing more cost-effective because more biomass is collected during each harvest. An organism that can efficiently use nutrients to reach these ideal growth parameters would be even more cost-effective because the culture would require reduced levels of nutrient supplementation.

Another important consideration for the cost-effective growth of algae for biofuels production is the culture's resistance to predators and pathogens. For large-scale production of biofuels, it is likely that algal cultures will be grown outside in huge open raceways prone to the introduction of contaminants through the air. Even if the algal culture can grow quickly to high density, this growth efficiency can be decimated by a predator, leading to the production of low amounts of biofuel. Many researchers have traveled around the world in the search for organisms with unique adaptations to protect themselves against predators and pathogens. Introduction of these traits into algal cultures used for biofuel production could protect them from infestation and secure the crop as a stable biofuel producer.

Finally, any organism or genetic trait that allows for a higher production of target molecules such as lipids or sugars enhances the commercial viability of biofuels. As you have learned, the production of lipids used to generate biodiesel and sugars used to produce bioethanol are controlled through an organism's metabolism, but all metabolisms are not created equally. Some organisms have the ability to produce larger quantities of these metabolites due to specific genetic traits. In some cases, an organism with "biofuel"-enhancing traits can be used directly as a producer, but in other cases complications in growth conditions or cost of growth may prevent further development. In these cases, individual genetic elements allowing organisms to produce higher levels of a biological metabolite can be extracted and bioengineered into an established biofuel-producing strain, which may lead to an increase in the production of ethanol or lipids within the producer. This process of genetic engineering will be discussed in more detail in the next section.

As you can see, identifying new organisms and novel traits can be hugely beneficial to the future production of biofuels, but how do scientists go about finding biofuel enhancers? Bioprospecting usually begins by identifying an environment that may have the potential to host organisms with some of these ideal bioenergy traits. For instance, when looking for drought tolerance adaptations, a scientist would likely travel to the desert or when looking for heat tolerance adaptations, a scientist would likely look within ecosystems surrounding hot springs or hydrothermal vents. Travel to these environments can be quite an adventure, whether scuba diving in the tropics or hiking in the mountains, and can sometimes be difficult, such as slugging through a salt marsh or snowshoeing through the arctic tundra. Regardless of the location, the goal is the same: to find and collect environmental samples. The collection of environmental samples can involve methods such as picking samples with one's hands, coring out a sample of soil from the ground, or filtering water. Each can yield a wide variety of life including bacteria, viruses, plants, algae, and invertebrates. This mass of organisms can be complex and present a problem in identifying individual characteristics, so once samples are collected they are generally brought back to a laboratory. In the laboratory, scientists focus on isolating individual organisms from the group, isolating genetic material and screening for traits of interest. Sometimes the isolated organisms may have a number of interesting traits related to bioenergy and can be directly tested for their productivity levels in biofuel production. However, since it is unlikely that a single native organism will have all of the ideal properties for biofuel production, in many cases it is the individual genetic trait found within this organism that is most valuable. Once a trait is identified, the target gene can be isolated and then spliced into the DNA of an established biofuels-producing organism like algae, giant miscanthus, or jatropha in a process known as **metabolic engineering** to give it an environmental or productive advantage.

Value of Metabolic Engineering

While breeding and traditional agricultural techniques will certainly be important in the future of bioenergy, many scientists believe that the massive quantities of biofuels needed to meaningfully supplant petroleum in a short time frame will require genetically manipulating biofuels-producing organisms. Genetic manipulation is likely to come in the form of recombinant DNA technologies and bioengineering techniques that can directly modify certain cellular metabolic functions and properties. Modifications may include the introduction and deletion of genes or perhaps modification of entire metabolic networks to create or enhance biofuel production. All of these genetic and metabolic processes can largely be manipulated through metabolic engineering.

The basis for metabolic engineering begins with the **Central Dogma of Molecular Biology** as diagrammed in figure 10.1A. This dogma describes the importance of DNA as a blueprint for metabolic functions within the cells of an organism (Crick, 1970). DNA is the genetic instruction book for any organism that is passed down from generation to generation. The primary structure of DNA, or deoxyribonucleic acid, consists of a linear sequence of nucleotides—adenine, guanine, cytosine, and thymine. The nucleotide sequence order of DNA defines genes that can be recognized by enzymes known as RNA polymerases to produce a transcript of messenger RNA in a process known as **transcription**. This RNA transcript is the messenger for the cell, taking information in the DNA and presenting it in a way that can be used to produce proteins. The synthesis of proteins from messenger RNA occurs in a process called **translation**. Proteins are what give a cell its metabolic functions and are the basic machinery operating cell function.

The use of DNA as the blueprint of life is what allows organisms to evolve and adapt to many different environmental and ecological circumstances and offers a range of opportunities for genetic manipulation. Adaptation occurs naturally when environmental change happens within an ecosystem, such as an organism being exposed to higher than normal levels of salt. At first, many individuals of that species may die, but a few individuals may have mutations within their DNA, allowing them to tolerate the excess salt. These individuals will persist and become the parents to all future individuals, thus passing on the specific DNA mutations that allowed them to survive in the salty environment.

Scientists can use genetic mutations to create an organism having desired characteristics. A few genetic mutations are shown in figure 10.1B. One example occurs when bacteria have genes responsible for an undesirable phenotype such as a red coloration. The genes involved in producing the red phenotype can be identified and "knocked out" or removed from within the DNA blueprint of the bacteria using genetic techniques. By removing the genes, the bacteria will no longer have the ability to turn red.

Another example of genetic manipulation is gene overexpression. Sometimes it may be desirable for an organism to produce more of a given product such as a biomolecule important in biofuels or medicine. Since these biomolecules

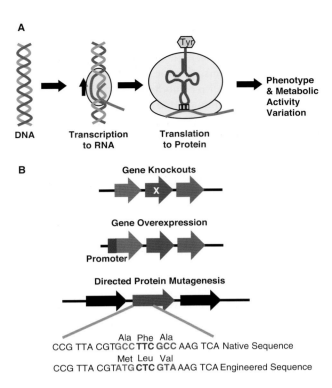

A

DNA

Transcription to RNA

Translation to Protein

Phenotype & Metabolic Activity Variation

B

Gene Knockouts

Gene Overexpression

Promoter

Directed Protein Mutagenesis

Ala Phe Ala
CCG TTA CGT**GCC TTC GCC** AAG TCA Native Sequence
Met Leu Val
CCG TTA CGT**ATG CTC GTA** AAG TCA Engineered Sequence

FIGURE 10.1 Diagram for the basis of genetic engineering.

A The central dogma of molecular biology showing how DNA is transcribed into RNA carrying the message that will eventually be translated into a protein resulting in the various phenotypes and metabolic reactions within cells and organisms.

B Three types of genetic manipulations that are important in biotechnology. These include gene knockouts where the function of a particular gene product is inhibited, gene overexpression where a promoter is added to a gene, resulting in an increased amount of the protein and the product of that protein's reactions, and directed protein engineering where individual nucleotides are changed, resulting in an alteration in the amino acids (shown as their three letter abbreviations) that make up a protein and ultimately a change in the structure and function of the protein.

are usually produced from enzymatic reactions involving proteins and because the amount of protein is based on the amount of messenger RNA, by increasing the amount of messenger RNA it is possible to increase the amount of the biomolecule. This increase in messenger RNA occurs by adding a short DNA sequence called a promoter into the metabolic pathway involved in producing the biomolecule. This promoter can then "promote" the increase in the amount of the messenger RNA, the amount of protein, and ultimately the quantity of the important biomolecule. This could be a very valuable tool in producing higher quantities of lipids for the production of biofuels!

Finally, metabolic engineering can also be used to modify the structure of a protein, thus changing its function. For instance, you may want to alter an enzyme to produce an 18-carbon lipid instead of a 10-carbon lipid when making

biodiesels. Since enzymes are proteins that catalyze or speed up a reaction within an organism, it may be possible to physically alter the structure of that enzyme to allow for the formation of the longer carbon chain. An enzyme is designed to specifically bind only a certain substrate similar to how a lock is fitted with only one key. In the case of the formation of a 10-carbon lipid, this substrate is acetyl-CoA. Once bound to the enzyme, the acetyl-CoA will undergo a series of reactions that will increase its length from 2 to 10 carbons. Once it reaches 10 carbons, it fills the binding pocket of the enzyme, triggering its release. However, if that binding pocket were to be expanded through a direct mutation to the structure of the enzyme, it may be possible that the lipid would continue to grow until it once again fills the pocket and instead of releasing the lipid at 10 carbons, the enzyme will now release the lipid when it reaches 18 carbons. In this case, metabolic engineering would create an organism capable of producing a biodiesel instead of a biogasoline.

In review, by understanding the Central Dogma, you will better understand the potential value of genetic manipulation. By altering any step within this Central Dogma, the ultimate biological result can change, resulting in the production of a more valuable product.

Introduction to Genetic Cloning for Metabolic Engineering

Metabolic engineering or genetic manipulation usually begins in a laboratory with the identification of a gene of interest often facilitated by bioprospecting. This gene of interest can be amplified from an organism's DNA through a process known as the **polymerase chain reaction** or PCR. PCR exponentially amplifies DNA from a host through a process that varies the temperature of the DNA and uses specific primers to recognize the section of DNA most valuable to the experiment. By using PCR, an individual gene of interest can be amplified to high enough quantities to be used in genetic experiments.

Once the gene is amplified, it must be cloned into a vector in preparation for metabolic engineering. Cloning is the process of inserting a gene of interest into a vector or a previously prepared circular fragment of DNA that contains all of the components needed to move that gene into another organism. The vector will generally contain a promoter to signal the cell to begin using the gene of interest and a resistance gene. The resistance gene is usually a gene linked to a selection factor that allows for the growth of only those organisms containing the gene of interest when grown in conditions containing that selection factor. Resistance to antibiotics is a selection factor commonly used during genetic manipulations. Once the gene of interest is cloned into a vector, it is ready to be inserted into a new organism in a process known as **transformation**. Methods for transformation are varied and range from heat shocking or electrically shocking the cell, which disrupts the cell membrane

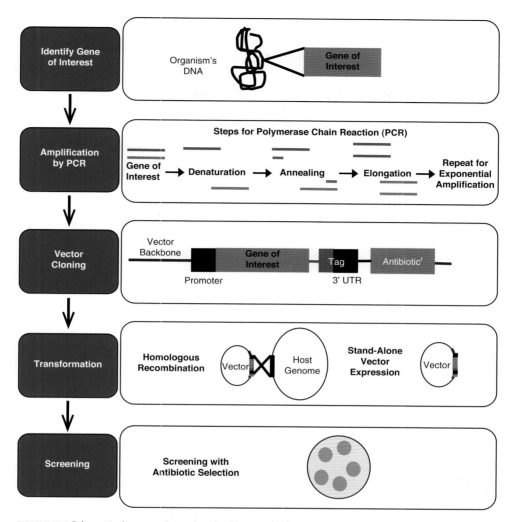

FIGURE 10.2 Schematic diagram of steps involved in genetic cloning. First, a gene of interest is identified and then amplified by the polymerase chain reaction (PCR). The amplified gene is then cloned into a vector that can be transformed into a host organism. Generally, this vector also gives the host a selective advantage, in this case antibiotic resistance. This advantage can be used to screen the host organisms to identify candidates that contain the new gene of interest.

to allow foreign DNA to enter, to physically "shooting" the cell with the DNA in a process called particle bombardment. In the end, the goal is to take a gene from one organism and place it into another organism to give that organism a specific function. Therefore, once an embedded gene is confirmed within the DNA of an organism of interest, the organism must be tested for the desired physical or phenotypic change. The process of genetic cloning as described above is outlined within figure 10.2.

Metabolic Engineering for the Production of Biofuels

The process of metabolic engineering is used for a number of applications ranging from biomedical research to bio-

technological advances. One of the easiest ways for us to understand the value of metabolic engineering in the context of bioenergy research is to study the principal pathways and metabolites relevant to this research and ask questions about how to manipulate them in a way that creates a more proficient bioenergy production system. Figure 10.3 outlines some of these important pathways and metabolites for bioenergy research including those involved in starch and triacylglyceride biosynthesis as well as ethanol production (Clarke, 2010).

The very first step toward productivity for every type of photosynthetic organism is the conversion of carbon dioxide (CO_2) into a 5- or 6-carbon sugar such as arabinose or glucose, respectively. This photosynthetic conversion sets the stage for the growth of the organism and providing

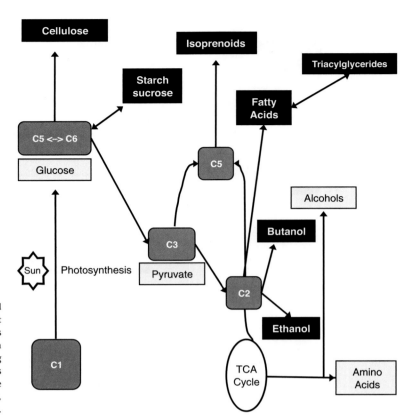

FIGURE 10.3 Outline of important pathways and metabolites for bioenergy research. Important metabolites for bioenergy are shown in black boxes following varying branches of metabolism beginning with the fixation of carbon during photosynthesis. Each branch of metabolism is marked by the number of carbons involved in the end product (C1, C2, C3, C5, or C6) (source: Clarke, 2010).

the metabolic building blocks for all of the metabolites produced by the organism. Thus, maximizing the photosynthetic efficiency within any photosynthetic organism could result in increased yields of any type of biofuel produced by that organism. In order to enhance growth and stimulate metabolite production through photosynthesis, it is necessary to initiate a careful study of the individual components and complex interactions involved in photosynthesis. By understanding the process of photosynthesis, it is possible to identify individual steps that may be more valuable in manipulating its productivity. For instance, a question that may arise is, "can chlorophyll (the light harvesting pigment involved in photosynthesis) be engineered to absorb more photons of light or a wider range of wavelengths during the light reactions?" or "can the electron transport chain be engineered to provide more energy (ATP) and more of the reducing equivalents (NADPH) for the fixation of carbon dioxide during the Calvin cycle?" A genetic engineer would seek to better understand how chlorophyll absorbs light in order to use that light to produce more energy for the cell. The production of more energy and reducing equivalents will then lead to the production of more sugars that can be converted into biofuels. By asking these questions, a scientist can begin to devise an experiment to test for these answers.

Take the manipulation of the electron transport chain as an example. Three molecules critically important in the function of the electron transport chain are plastoquinone, plastocyanin, and ferredoxin. The names of these molecules are not important, but the functions are vital for photosynthesis to be successful. Each of these molecules is known to be rate limiting, or the slowest part of the chain, when it comes to the transfer of electrons. By slowing down electron transfer, these molecules slow down the conversion of energy and reducing equivalents important in photosynthesis. By introducing more of these electron transport chain molecules, it may be possible for a scientist to overcome this slow stage of the electron transport chain and produce more energy and reducing equivalents. To test this idea, the genes encoding one of these molecules, for example, plastoquinone, can be promoted to express at higher levels within a cell. By "overexpressing" these genes, it may be possible to produce (1) more of the enzymes used in the production of plastoquinone, (2) more plastoquinone, and (3) a higher quantity of energy in the form of ATP and reducing equivalents in the form of NADPH. Thus, through genetic manipulation, the process of photosynthesis can become more efficient (Peterhansel et al., 2008).

After considering the enhancement of plant growth through metabolic engineering of the photosynthetic process, another question to consider is how an organism partitions its carbon. For instance, does the organism maintain stores of carbon as cellulose, starch, or lipids? In some cases, such as in algae, where the desired product is often high lipid yields, it may be necessary to find a mechanism for repartitioning carbon accumulation from starch storage to lipid storage. Algae are known to have the most productive growth in conditions where nutrients are abundant. When algae are grown under these conditions, they typically partition most of their carbon from carbon dioxide into carbohydrate biosynthesis, leaving only a limited amount for lipid biosynthesis. Yet, if an algae cell is switched to growth conditions containing low nutrients, then the growth rate will decrease, but the amount of carbon being partitioned into lipid biosynthesis will increase. This natural switch known to happen in algae indicates there is some type of regulatory control for how algae partition their carbon. If this control is identified, perhaps metabolic engineering can be used to regulate carbon partitioning so more carbon is partitioned into lipid biosynthesis to produce a better lipid yield. This switch is in fact possible because if starch biosynthesis in the green algae *Chlamydomonas reinhardtii* is knocked out through metabolic engineering methodologies, then the amount of lipids that accumulate are increased (Li et al., 2010).

Metabolic engineering can also be used to generate an organism that produces a better bioenergy product more efficiently. We can study this idea through three examples: engineering better fuel properties for biodiesels, advancing cellulosic processing, and creating alcohols with a higher energy content than ethanol.

Making improvements in fuel properties can be envisioned in two ways: creating a different fuel or obtaining that fuel more efficiently. Let us look at a recent example. A cyanobacterium, *Synechocystis* sp. PCC6803, was shown to produce mostly saturated lipids containing 16 carbons. These lipids would represent more of a diesel fuel type molecule due to the length of the carbon chain; however, the researchers were really interested in trying to generate a biofuel that could be used as a jet fuel. Jet fuels are usually slightly shorter than typical diesels, only containing 12–14 carbons. In order to try to engineer the cyanobacterium to produce these shorter lipids, they looked at the biosynthetic pathway involved in the production of lipids. Within this pathway, it is known that the length of a given carbon chain is generally controlled by the very last enzymatic function in the pathway known as the thioesterase. In this case, the native thioesterase has a long binding pocket that allows the carbon chain of the lipid to continue to grow until it reaches the end of the pocket, utilizing exactly 16 carbons; thus, the length of the pocket determines the length of the chain. Therefore, these scientists set out to replace this original thioesterase with an engineered thioesterase that had the potential to produce more of the 12 and 14 carbon length lipids. Through a series of steps

involving directed mutagenesis of a thioesterase from *Escherichia coli*, another bacterium, they were able to increase the production of these shorter lipids while decreasing the amount of 16 carbon length lipids produced. Thus, a cyanobacterium was genetically engineered to generate a higher quantity of the desired shorter lipid in a similar fashion to what was discussed earlier for creating longer carbon chains (Liu et al., 2011).

While working with this *E. coli* thioesterase, the authors also found that it contained a signaling peptide instructing the *E. coli* to keep the lipids inside the cell. However, if that peptide fragment was removed, then the lipids would be excreted. When this thioesterase from *E. coli* was introduced into the *Synechocystis* strain of cyanobacteria without the signal peptide, the cyanobacterium also began excreting lipids. Therefore, not only could they produce a lipid with a specific desired length, but they were also able to have the organism excrete the lipid, making extraction easier. This experiment is an excellent demonstration of how identifying a single gene within an organism and using metabolic engineering technologies can make a big impact on the production and control of biofuel molecules (Liu et al., 2011).

As covered in Chapter 6, cellulosic ethanol has growing potential to play a role in the replacement of liquid transportation fuels in the future; however, this role is dampened by the high cost and inefficiency of the enzyme cellulase when it comes to efficiently breaking down the cellulose from these plants for fermentation into ethanol. However, there are three areas of cellulose processing that could greatly benefit from metabolic engineering pursuits.

First, the efficiency of the conversion of cellulose into sugars could be enhanced through engineering the enzyme involved in this conversion, cellulase. This enhancement could occur by creating a more stable enzyme that can withstand the higher levels of heat required to prevent microbial contamination or by allowing the enzyme to function efficiently in a thicker mixture so that larger quantities of plant materials can be digested in a shorter period of time (Heinzelman et al., 2009). This efficiency could also be enhanced by creating a "cellulosome" that has a number of different cellulase enzymes docked together, resulting in an ability to break down many different types of sugar substrates. By doing this, the cellulosome could break down not only cellulose but also other cellular components including hemicellulose. This development would ultimately provide more substrate for the fermentation process (Wilson, 2009).

In a perfect world for cellulosic ethanol, a plant would be made of only cellulose that could be fully broken down for fermentation; however, as already discussed in Chapter 6, a significant amount of the polysaccharides in plants are hemicelluloses. Hemicellulose is largely made of 5-carbon sugars such as xylose and arabinose that are not fermentable by the traditional species of yeast used in

METABOLIC ENGINEERING FATTY ACIDS IN ALGAE

Travis L. Johnson

As discussed in this book, algae are being researched to produce biodiesel, which can be used in our existing infrastructure without replacing current refineries or engines. The molecules that make up algal biodiesel come from the organisms' fatty acids that are created through carbon metabolism. These fatty acids are then converted in algae to many kinds of lipids, all of which can then be converted into liquid fuel. Although algae naturally produce lipids, metabolic engineers have explored ways to make the molecules more ideal for biofuel production. One of these projects was to engineer algae to produce medium chain fatty acids that have properties closer to that of petrodiesel, including better cold flow properties, ignition properties, and oxidative stability. The research team knew that there was an enzyme in vascular plant called thioesterase that produced medium chain fatty acids for plant seeds, so they inserted this gene from a number of different plants into the algae's genome. The group also overexpressed the endogenous version of this gene to see if it would have an effect. To their surprise, none of the vascular plant thiosterases led to increased medium fatty acid accumulation, but the overexpressed algal gene did. Their hypothesis is that the vascular plant thiosesterases are not interacting with the algae's native fatty acid synthetic enzymes, and thus did not create the desired medium chain fatty acids. The idea is that through evolution, algal thioesterase has evolved to be different from vascular plant thioesterase, making these enzymes incompatible with the opposite host's metabolism.

SOURCES: Burkart (2014); Blatti et al. (2012).

---

fermentation. Thus, a second way to enhance the efficiency of the conversion of cellulosic plant sources into sugars is through the metabolic engineering of this traditional yeast to allow it to utilize 5-carbon sugars efficiently during fermentation. This would lead to a significant increase in the amount of sugar substrate available for the fermentation process and raise ethanol yields (Alper and Stephanopoulos, 2009).

Finally, cellulosic ethanol production is also hindered by the presence of lignin. Lignin is a very versatile molecule that gives many plants including tree trunks their rigidity. Lignin also prevents cellulases and other enzymes from being able to access the cellulose and hemicellulose to break down in the fermentation process. Thus, plants with higher levels of lignin are much more difficult to use as a feedstock for cellulosic biofuels. Metabolic engineering of the lignin biosynthetic pathway, particularly the knockout of key biosynthetic enzymes involved in the formation of this molecule, can be used to lower the lignin content of a plant or tree. A plant or tree specifically designed to have lower lignin content would be beneficial when used specifically for the production of cellulosic ethanol (Wagner et al., 2009). Overall, these three examples—better cellulosic conversion enzymes, genetically engineered yeast for hemicellulosic fermentation, and knocking out biosynthetic enzymes in lignin formation—represent a small number of the many metabolic engineering experiments that could advance the production of cellulosic ethanol and ultimately influence the future of bioenergy.

Metabolic engineering can also be used for the production of new and better biofuels. One example is the production of biobutanol, rather than bioethanol, during fermentation. Biobutanol has a higher energy content by volume than ethanol, is less likely to absorb water during storage, and has a better blending ability; however, biobutanol is not normally produced by the commonly used yeast species for fermentation. That said, engineering key enzymes from the natural butanol biosynthetic pathway from the bacterium *Clostridium acetobutylicum* into these yeast species has been shown to allow for the production of this fuel during the process of fermentation (Steen et al., 2008).

All of these examples of metabolic engineering for the advancement of biofuels stem from one source: nature. Nature provides a host of resources for the production of these fuels, but these resources must be used responsibly and safely. In general, metabolic engineering results in a **genetically modified organism** or GMO. To be considered a GMO, an organism must contain DNA from a different species and have been manipulated through recombinant DNA technologies such as metabolic engineering. GMOs have recently resulted in a significant level of controversy. While many people are concerned with the spread of altered genetic materials between crops and the effects of genetically manipulated crops on human health, the reality is that genetically modified foods have existed for a long time. But this does not dismiss the need for careful regulation of the use of these technologies and testing and research on the potential for negative consequence to humans and the environment in the future. It will be critical to establish responsible practices so that GMOs are not allowed to contaminate native species or alter established ecosystems.

"Omics": Advancing Bioenergy through Unseen Organisms

While much of the discussion thus far has been based on bioprospecting where scientists use organisms and traits they can physically isolate from nature, the reality is that environmental scientists predict only 2% of most bacterial species can be cultured in the laboratory (Wade, 2002). Therefore, 98% of the microbial resources on Earth and their many adaptations and genetic traits might never be recovered from these traditional collection methods. Does this mean these unculturable organisms are simply unavailable for discovery? No, fortunately, due to advancing technologies, it is possible to identify many of these organisms and traits using only their DNA. New methods in hybridization technologies and genetic sequencing are opening doors to the identification of many organisms that are simply not culturable within a laboratory.

In order to take advantage of these new identification technologies, a realm of science has developed known as the "omics" technologies. The "omics" technologies focus on the understanding of each step of the Central Dogma using a combination of laboratory experiments and computer programs. There are four major categories of "omics" technologies as summarized in Table 15 including **genomics**, **transcriptomics**, **proteomics**, and **metabolomics**. Genomics looks carefully at DNA, transcriptomics focuses on messenger RNA, proteomics looks at translation and the generation of proteins, and metabolomics focuses on understanding how the metabolites produced by an organism change under varying conditions.

One great example of the power of the "omics" technologies can be found in the use of metagenomics. In a metagenomic approach, rather than focusing on a single organism's genome (DNA), an environmental sample is taken, such as a soil sample, and all of the DNA from all of the organisms within this sample are isolated and combined. The isolated DNA is then submitted to a process known as

TABLE 15
Description and types of "omics" technologies

"Omics" Technology	Metabolite Utilized	Description
Genomics	DNA	Study of the genomes of organisms
Transcriptomics	RNA	Study of the expression of genes within a cell or organism
Proteomics	Proteins	Study of the structure and function of proteins
Metabolomics	Small molecules	Study of the collection of metabolites produced by an organism

sequencing where the actual nucleotides making up that DNA can be read. Once all of the nucleotides are identified, sequences are compared with existing information in databases in a process known as **annotation** to predict the most likely function of each gene. In doing this annotation, novel organisms, novel traits, and even entirely novel metabolic pathways can be identified without ever having to isolate the organism. "Omics" approaches are especially valuable in samples from environments where mimicking the conditions can be very difficult in the laboratory such as in anaerobic methane communities, acid mines, or a hypersaline mat. Having a method to find novel traits from these environments opens the door to the discovery of potential genetic traits that can give a biofuel producer a competitive advantage and change the future of bioenergy.

The Future of Biotechnology in Biofuels

While much of what has been discussed in this chapter including metabolic engineering and the "omics" technologies has really only been highly used over the last few decades, biotechnology is advancing so quickly that it already appears that the future of this field lies beyond using only what nature provides. In recent years, a field of science has emerged known as **synthetic biology**. Synthetic biology is the complete design, redesign, or construction of biological parts or systems. This field of study offers a tremendous amount of potential in the future due to the ability of scientists to completely design and chemically synthesize genetic sequences that can then be imported into a host. Thus, scientists can now design individual genes, sequences of genes, and even metabolic pathways that can simply be transformed into an organism to give it a completely different phenotype including the specific

BREEDING BIOFUELS ORGANISMS

Travis L. Johnson

Breeding is a genetic modification technology that has been around for thousands of years. Animal husbandry of livestock and pets, and selective breeding of plants for agriculture has shaped all of the domesticated organisms that benefit humans. Although not as new and high-tech as genetic engineering or synthetic biology, breeding is still a powerful tool that can be used for the development of algae biofuel sources. When organisms have sex, their offspring get a combination of each parent's genes. Through selective breeding—breeding organisms that have a desired trait—beneficial genes can be concentrated and put into genetic backgrounds, and detrimental genes can be taken out. An example of this is the breeding of *Jatropha* for biodiesel production. Researchers collected different cultivars of *Jatropha* and built up a large diverse germplasm of all the genetic resources available for trait selection. With a collection of over 12,000 genotypes, researchers successfully bred *Jatropha* plants to have desired traits such as high oil content, better fatty acid profiles, and high fruit yields.

Breeding can also be used to enhance the production of algae biofuels. A challenge though is that it is often hard to know the conditions that produce the algae's gametes for sexual reproduction. Many algae strains are facultatively sexual, meaning that in most conditions they will reproduce asexually, but in some unique circumstances, gametes may be induced and breeding can occur. It is difficult to know what those unique circumstances would be for the numerous species of algae.

SOURCES: McBride (2014); Schmidt (2014).

and high-yielding production of biofuels. While this technology is certainly exciting and offers an amazing amount of potential for the future of biofuels, it is important to once again remember that science must be viewed with a high level of responsibility. Sometimes the function of unknown genes and proteins within a new system cannot be entirely predicted; therefore, it is important to carefully monitor and regulate the use of these technologies in order to preserve the natural world while advancing biotechnology.

Overall, bioprospecting, metabolic engineering, "omics" technologies, and synthetic biology have created a new level of potential for the future of biofuels. The opportunities from these advancing fields are nearly endless, and as scientists continue to discover and create organisms ideally suited for biofuels production the productivity and the competitiveness of biofuels will continue to grow.

STUDY QUESTIONS

1. What role will bioprospecting likely play in the future of bioenergy?
2. Explain how the Central Dogma of Molecular Biology plays a role in metabolic engineering. Why is this significant for biofuels development?
3. Briefly describe the metabolic pathways most relevant to metabolic engineering for the production of biofuels. Explain how each of these pathways can be enhanced to improve the future of biofuels.
4. What is a GMO?
5. Explain how synthetic biology and "omics" technologies could shape the future of bioenergy.

CHAPTER ELEVEN

Thermochemical Conversion Technologies

Today renewable replacements for fossil fuels come mainly from sources that are well developed, like corn ethanol and biodiesel from oilseeds. Future biofuels will need to come from nonfood sources, and this development will require an aggressive research plan to optimize the productivity of new biofuels-related feedstocks through both traditional and modern agricultural methods including a significant contribution from biotechnology. With reserve-to-production ratios predicting a supply of less than 60 years remaining for petroleum-based resources, it seems likely that the production of liquid biofuels from biomass feedstocks will need to advance rapidly to meet the rising demand for these liquid fuels, at least in the short term. Even before these petroleum resources are completely exhausted, the price of these resources is likely to skyrocket, resulting in extreme economic and societal stress for many developed nations. Thus, we must also consider technologies that are not focused as closely on existing biofuels processes, but rather processes that leverage the chemical and engineering progress that has occurred over the past 200 years as well. One example of this is seen in the development of thermochemical technologies and their ability to utilize a wide array of feedstocks ranging from waste materials including municipal and plant waste to dedicated plant materials for the production of liquid fuels. A combination of these thermochemical technologies with the increased production of agriculture-based biomass over the coming decades could undoubtedly result in a significant production of liquid fuels, and with that an extension of the supply of liquid fuels much past the predicted 60 years.

Introduction to Thermochemical Processing

From a business perspective, the goal of the use of organic material such as biomass as a resource is the production of a finished product with the highest value possible. That product can range from fuels to food to chemicals and even to renewable power sources. The road from the growth of a biomass resource to the end product is full of alternative paths. Some of these paths, such as the fossil-fuel-based conversion technologies of converting coal to liquid fuel and natural gas to liquid fuel, and the biochemical-based conversion technologies such as sugar and starch fermentation to ethanol, have already been commercialized. However, other pathways such as the direct use of methane for the production of chemicals or the conversion of lignocellulose and/or algae into liquid fuels and chemicals are just now beginning commercial scale-up. Regardless of developmental stage, each of these paths can be considered just one method to convert a single type of biomass into more or less a single product. These specific conversion pathways can be valuable by themselves, but the reality is that the highest potential for the use of biomass comes when a single method can convert many types of biomass into many types of finished products. Such flexible conversion processes are not common to biology unless multiple processes are linked together. Nevertheless, a combination of chemistry and engineering offer a conversion method called **thermochemical** processing, which is capable of producing a variety of products from almost any type of biomass resource.

Thermochemical processing converts any carbon-containing material into an energetic product by taking advantage of the principals of thermochemistry, or the study of the energy and heat released during chemical reactions. The simplest example of thermochemical processing is one that has already been discussed, combustion. During combustion, organic materials are combined with oxygen to release carbon dioxide, water, and heat. You are well aware of this process, as this is what is happening when you are burning wood in a campfire or fireplace. Of course

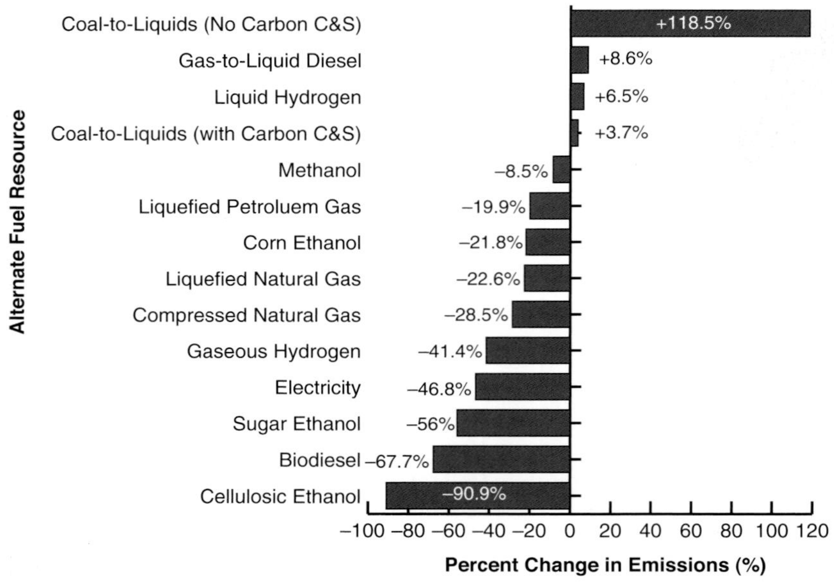

FIGURE 11.1 Percent change in greenhouse gas emissions over the life cycle of alternative fuel resources compared on an energy equivalent basis to the replacement of petroleum. Plant-based biofuels have the largest reduction in emissions, while thermochemical technologies converting coal and gas to liquid fuel have the highest relative emission. The conversion of coal to liquids can actually significantly increase emissions if carbon capture and sequestration (C&S) are not utilized (data from EPA, 2007).

exactly the same process is used to generate steam to produce electricity with the burning of coal or natural gas. Biomass can also be used to generate electricity resulting renewable power.

The production of heat or electricity from biomass is an important renewable energy process, and one that is a considerable part of the total energy picture. The problem with this use of biomass is that while it is certainly a renewable energy, it is not the highest value product that can be made, which are liquid fuels. We certainly cannot run cars or airplanes by burning wood. The combustion processes, if controlled carefully, can help in the production of desired products such as liquid fuels from many sources of biomass. This idea relies on the basic thermochemical principal that most sources of biomass can be broken down into individual molecules in the presence of high heat and pressure (in a low oxygen environment to avoid complete combustion) and then these molecules can be allowed to recombine to form desired products. One example of this is when biomass is exposed to high temperatures in low oxygen conditions to produce carbon monoxide and hydrogen. Hydrogen gas, as you have learned, can be a valuable fuel, but carbon monoxide, although an environmental pollutant, also contains useable energy. So in a second reaction, carbon monoxide and hydrogen can be combined to produce a valuable product, liquid fuels. The equations for both processes are given below:

$$C_xH_yO_z + \text{reduced } O_2 + \text{High Temp} \rightarrow CO + H_2$$

$$CO + H_2 \rightarrow \text{Liquid Fuels}$$

Carbon monoxide is an example of a **product gas** or **synthesis gas** that can then be used as a feedstock for the production of many products including fuels, electricity, and chemicals.

Aside from combustion, the commercial use of thermochemical processes is limited mainly to the fossil fuel conversion technologies coal-to-liquid (CTL) fuel and gas-to-liquid (GTL) fuel production mentioned earlier. These types of conversions began in 1923 with the development of the Fischer–Tropsch process by Franz Fischer and Hans Tropsch. **Fischer–Tropsch synthesis** allows for the conversion of synthesis gas, like carbon monoxide, created from either natural gas or coal, to be converted into liquid fuels (DOE, 2012). The Fischer–Tropsch process revolutionized thermochemical processing and has played an important role in the availability of liquid fuels to many countries at various times throughout history. This process is still commercially important today particularly in places such as China where coal is much more abundant than petroleum, and in the Middle East where large stocks of methane are available for conversion to the more easily transported liquid fuels. While both CTL and GTL are important, these technologies rely on the use of traditional, finite fossil fuel resources. Although coal is supposed to last significantly longer than either natural gas or petroleum, the widespread use of coal to create liquid fuels would ultimately decimate the coal reserves and lead to a much quicker depletion of proven reserves. Reliance on CTL and GTL to replace a significant amount of petroleum-based fuels would also significantly increase greenhouse gas emissions, particularly when coal is used as the initial feedstock. Figure 11.1 compares the change in greenhouse gas emissions if petroleum fuels were replaced by various

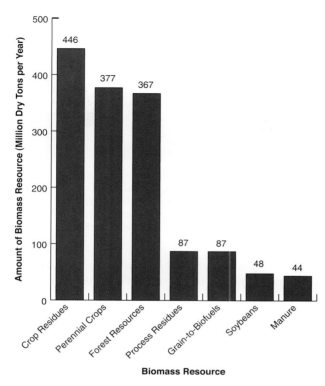

FIGURE 11.2 Comparison of the amounts of various biomass resources available in the United States where crop residues, perennial crops, and forest resources offer significant quantities of potential biomass for biofuels production (data from USDA and DOE, 2005).

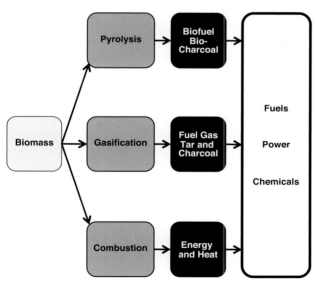

FIGURE 11.3 Flow diagram of the conversion of biomass into products via three primary conversion technologies: pyrolysis, gasification, and combustion.

technologies. The graph shows how adoption of a CTL process would greatly increase the level of emissions, especially if carbon capture and sequestration were not adopted. Other fossil-fuel-based conversions including GTL would also result in an increase in greenhouse gas emissions, while some of the other technologies including liquefied natural gas, corn-based ethanol production, cellulosic ethanol production, and biodiesel production could decrease overall emissions. Thus, the widespread use of thermochemical technologies with fossil fuels would likely hasten global climate change.

Luckily, the Fischer–Tropsch process requires carbon monoxide and H_2 gas, but it does not matter where this synthesis gas comes from, so the process itself does not require the use of coal or natural gas. In fact, this process is quite effective with a range of biomass sources. As shown in figure 11.2, the United States has a large quantity of biomass predicted to be available for harvest including crop residues, perennial crops, and forest resources. These vast sources, representing over a billion tons of available biomass that is spread over a wide area of the country, offer a significantly valuable opportunity for biofuels production with the proper conversion technology. Since thermochemical processes can be used to convert almost any type of biomass into liquid fuels, these processes could also prove valuable in meeting rising future demand for liquid fuel.

Primary Conversion Technologies

One of the greatest benefits of thermochemical processing, as mentioned previously, is its ability to use a variety of biomass feedstocks and produce a wide array of different products. This flexibility allows feedstocks ranging from forestry waste to dedicated energy crops to municipal solid wastes to be used in a single series of processes to produce fuels, power, and chemicals. Once harvested, dried, and prepared, these feedstocks can all be submitted to one of three primary thermochemical conversion technologies, combustion, **pyrolysis**, or **gasification**, to produce valuable fuel products as shown in figure 11.3.

Combustion, as already discussed, is one of the first methods of energy consumption by humans who relied on dried wood and vegetation to start fires for heating and cooking. Combustion is simply the burning of a fuel. This fuel can be anything that will burn ranging from traditional biomass sources like sugarcane bagasse for heating to coal and charcoal for electricity generation and even the burning of gasoline in a car's engine. Combustion is a chemical reaction where a form of chemical energy like wood, methane, or gasoline is combined with oxygen and a little added energy, such as a spark, to initiate the process. The reaction uses oxygen in the air to convert the chemical energy from the original source into heat while producing water and other by-products, the most common of which is carbon dioxide. Below is an equation for the combustion of methane gas:

$$CH_4 + 2O_2 \rightarrow CO_2 + 2H_2O + Heat$$

Clearly, with its versatility and its production of heat energy, combustion is still a valuable primary conversion technology used in many applications every day. The equation above and the term combustion generally refer to complete thermochemical decomposition of a material. This means that the original material—such as sugarcane bagasse or gasoline—is completely burned in the presence of oxygen, ultimately producing just three by-products: carbon dioxide, water, and heat energy. Pyrolysis and gasification are two other primary conversion technologies that can be used to create products from biomass that are normally created from fossil fuels today. Both of these technologies rely on thermochemical decomposition except that, unlike combustion, they result in an incomplete combustion process.

Pyrolysis occurs when organic material is exposed to high temperatures in the absence of oxygen. This process can happen in a regulated oxygen-free environment such as occurs in commercial facilities for the production of chemicals but can also occur naturally. Pyrolysis can take place in any situation where a fire uses up all of the oxygen but continues to burn, as in extremely hot structure fires or in situations where vegetation comes in contact with lava during volcanic eruptions. Pyrolysis can be used to convert most organic materials into solid, liquid, or gas. The end product of pyrolysis depends in part on the temperature that the organic material is exposed to, with lower temperatures around 400 degrees Celsius tending to produce larger amounts of solid char, while slightly higher temperatures around 500 degrees Celsius producing more gas and liquid oil known as **bio-oil** (Verma, 2012).

Pyrolysis is also common in cooking, particularly baking, frying, grilling, caramelizing, and any other food preparation technique that requires cooking food in a vessel with limited oxygen exposure to avoid burning. A good way to visualize pyrolysis is in cooking over a wood-fired grill. The flame burns away all of the oxygen but continues to heat the food, resulting in a golden brown coating on meats from the pyrolysis of the sugars and fats on the surface. If using wood for cooking, the wood also undergoes pyrolysis until the wood itself becomes a blackened solid material similar to charcoal, before eventually turning to ash. In fact, pyrolysis is the main thermochemical technology used for the production of charcoal and other solid residues rich in carbon, known collectively as char (Overend, 2005). Applications of pyrolysis are widespread not only because this process produces solid char, but also because it can produce both gaseous and liquid products.

The main applications for pyrolysis include creating a denser, more easily transportable material than biomass, such as charcoal and pyrolysis oils (Biomass Energy Centre

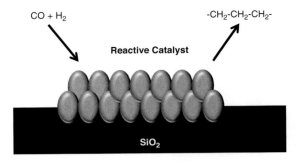

FIGURE 11.4 Diagram of Fischer–Tropsch synthesis showing the conversion of carbon monoxide gas into hydrocarbon chains using a reactive catalyst. This process represents an important conversion technology for the future of liquid hydrocarbon fuel production (source: Sun et al., 2000).

2011). Pyrolysis oil, or bio-oil, is a dark-brown liquid with a heat index about half that of conventional fuel oil, but generally much higher than the starting biomass. It is important both for co-firing technologies and as a feedstock for gasification and upgrading to other fuels. Bio-oil is produced in a fast pyrolysis method where the biomass material is exposed to high temperatures with a high heat transfer and then quickly cooled to form a liquid containing a variety of organic acids and hydrocarbon, plus oxygen (DOE, 2005). Bio-oil is different than petroleum because it is composed of a high quantity of acids, alcohols, phenolic, furans, and water rather than the traditional long chain hydrocarbons, contains a higher amount of oxygen (50%), and is very acidic (pH 3). It is, therefore, not an ideal substitute for liquid fuels by itself. One option for using bio-oil as a replacement for fossil fuels is through upgrading. Upgrading bio-oil takes this viscous liquid with a high oxygen and low hydrogen and carbon content and converts it into a liquid with much less oxygen and a higher ratio of hydrogen and carbon. Upgrading occurs by exposing the water-soluble fraction of the original bio-oil to steam processing to produce hydrogen. This hydrogen is then reintroduced to the original bio-oil in order to deoxygenate the oil essentially replacing the oxygen with hydrogen. Once complete, the oil is exposed to a catalyst such as zeolite to remove any remaining oxygen and water. This upgraded bio-oil is usually composed of molecules now much more similar to classical liquid fuels like benzene and can be fungible in many fuel applications (Verma, 2012). As you might imagine upgrading bio-oil is an energy-intensive process that releases a fair amount of carbon dioxide as well, so this process will need significant improvements before it will be used for any commercial fuel applications.

The other option for using non-upgraded bio-oil is as a feedstock for gasification. Gasification is the third primary conversion technology that utilizes not just bio-oil but also any type of organic feedstock including biomass and municipal wastes. Gasification is similar to pyrolysis except that, during this process, organic material is generally exposed to a much higher temperature at 800–900 degrees

BIOMASS GASIFICATION PROCESS

Travis L. Johnson

Let us take a look at the steps of producing liquid fuels using the biomass gasification process.

Step 1: pretreatment   First, the biomass must be dried, shredded, and grinded to small enough pieces to feed into the gasifier. It is important to dry the biomass before it starts the gasification process because drying during gasification could take a significant amount of energy.

Step 2: gasification   Once the biomass is dried and ground, it moves to the gasification reaction chamber. There are a few types of reaction chambers including moving bed and fluidized bed. A moving bed gasifier is a single chamber in which biomass and air are continually fed, while gas and char are produced. As the name implies, there is a moving bed of biomass in the chamber that undergoes different temperature and reaction zones. A fluidized bed gasifier is filled with biomass particles and ceramic particles similar to sand. Steam is injected into the bottom of the gasifier and as the vapor moves up the reactor, it lifts the biomass and ceramic particles up, creating a fluidized state. The produced gas and steam move up to the top of the reactor and are captured, while solid char and ceramic particles are captured.

Step 3: gas clean-up   The gas that comes out of the gasifier needs to be cleaned because it contains sulfur, minerals, and heavy hydrocarbon tar that could deactivate the catalysts used in the gas-to-liquid step. Tar can be removed by condensing the gas and discarding the contaminants, absorbing the tar vapors in liquid oil (stripping), or reacting the gas with steam and a catalyst to create more gas (reforming).

Step 4: gas-to-liquid   Once the produced gas has been cleaned, it can now undergo the gas-to-liquid step. This gas is made up of carbon monoxide (CO) and hydrogen (H_2) known as syngas. Different liquid fuels require different amounts of CO and H_2 molecules and different solid catalysts. Methanol synthesis requires one CO molecule and two H_2 molecules. If two molecules of CO and four molecules of H_2 are reacted with a catalyst, ethanol and water are produced. Reacting seven CO molecules with 15 H_2 molecules will produce diesel and water, which is known as the Fischer–Tropsch process. Since all of these reactions require different ratios of CO and H_2, this ratio can be adjusted by reacting CO with H_2O to create H_2 and CO_2, which would increase the number of H_2 molecules and decrease the CO molecules.

SOURCE: Herz (2014).

Celsius and a limited amount of oxygen. In fact, pyrolysis will occur as the apparatus housing the organic material warms up to these extremely high temperatures. These high temperatures will actually melt organic materials all the way down to a gaseous mixture of carbon monoxide and hydrogen known as synthesis gas or syngas, as mentioned earlier. **Syngas** can be used directly as a gaseous fuel in furnaces, but because of the high heat required to produce this gas, thus requiring large amounts of energy input, there is a significant energy cost to this technology. Thus, the value of syngas may be in the higher value products created after upgrading in a similar manner to bio-oil in producing liquid fuels (Damartzis and Zabaniotou, 2011).

Upgrading syngas occurs through the Fischer–Tropsch synthesis discussed earlier in the chapter and depicted in figure 11.4. As a reminder, this synthesis is a chemical process that converts carbon monoxide and hydrogen into hydrocarbon chains using a reactive catalyst. The process actually leverages the partial oxidation of the carbon in carbon monoxide as this partial oxygenation makes carbon monoxide a very reactive molecule. When two molecules of carbon monoxide react with a catalyst and one molecule of hydrogen gas, they form one molecule of carbon dioxide and a reduced carbon species (CH_2). As the reaction continues, these CH_2 species react to cross-link and form large hydrocarbon molecules, eventually attaining a length that makes them liquid fuel molecules (Damartzis and Zabaniotou, 2011).

This process of pyrolysis, gasification, and Fischer–Tropsch synthesis may bring to mind diagenesis from Chapter 2. To reiterate, diagenesis is the geologic process when organic material is buried deeper and deeper within the Earth and subjected to high temperatures and pressures until the organic material degrades and restructures into petroleum and natural gas, known, respectively, as catagenesis and metagenesis. Diagenesis is similar to the Fischer–Tropsch synthetic conversion process, in that cross-linking is free to occur randomly, and hence there is a limited amount of control over what molecules come out. The molecules produced are often dependent on how quickly the temperatures are changed and the pressure of the system, and on the catalyst used. Sometimes synthesis reactions only produce really long hydrocarbon chains in the form of waxes. These waxes must then be submitted to a round of energy-intensive treatment known as hydrocracking, exactly as done with petroleum, to produce useable molecules including jet and diesel fuels. Thus, the focus of some research related to these synthetic fuel technologies is in the development of catalytic conversion protocols that allow for better control of the final product. Since the production of synthetic gasoline and jet fuels is in high demand, a catalyst that produces hydrocarbons ranging in size from 4 to 12 carbons in length would alter the future energy landscape.

Future of Thermochemical Technologies

The thermochemical processes described above can be used to convert nearly any type of biological material—be it biomass, plastics, or municipal waste—into a variety of liquid fuels and chemicals, similar to what is presently derived from petroleum. Already, conversion technologies including pyrolysis and gasification are used commercially around the world using fossil fuels as the main feedstock. The fuels produced are often fungible with fossil fuels, meaning they can be transported and used in existing infrastructure and distribution networks around the world. This advantage is likely to prove crucial in leveraging the trillions of dollars already invested in energy infrastructure. Given all of these benefits, why are biomass thermochemical conversion processes not used commercially today? A close look at the primary reaction tells much of the story. First, significant amounts of energy input must be used to reach the temperatures required for any of these processes to occur. Second, oxygen is introduced into the system, meaning that before any liquid fuels are even created the biomass is already partially consumed, and this energy cannot be recaptured from the process. These are two significant energy components of the process that simply cannot be changed, meaning that all other processes must be optimized to overcome these basic energy restraints.

There remain a number of other challenges to the thermochemical conversion of biomass into synthetic fuels. One important issue is the implementation of this technology at large scale. It will be necessary to build large-scale plants that are investment capital intensive. By increasing the size of the plant, the overall cost of the product typically decreases, known as **economies of scale**. However, with the need for large-scale plants comes the need for large quantities of biomass feedstock. Additionally, thermochemical conversion technologies often produce caustic and toxic side reactions requiring the development of methods and practices to prevent these reaction products from entering the environment. Finally, biomass feedstocks have low energy density, meaning they cannot be transported long distances and remain energy competitive, so thermochemical processing facilities will need to be built near the biomass feedstock sources.

With further research into catalysts and methods for controlling end products, thermochemical technologies could prove highly valuable in sourcing synthetic fuels from biomass under many conditions. In addition, the utilization of these conversion technologies due to their flexibility in feedstock use, potential to reduce carbon dioxide emissions, and capacity to meaningfully augment liquid transportation fuels, suggests that we continue to develop these technologies to make them more energy efficient, providing another important technology in our quest for fossil fuel replacements.

STUDY QUESTIONS

1. How are the thermochemical conversion technologies different than the use of previously described biofuel technologies for the production of fuels?
2. What is the simplest thermochemical conversion technology? Why can't it be used for the production of fuels?
3. Briefly describe the role of pyrolysis in producing bio-oil.
4. How can gasification be used to produce fuel? What role does syngas play?
5. What are the benefits and challenges with thermochemical conversion technologies for the future?

CHAPTER TWELVE

Environmental Impacts of Biofuels: Water, Land, and Nutrients

The future of bioenergy is promising because of the wide diversity of feedstocks and technologies being developed and implemented. Such a large canvas of valuable biofuel options means there is an incredible opportunity to impact the fuel market, lower the consumption of fossil fuel, reduce environmental effects, and provide energy stability and security. As the development and production of biofuels from different feedstocks increases, it is important to begin considering the results of commercializing these products. For some feedstocks, commercialization will require huge investments in land and other natural resources in addition to economic and political support. While the economics and politics of biofuels will be discussed in Chapter 14, the next two chapters will focus on the environmental impacts of biofuels.

Biofuels and the Environment

With transportation accounting for nearly 30% of energy consumption in the United States and most of this energy being derived from liquid transportation fuels, the development of renewable, transportable, and storable liquid fuels has become a major focus of bioenergy research (EIA, 2014a). Liquid biofuels including ethanol and biodiesel have played a role in energy consumption since the founding of the automobile at the end of the nineteenth century, but this role has wavered in importance depending on the fossil energy climate. Recently, fluctuating and higher prices for petroleum and increasing concerns over petroleum shortages, national energy security, and environmental degradation have led to a renewed focus on the development of renewable fuel sources. Today, economists, scientists, and engineers believe the increasing demand for petroleum-based fuels cannot easily and cheaply be met with available supplies for much longer. This realization has led to the establishment of the Renewable Fuel Standard

(RFS) program under the Energy Policy Act in 2005 and later revised under the US Energy Independence and Security Act (EISA) in 2007. The RFS now calls for the production of 36 billion gallons of renewable fuels by 2022 with an emphasis on volume requirements for individual biofuel sources to limit the overall production of corn-based ethanol (EPA, 2012a).

In 2013, the United States produced about 14.5 billion gallons of biofuels composed primarily of corn-based fuel ethanol (EIA, 2014b). While this amounts to less than half of the 36 billion gallon goal for 2022, this production level of ethanol is already placing stress on the current US agricultural system and food production. The current RFS now mandates most of the remaining yield be met through the production of larger quantities of cellulosic and advanced biomass feedstocks such as miscanthus and algae. To qualify, the RFS requires the production of advanced biofuels, biomass-based diesels, and cellulosic biofuel that result in a reduction of greenhouse gas emission of between 50% and 60% over the life cycle of the fuel's production. The RFS also says the growth of feedstocks to produce these fuels cannot use virgin agricultural lands cultivated after 2007 or trees and tree crops from federal lands. These restrictions are designed to lower the environmental impact of fuel production and protect the land needed for additional food production (Schnepf and Yacobucci, 2012). Predicting and avoiding environmental impacts including emissions, land use changes, and natural resource stress will be important for the long-term success of renewable fuel sources.

Carbon Neutrality and Biofuels

As mentioned in the previous paragraph, to meet the requirements of the RFS, most biofuels must reduce greenhouse gas emissions by 50–60% during the entire life cycle of fuel production and processing (Schnepf and Yacobucci,

2012). This reduction is an important environmental component of biofuels because the combustion of fossil fuels over the past century has placed the entire planet under environmental stress. As you read in Chapter 3, the rise in greenhouse gases, particularly carbon dioxide, from the combustion of fossil fuels has resulted in an increase in positive radiative forcing in the atmosphere and a rise in average global temperature by about 0.85 degrees Celsius. It is expected that continued release of greenhouse gases could lead to an increase in global temperature of 2 to 4 degrees Celsius. Although temperature changes are not uniform across the planet, they can adversely affect precipitation, seasonal agricultural growth patterns, and animal migrations (Pachauri et al., 2014) and have direct impacts on productivity. Thus, it is important to consider the differences in greenhouse gas emissions between biofuels and fossil fuels.

The term biofuel is often associated with a clean-burning, carbon-neutral fuel that is sustainably derived from a biomass feedstock. While partly true, this definition can also lead to some misconceptions particularly in understanding **carbon neutrality**. A carbon-neutral fuel refers to a fuel whose net intake and output of carbon is zero. While this may seem simple, many people confuse a carbon-neutral fuel with a fuel that does not produce any carbon dioxide. The combustion of a biofuel still produces carbon dioxide, and in the case of some fuels such as biodiesel or biogasoline, the amount of carbon dioxide produced upon combustion is equivalent to that of the combustion of the same petroleum-based fuels. Think back to the previous diagrams showing the different fuel structures—once extracted and purified fossil-derived gasoline and biogasoline are almost exactly the same molecules. They burn exactly the same way in a car's engine; therefore, they are going to release exactly the same amount of carbon dioxide during combustion. The major difference between these two molecules is the original source of the fuel. Fossil-based fuels come from plant remnants that have been buried below the surface of the planet for millions of years; thus, when we remove the crude oil from the Earth to produce gasoline and then burn the gasoline in a car's engine, we are only adding greenhouse gases to the atmosphere. However, a bio-based fuel is derived from a plant grown for the purpose of producing that fuel. Since this plant must use photosynthesis to grow, which we know is a process that results in taking up or sequestering carbon dioxide from the atmosphere, the plant has the potential to take up at least as much carbon dioxide as is released when the biofuel is produced and burned in an engine. Therefore, the ideal biofuel will balance the use of carbon dioxide by the plant source with the output of carbon dioxide emissions during combustion. For this reason, it is important to remember that not all biomass sources are created equal. Two biomass sources could both sequester an equal amount of carbon dioxide, but the processing of these two sources to produce the biofuel could be very different. One source could require a lot of processing and result in the production of more greenhouse gases, while the other source could require less processing and release fewer emissions. In another example, some sources such as switchgrass, a cellulosic ethanol source, can sequester much higher volumes of carbon and lead to a higher net reduction of carbon dioxide than other bioethanol biomass sources such as corn (Qin et al., 2012). Figure 12.1 depicts the potential of cellulosic feedstocks used in fuel production to reduce carbon emissions due to their higher levels of soil sequestration. In figure 12.1A, the cellulosic feedstocks switchgrass, miscanthus, and corn stover are compared with the common corn kernel feedstock for ethanol production. Here it is apparent the use of cellulosic feedstocks for fuel production decreases carbon emissions due to the higher level of soil sequestration by these plants. Switchgrass is predicted to have the highest level of reduction due to soil sequestration. While it is convenient to see cellulosic feedstocks reduce emissions to a higher extent than corn kernel-based ethanol production, how do these feedstocks really compare to emissions from gasoline? Figure 12.1B shows that the use of corn for ethanol production reduces total carbon emissions by about 37% compared to gasoline emissions, while the cellulosic feedstocks including corn stover, miscanthus, and switchgrass are predicted to reduce total carbon emissions by 94%, 130.5%, and 175.1%, respectively (Khanna, 2008).

Finding a biomass source that can sequester high concentrations of carbon is the first step toward reaching carbon neutrality and possibly even a carbon negative fuel source. However, this benefit can be negated if the production and processing of fuel from this biomass has a high-energy requirement. For instance, the harvesting and separation of algal cells from water during biodiesel production may have higher energetic costs than the harvest of soybeans, particularly because the era of industrial agriculture has improved the efficiency of field-based harvesting methods. Sometimes these energetic needs can be offset by hybrid systems using various forms of biofuels such as biogas where the biogas is used to drive the processes needed for harvest, separation, and extraction. However, in most cases, these energetic needs are still acquired from fossil fuels, resulting in a significant addition of carbon emissions to the overall footprint of the production of the final fuel product. To truly understand the environmental effects of a fuel regardless of an individual parameter such as harvesting or separation, the production and processing of this fuel throughout its entire life cycle must be considered. This comparison is known as a life cycle assessment and will be discussed in Chapter 13. Ultimately, in order to achieve the maximum level of energy sustainability in the future, it will be critical that the biomass sources used as biofuels platforms are engineered and cultivated in a way to maximize fuel volume while minimizing emissions.

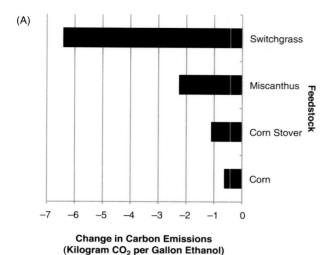

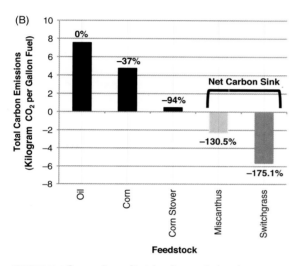

FIGURE 12.1 Comparison of total carbon emissions between cellulosic feedstocks used in ethanol production, corn-kernel-based ethanol production, and gasoline production.

A Comparison of the impact of soil sequestration for cellulosic feedstocks corn stover, miscanthus, and switchgrass with corn-kernel-based (corn) feedstocks for ethanol production. The cellulosic feedstocks sequester a higher level of carbon than the corn kernel feedstock; therefore, they reduce the level of overall carbon emissions to a higher extent. Switchgrass is predicted to have the largest impact on carbon emissions due to soil sequestration with a reduction of 6.4 kilogram CO_2 per gallon ethanol.

B Comparison of total carbon dioxide emissions due to soil sequestration between ethanol-producing feedstocks and gasoline. The use of corn-kernel-based ethanol reduces carbon emissions by 37% compared to gasoline, but the use of cellulosic feedstocks corn stover, miscanthus, and switchgrass used for ethanol production are predicted to reduce carbon emissions by 94%, 130.5%, and 175.1%, respectively.

DATA FROM: Khanna (2008).

Land Use Changes

Another important consideration for greenhouse gas emissions during biofuel production is the growth of the feedstocks themselves due to land use changes. As mentioned previously, unlike fossil fuels which have been buried for millions of years and simply release greenhouse gases like carbon dioxide into the atmosphere when combusted, biofuels are generally produced from a living product like crop plants or algae and have the potential to also sequester carbon dioxide from the atmosphere. The ability to sequester carbon dioxide by plants and algae through photosynthesis gives these feedstocks a carbon uptake credit that can often be large enough to counterbalance the release of carbon dioxide through the combustion of the biofuel itself. While this carbon uptake credit may seem ideal, in many cases, the carbon benefits associated with biofuels do not take into consideration the direct and indirect land use changes that occur when a plot of land is switched from either its natural state or agricultural state to a crop dedicated to biofuel production (Searchinger et al., 2008).

Direct land use changes are most often associated with the loss of natural ecosystems and their evolutionary balance of carbon. For instance, a farmer may destroy a forest or grassland area that has existed for thousands of years. In doing this, the farmer releases huge amounts of carbon that have been stored in the plants and soils through either decomposition or fire. This release of carbon can far outweigh the amount of carbon that will be sequestered by the replacement of these natural ecosystems with an agricultural system for biofuel production. And this replacement agricultural system can lead to an increase in overall emissions associated with the final production for many years (Searchinger et al., 2008; Ahlgren and Di Lucia, 2014).

Indirect land use changes are a little harder to quantify but also play a significant role in the final emissions value for biofuel production. Indirect land use changes occur when a farmer who already uses land for agricultural purposes decides to stop using the land to produce food and rather use this land to produce a dedicated biofuel crop. In this case, the actual emissions between the two crops are likely to be very similar, but the difference comes in the need to replace that food crop somewhere else. By diverting an agricultural food crop to a biofuel feedstock crop, a farmer can influence the price and demand for a food crop that leads to other farmers in other areas clearing more land to produce the food crop. While this change directly influences carbon emissions associated with food production, it is indirectly derived from the use of cropland for biofuels production (Searchinger et al., 2008; Ahlgren and Di Lucia, 2014).

Both direct and indirect land use changes are important when considering overall environmental impacts during the production of biofuels, particularly in relation to greenhouse gas emissions. As shown in figure 12.1, every crop is

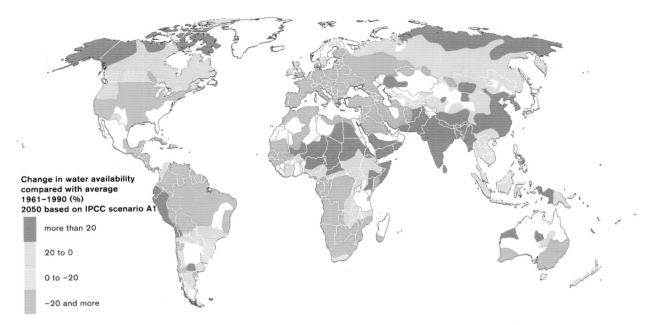

FIGURE 12.2 Predicted change in global water availability by 2050 compared to the average between 1961 and 1990. Based on these predictions, important agricultural regions including the United States, Europe, and South America are expected to see a decrease in water availability of over 20% (map created by Rekacewicz, 2009).

**Change in water availability
compared with average
1961–1990 (%)
2050 based on IPCC scenario A1**

more than 20

20 to 0

0 to −20

−20 and more

slightly different, and it is possible that one crop will be able to sequester more carbon than another; therefore, research is needed to better understand the growth characteristics of an individual feedstock as well as the environmental and economic impacts of using this feedstock. Generally, over time land use change emissions will balance out for most crops because the amount of carbon being sequestered will ultimately be more than the one time land use change deficit. However, in a time when greenhouse gas emissions, particularly carbon dioxide, are becoming a clear environmental, social, and economic concern, we must consider these land use changes now to help reduce greenhouse gases and avoid further negative climatic impacts in the near future.

Importance of Water on Earth

The production of enough biomass for food and fuel at scales to maintain the current levels of agricultural production and continue to feed a growing population will likely require a higher demand for two important natural resources, water and nutrients.

With over 70% of the Earth covered with water, it may seem that water should not be a significant limiting factor with regard to biofuels. However, the limiting factor is not necessarily in quantity of the water but rather in quality. Over 97% of water on the Earth is saline including the oceans. This leaves less than 3% of Earth's water in the form of freshwater such as glaciers, aquifers, rivers, and lakes. Saline water and many freshwater sources like glaciers are

considered largely unusable for agricultural or human consumption; thus, only about 1% of the total volume of water on Earth is actually available for human repurposing (Theis and Tomkin, 2012). Agricultural and industrial applications and human consumption all depend on the efficient allocation of this 1%.

Freshwater is one of the most valuable substances on the planet, and we continually consider it a renewable resource instead of a finite resource. Why? The answer lies in one of Earth's most important cycles, the **hydrological cycle**. As outlined in figure 4.11, the hydrological cycle is the continuous cycle of water in its various forms on Earth. During this cycle, water is evaporated from the oceans to form clouds that blow over land where precipitation will fall into the mountains and travel back toward the ocean either directly through rivers, lakes, and streams or seeping slowly through groundwater discharge. This constant cycle of evaporation and precipitation is what replenishes the freshwater supply, allowing 1% of the Earth's water supply to support so much life.

Although water is renewable, freshwater is not always available. Precipitation in some regions such as the tropics is much higher than in other regions like the temperate zones. Climate change and growing populations in many of the regions already experiencing lower levels of precipitation are exacerbating the issue and leading to water scarcity. Figure 12.2 shows the predicted change in water availability by 2050 for different regions around the planet. Here you can see that in many regions including the United States, Europe, and South America (all important agricultural hubs) there are predicted to be decreases in water availability of more than 20% (Rekacewicz, 2009). With the United Nations estimating nearly 11% of the global population already lives under water stress conditions, a decrease in water availability to this level could have drastic consequences to agriculture and human health (Theis and Tomkin, 2012). Water stress conditions will likely contribute to less food availability, less drinking water availability, and poorer sanitary situations that can lead to a higher rate of disease and even premature death, particularly in developing regions.

While it is easy to think of water stress when visualizing the desert climates of the Middle East and North Africa, water stress is not limited to these regions; many US states, particularly those in the south, are already facing water shortages due to increases in heat and drought conditions. For instance, the Ogallala Aquifer that supplies most of the water used in North Texas to support cotton and corn growth is decreasing at least 1 foot per year and may only meet demand for the next 10–20 years (SARE, 2010). As the Ogallala Aquifer is a major agricultural water source, this decrease could place a significant strain on the availability of corn to support the ethanol and cattle industries in this region. With water shortages spreading and the human population increasing, there is a definite need to conserve water resources wherever possible, which includes in the production of our energy resources.

Balancing Water Stress and Biofuels Production

Water plays an important role in energy availability. There are obvious sources of energy that use water like hydropower; however, hydropower is generally considered a nonconsumptive use of water. This means that the water is never completely removed or depleted from the natural environment where it is taken. However, other uses of water for energy are consumptive and make the water withdrawn unavailable to the environment in the region. The production of both fossil fuels and biofuels requires the consumptive use of water. For example, the production of diesel or gasoline from crude oil consumes about 3.4–6.6 gallons of water per gallon of fuel (Wu et al., 2009). Considering that the United States alone produces 10 million barrels of oil daily (approximately 420 million gallons), the production of diesel and gasoline in the United States consumes over 1.4 billion gallons of water every day (British Petroleum, 2014). While this volume of water may seem high, it is actually much less than the water consumed in the production of ethanol from cornstarch. Depending on the amount of precipitation in the area where the corn crop is grown, water consumption for cornstarch-derived ethanol can range from 10 to 324 gallons of water per gallon of ethanol (GAO, 2009). Even with low levels of irrigation, 10 gallons of water per gallon of ethanol is still double that of the volume of water used in the production of gasoline and diesel.

Taking into consideration the lower fuel density of ethanol, the estimated 1.7 million additional acres of corn that will need to be planted to reach the 15 billion gallons of ethanol from cornstarch goal set by the RFS mandate seems unrealistic. In addition, other biomass crops being researched such as cellulosics are likely to have substantial water requirements of their own. Thus, reaching the 60 billion gallons of ethanol that the USDA and DOE predict can be produced in the United States to replace 30% of US gasoline demand by 2030 may be an unreasonable goal (NRC, 2008; GAO, 2009).

A significant amount of the water used in producing ethanol results from irrigation of crops; however, there is also a large volume of water required in the actual refining of ethanol within processing plants. By improving equipment and becoming more energy efficient, many plants have been able to reduce their water usage from 5.8 to 3.0 gallons of water per gallon of ethanol in the last 10 years; however, this still results in a corn-based ethanol production plant requiring nearly 300 million gallons of water to produce 100 million gallons of ethanol, an amount equivalent to the quantity of water used by a small town. In addition, with the increasing demand for the production of ethanol from either corn or cellulosic sources, many of the new ethanol-processing facilities are planned for the regions in the United States where water withdrawals from the existing aquifers already outpace the recharge rate of those aquifers, setting up for a future of extreme water stress (GAO, 2009).

Biofuels, Agriculture, and the Water Supply

Biofuels are derived from biomass feedstocks produced in basically the same way that food has been produced for thousands of years. Many of these production methods even now are not sustainable in the sense that they require water in quantities that cannot be guaranteed into the future. However, pursuing RFS biofuel targets using current agricultural practices could accelerate water overuse even faster. An understanding of the importance of water quantity, water quality, nutrient availability, and nutrient recycling—both to agriculture and to human life—is crucial in developing practices that will allow the production of biofuels to be sustained for many generations.

Regardless of feedstock, all plants require water and nutrients. Water can be considered in terms of both quantity and quality: quantity referring to an amount of water that can support agricultural practices, human consumption, and the native environment, and quality referring to the degree to which water sources are free of contamination.

To understand the quantity of water used in agricultural practices, we must first consider water flows involved in the growth of crops. Water flows in agriculture are a balance of inputs and outputs as shown in figure 12.3. A crop can be supported by the input of water naturally from precipitation or supplemented with irrigation practices that use

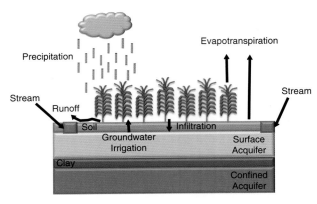

FIGURE 12.3 Diagram of agricultural water flows showing how precipitation and groundwater can provide irrigation to crops while much of that water is recycled through transpiration, runoff, or infiltration (source: NRC, 2008).

surface water and groundwater resources. The crop itself will uptake a portion of this water and then lose some to **evapotranspiration**, with the remaining water left to evaporate from the ground, infiltrate the soil, or run off into a nearby stream or river. Precipitation, infiltration, and evaporation are all natural processes that can often support plant life in their native environments without a problem. However, often agriculture has taken nonnative plants and grown them in regions whose natural water flows do not support high productivity levels in these crops. In these cases, irrigation is used as a supplement to natural water input.

Irrigation is any type of water added to land from an artificial source such as a sprinkler or a water drip system. This artificial redistribution of water was one of the main factors allowing humans several thousand years ago to develop agricultural practices. Today, irrigation is still a critical practice that grants humans the ability to produce such a wide variety of foods in amazing abundance. In fact, nearly 60% of the global consumption of water is due to irrigation (Perlman, 2012). In 2007, 414 million acres of land were used as cropland, equaling approximately 18% of the total land area in the United States (Nickerson et al., 2011). The 2007 US Census of Agriculture found 54.9 million acres of farmland were irrigated, which translates into a consumption level of about 29.8 billion gallons of water. Although this irrigated land only represents 13% of total US cropland, a significant quantity is located in low precipitation states such as California and Texas that are already facing water shortages and drought conditions (USDA, 2008).

Corn is one of the most important crops grown in the United States, supporting food and animal feed production as well as ethanol generation. Corn is also the most highly irrigated crop (Nickerson et al., 2011) and in 2007 accounted for about 30% of the total US crop harvest. As total rainfall among states varies widely, corn correspondingly requires differing amounts of irrigation depending on location. For

instance, the corn belt (Iowa, Indiana, Illinois, Ohio, and Missouri) uses about 7.1 gallons of irrigated water to produce 1 gallon of ethanol and the Great Lakes (Minnesota, Wisconsin, and Michigan) use about 13.9 gallons of irrigated water per gallon of ethanol. However, states within the Northern Plains (North Dakota, South Dakota, Nebraska, and Kansas) require the use of about 320.6 gallons of irrigated water per gallon of ethanol, with much of this water taken from ground sources (GAO, 2009). Such consumption rates can result in lowering of the water table and water shortages for nonagricultural needs.

Water used in irrigation largely comes from two sources: surface waters such as lakes, streams, and rivers, and groundwaters such as aquifers. Groundwater sources are generally replenished through seepage of water through the soil back to the water table, while surface waters are primarily renewed through direct precipitation. Depletion of water sources occurs when usage rates exceed replacement rates, although rate of depletion is also water source dependent. Surface waters will be depleted in part by human consumption, but depletion can be accelerated based on climate change related effects on precipitation. In the United States, northern states are expected to become wetter, while southern states will continue to become drier (EPA, 2012b). Precipitation changes will undoubtedly affect the flow of streams and rivers and lower the volume of water found in lakes in southern regions.

While surface waters play an important role where available, many regions are wholly dependent on groundwater. Groundwater fulfills the water needs of about half of the US population and also serves as a source of agricultural water in irrigation (Perlman, 2012). Although groundwater can be replenished through seepage from precipitation, rivers, lakes, and municipal discharge through the ground into the water table, these water resources can also be easily depleted through overuse. Depletion can be localized (a well dug for an individual home runs dry because the water table has dropped below the level of the well) or more widespread (a city reduces its water table to the point of creating a cone of depression). Overpumping can scale to huge regions, affecting multiple states when large aquifers are depleted (Perlman, 2012). In the northeastern United States, the depletion of groundwater is due to the consumption of water for domestic use in highly populated urban areas; however, in other areas such as the Midwest and West, the majority of the depletion is more likely because of irrigation. As mentioned earlier, one of the world's largest aquifers, the Ogallala Aquifer, is located in the US high plains. This aquifer underlies eight states, including Colorado, Kansas, Nebraska, New Mexico, Oklahoma, South Dakota, Texas, and Wyoming, and is responsible for providing 30% of US groundwater used for irrigation. Withdrawal of large volumes of water needed for irrigation from Ogallala have far outpaced the replenishment (USGS, 2011), and water levels have been reduced by 100 feet in some areas (Perlman, 2012). Unfortunately, the strain on this aquifer will continue as

LAND SUBSIDENCE

Travis L. Johnson

California is one of the largest agricultural producers in the United States and that is thanks in great part to the San Joaquin Valley. This area, along with the rest of California's Central Valley, produces roughly 25% of the country's agriculture on only 1% of its farmland. In order to sustain this enormous production, irrigation was put into place taking water from rivers and diverting them to farms through canals and ditches. But farmers were restricted by the natural surface water flow, and as crop demand grew, they started to develop groundwater resources in the early twentieth century. This new water source proved to be economic and stable, providing California farmers enough water to grow crops to feed the nation. It was not until the mid-twentieth century that people became concerned that pumping groundwater had resulted in the land subsidence that had occurred in the San Joaquin Valley. Land subsidence occurs when the ground elevation decreases because large amounts of groundwater have been withdrawn,

Image of land subsidence in San Joaquin Valley, CA

resulting in ground compaction. The largest measured land subsidence occurred in the town of Mendota, CA, where the ground's altitude had decreased by more than 28 from 1925 to 1977. Over 5,200 square miles, or more than half of the entire San Joaquin Valley, had at least 1 foot of land subsidence, making this one of the largest human-caused Earth surface alteration.

SOURCE: Cone (1997). Image provided by the United States Geological Survey.

---

the Northern Plains are expected to see the largest increases in corn production in the coming years (GAO, 2009).

Clearly, the quantity of water in the United States is already under threat from current domestic and agricultural usage. Yet, with the increase in biomass required to meet the potential demand for biofuels in the future, the consumption of water may increase significantly. When considering the various biomass feedstocks being used, studied, and developed, water usage is a criterion that should be carefully considered. Native crops can often survive only using the natural water supply, while introduced crops may require additional irrigation. In this case, native crops may represent a better resource as a feedstock to avoid overuse of water. In addition, many cellulosic crops are being studied that do not have the same water requirements as other feedstocks. For instance, corn stover (stalk) can be harvested as a cellulosic feedstock from corn already being grown and irrigated for food and feed.

Nutrients, Fertilizers, and Water Quality

Consumption of water for domestic and agricultural uses is a continued threat to water supply; however, in some cases, even a large water supply can be ruined by human practices affecting water quality. The most obvious threat to water quality is pollution. Dumping or leaking toxic materials into the water supply or throwing solid trash into waterways can cause water pollution, but pollution in water sources can also come from less obvious and more organic sources like agriculture. Nutrient enrichment and sedimentation within surface waters from agricultural runoff and contamination from agricultural water infiltration all play a significant role in polluting the water supply in the United States.

Sedimentation occurs when particulate matter is suspended in waterways. Sedimentation often results from soil erosion, a form of soil degradation that occurs when water and wind move soil from one location to another, thus often depleting the quality of the soil in the original location. Soil erosion can be increased on croplands due to the replacement of native plants with crops that are unable to sustain the integrity of the soil and in conditions where large volumes of irrigation water washes the soil away from the crop. The combination of these two factors can result in a significant amount of sediment being washed away from cropland and entering waterways. Increases in sedimentation can significantly affect the topology of a waterway and impact its ecology, causing habitat loss and reducing fish and plant life (Victoria, 2011). Maintaining ecosystem integrity within waterways is critical for water quality because many organisms play a role in filtering, clarifying, and upholding water source health.

Runoff and infiltration can also contribute to a decrease in water quality. While neither of these processes is a serious detriment when occurring under natural conditions, both are capable of harming waterways when contaminated with pollutants. Fertilizers are a major source of water contamination through both runoff and infiltration. Applying fertilizers to crops is a foundational agricultural practice that has played an important role in the large increase in crop productivity over the past century. Native plants are usually accustomed to the natural nutrient balance found within their native soil; however, many crop plants are both unaccustomed to this balance and tend to deplete the nutrient resources in the soil due to soil reuse. Therefore, these crop plants often require the addition of a number of major, minor, and micronutrients to survive and thrive. As shown in figure 12.4A, nitrogen, phosphorus, and potassium are considered some of the most important macronutrients needed by crop plants. These macronutrients combined with lower quantities of micronutrients also shown in the figure are usually important components of fertilizers. Fertilizers allow for better plant growth and productivity when considering maximizing food and fuel production. Compare the changes in corn biomass associated with varying levels of nitrogen fertilization as shown in figure 12.4b (Gallagher et al., 2011). As shown, applying higher fertilizer quantities to the corn crop increases total biomass associated with the corn plant including the grain, leaf, and stem, and the crop residue. The findings are important because the grain is the major part of the plant used for both food and the production of ethanol. Increasing the level of grain production can ultimately lead to higher quantities of food and fuel with the potential for cheaper prices. While it is clear that fertilizers are useful in agriculture, a balance must be struck between the overall environmental impact and value of the end product.

The total nutrient requirement for a crop depends on the plant. Corn crops require the largest quantity of total nutrients when considering nitrogen, phosphate, potassium, magnesium, and sulfur. Other crops such as soybeans, wheat, cotton, and rice have lower total nutrient requirements. The nutrient requirements of the plant in turn dictate the amount of fertilizer needed. The United States consumed about 20.8 million tons of nitrogen-, phosphorus-, and potassium-based

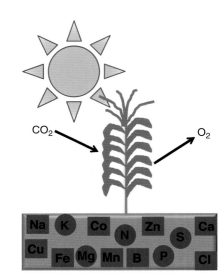

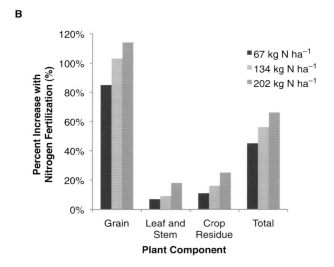

FIGURE 12.4 Fertilizer use in crops.

A Diagram of important nutrients required by plants for growth including the conversion of carbon dioxide (CO_2) into carbohydrates and oxygen (O_2). Major and minor nutrients shown in red circles (K – potassium; Mg – magnesium; N – nitrogen; P – phosphate; S – sulfur) and micronutrients shown in brown squares (Na – sodium; Cu – copper; Fe – iron; Co – cobalt; Mn – manganese; B – boron; Zn – zinc; Cl – chlorine; Ca – calcium).

B Effect of nitrogen fertilization on levels of corn biomass. Notice with the red bars that the initial use of the lowest amount of nitrogen fertilizer has a large effect on both grain and total biomass and the increase in biomass continues as the level of nitrogen fertilization increases.

DATA FROM: Gallagher et al. (2011); Jasinski et al. (1999).

fertilizers in 2010 compared to only 7.4 million tons in 1960. A substantial amount of this change is due to increases in the production of corn, where on average 97% of cornfields receive at least nitrogen fertilization (USDA, 2012).

While the application of fertilizers has increased agricultural productivity, fertilizers are not always completely used by the crops. In many cases, excess nutrients either infiltrate deep into the soils to the level of the groundwater or are

COMPOSTING

Travis L. Johnson

Nutrient recycling is not just for biomass production systems; it can also be done at home with compost. Composting is the process of taking organic waste and converting it into fertilizer and soil conditioner for gardens. Green waste, such as food scraps and fresh garden waste, is wetted and combined with brown waste, such as dry paper, cardboard, leaves, and sawdust. This combination provides an environment and the raw materials for microbes to break down complex compounds into simpler molecules they can absorb into their cells. The composting process produces heat which can rise to over 40 degrees Celsius for anywhere from a few days to a few months depending on the size of the compost. The microbes, including fungi and bacteria, secrete enzymes that degrade cellulose, sugars, starches, proteins, and other organic compounds, and create carbon dioxide, water, energy, nutrients, and humus. Soluble sugars are immediately consumed by the microbes, causing a population boom and increased temperature. Proteins decompose into amino acids, which are further decomposed into inorganic ammonium, nitrate, and sulfate that plants can use as nutrients. Anything else that is resistant to decomposition results in a complex organic mixture called humus, which can be added to soil for moisture and nutrient retention.

SOURCE: Cornell Waste Management Institute (2015).

washed through irrigation and rain as runoff into nearby surface waters. The addition of these artificial nutrients into aquatic systems is known as **eutrophication**. While agriculture is a significant contributor to eutrophication, the source of these nutrients can include residential street and construction runoff, industrial waste release, municipal waste discharge, and atmospheric nitrogen compounds (Paerl, 2006). While at first it may seem the addition of added nutrients to a waterway would benefit plant life similar to what is observed in agricultural croplands, this is not usually the case. Many natural water ecosystems have a balanced nutrient cycle; when components of fertilizers such as nitrogen and phosphorus are added to this ecosystem, the balance is tipped. In these cases, the over-enrichment of an ecosystem can result in a massive growth of algae at the surface of a water system, be it a lake, river, or ocean. Sometimes these algae are toxic forming harmful algal blooms (HABs) and causing the death of fish, wildlife, and domestic animals. At other times, the overgrowth of algae can reduce the clarity of the water, impacting photosynthetic productivity for native plants. Another harmful outcome, known as **hypoxia**, results when there is a reduced amount of oxygen in the water. Hypoxia occurs when the influx of excessive nutrients to an aquatic system causes a massive algal bloom. Once the nutrients are consumed, the algae will die and sink toward the bottom. As they sink, the algae are decomposed by other microorganisms that use up all of the available oxygen in the water column. This forms an oxygen-deprived layer in the water column incapable of supporting life, forming a **dead zone**. Dead zones are found at certain times of the year in many places around the world. The largest dead zone in the United States occurs in the Gulf of Mexico at the base of the Mississippi River. This dead zone is due largely to agriculture runoff occurring around the Mississippi River Basin.

The Mississippi River is responsible for most of the freshwater that flows into the Gulf of Mexico (LUMCON, 2012). In 2007, the USGS suggested nearly 52% of the nitrogen and 25% of the phosphorus that entered the Gulf of Mexico were derived from corn and soybean growth. Since it is likely additional crops, particularly corn, will be grown within the Mississippi River Basin zone to meet the demand for 15 billion gallons of corn ethanol, there will also be an increase in nitrogen runoff into this waterway, potentially exceeding 10% (GAO, 2009). Increasing amounts of nitrogen and runoff from other crops will undoubtedly continue to grow the size of the Gulf of Mexico dead zone and negatively affect an important area that supports a wide array of marine life and fisheries.

Eutrophication that causes massive dead zones is an easily recognizable environmental impact of the use of nitrogen-based fertilizers for agriculture. However, the effects of these fertilizers can also be found in a less obvious way due to their ability to infiltrate into the ground and enter the groundwater. Nitrates found in fertilizers have a low toxicity; however, the chemical reduction of these nitrates can produce more toxic nitrites. The Environmental Protection Agency monitors groundwater sources to make sure nitrate levels remain below 10 parts per million to avoid adverse human health effects (GAO, 2009). Increases in fertilizer use in agriculture will need strong safeguards against higher quantities of nitrates and nitrites entering groundwater sources to protect humans and wildlife.

Nutrient Recycling and the Future of Biofuels

Agricultural practices for common crops like corn and soybean are already having significant negative effects on the quantity and quality of water in the United States. Unfortunately, this trend is likely to continue if current bioenergy goals are pursued using traditional biomass feedstocks. The US Department of Energy's Biomass Research and Development Advisory Committee suggests biomass in the United States will supply 5% of power, 20% of transportation fuels, and 25% of chemicals by 2030, an energy equivalent to 30% of current petroleum consumption (Perlack et al., 2005). To meet this objective, in addition to the already 15 billion tons of corn grown for ethanol, nearly 1 billion dry tons of ligno-cellulosic feedstock will likely be needed annually. This will require an increase in crop fertilization by about 5.5 times the current application levels (Han et al., 2011). An increase of this magnitude will undoubtedly exacerbate the growing concern for the availability of clean freshwater for human consumption and food-based agriculture. However, implementing more sustainable practices such as nutrient recycling and developing alternative feedstocks requiring fewer synthetic fertilizers could greatly diminish the environmental impact of this fuel production. For example, the development of alternative and less nutritionally demanding ligno-cellulosic feedstocks such as miscanthus, switchgrass, and trees may reduce the need for both irrigation and fertilization. In addition, crops could be grown in association with rhizobacteria. Rhizobacteria are bacteria capable of converting inorganic sources of nitrogen such as gaseous nitrogen found in abundance in the air into organic sources that can be used directly by plants to enhance growth.

Finally, while much of the discussion has been focused on eutrophication from agricultural runoff and infiltration, there is also a huge supply of nutrients available in municipal wastewater sources. Although wastewater is not likely to be used in the production of agricultural crops for food, this water source could be used for other nonfood-based biofuel feedstocks such as algae. Nutrients can be recycled from wastewater in the production of algae biomass to generate biofuels and in **bioremediation**. Bioremediation occurs when the algae filter out pollutants such as nitrates from the wastewater, resulting in a much cleaner water that can be recycled back into the environment. Bioremediation is one strategy being used to restore waterways susceptible to wastewater eutrophication back to natural conditions.

Biofuel development must be pursued keeping in mind both emissions in the air and impacts on natural resources, including water. By focusing the future of bioenergy on feedstocks and technologies that can reduce energy consumption and environmental stress, the probability of biofuels becoming a reliable and sustainable way forward in our energy future will be significantly improved.

STUDY QUESTIONS

1. Explain carbon neutrality. Does the combustion of biofuels result in the emission of carbon dioxide? If so, why are biofuels considered carbon neutral?
2. With over 70% of the planet covered with water, why is freshwater so valuable? What role does the Earth's hydrological cycle play in the availability of freshwater?
3. Briefly explain how the production of gasoline and the production of corn-based bioethanol compare in water usage.
4. How do surface waters and groundwaters differ when considering depletion? What impact could these differences have on the future of agriculture and biofuels?
5. Briefly explain how sedimentation, runoff, and infiltration impact water quality.
6. What role do fertilizers play in water quality? Explain eutrophication and its impact on the environment.

CHAPTER THIRTEEN

Life Cycle Assessments for Evaluating Biofuels Production

The wide assortment of feedstocks and technologies being considered within the energy market hold great potential for the future generation of biofuels. Even though these biofuel feedstocks will be derived from plants or recycled materials, thus reducing their overall environmental impact compared with fossil fuels, the level of sustainability for each feedstock and fuel type remains highly variable. Clearly, to achieve a sustainable energy future, it will be critically important to consider three important characteristics of each biofuel source and process: **energy return on energy investment** (EROEI), greenhouse gas emissions, and finally the broader implications on the environment when a biofuel-generating process is scaled to the level required for fuel production on a world scale.

With seemingly endless feedstock possibilities, fuel types, and various conversion technologies, the overall impact of biofuel production is not easy to decipher. The variety of options in biofuels production leads to a plethora of individual steps within the production process where each could significantly contribute to the overall energy return and environmental impacts of the final products. In order to truly assess the impact of an individual biofuel type, a "cradle-to-grave" understanding must be obtained where all inputs, including energy and raw materials, are balanced with the outputs of fuels and materials that are recycled or returned to the Earth. This type of approach in understanding a product is known as a **life cycle assessment (LCA)**. This comprehensive overview of a biofuel's production allows companies, consumers, and policy-makers to select biofuel products that can most efficiently offset fossil fuel use, boost the economy, and protect the environment.

Why Do We Need Life Cycle Assessments?

An LCA is a technique used to assess the environmental impacts of a product through all stages of its "life" begin-ning with the extraction and utilization of natural resources used in production; followed by the transport, manufacturing, and packaging of the product; and eventually its final use and disposal. By considering this entire cycle, unintended consequences, either direct or indirect, can be avoided. Let us consider two examples that illustrate this concept in the energy market. First, the electric car. Many advertisements lead the consumer to the conclusion that electric vehicles are "green" vehicles because they run off of electricity and do not use gasoline that when burned emits carbon dioxide into the atmosphere. While this may seem true for the actual power used in the car, is it true for the car overall? In most cases the answer is no. The component parts including the steel, rubber, and fabric used to construct the car were all likely produced through a process involving the use of fossil fuels and resulted in the emission of carbon dioxide. Additionally, in most parts of the United States, the electricity used to charge the car's batteries is derived from electricity generated by coal or natural gas, both of which produce their own carbon dioxide emissions. Thus, by not considering this entire picture of inputs and outputs, it is easy to misinterpret a vehicle's net effect on the environment. At the same time, by understanding where such emissions occur during the production and use of an electric car, adjustments can be made to lower overall emissions. Instead of using coal (the dirtiest fossil fuel emitter) as a source of electricity, perhaps wind or solar energy can be used. Wind and solar energy represent a truly renewable way to power an electric vehicle and significantly lower the resulting emissions previously involved in electricity generation.

The electric car above is an example of unintended consequences related to its production and use. For the second example, consider a vehicle powered entirely by cellulosic ethanol. While this vehicle emits carbon dioxide from internal combustion in the same way as gasoline, the emissions from cellulosic ethanol are offset by carbon dioxide sequestered by photosynthesis in the agricultural production process. Does this mean that the use of cellulosic ethanol is carbon neutral? Probably not. Two reasons why cellulosic ethanol may emit more carbon dioxide than initially anticipated is because the distillation of ethanol requires energy and that energy most likely comes from either coal or natural gas. In addition, the feedstock used for its production could have been grown on land that was cleared for agricultural use. In clearing this land, trees and other native plants naturally involved in sequestering carbon dioxide are displaced. In some cases, this can lead to a net loss of carbon dioxide being sequestered for this land. This indirect land use has implications on the total GHG emissions from biofuel production, and serves as another example of a common yet oft-overlooked component of life cycle assessments for biofuels.

Types of Life Cycle Assessments

There are three types of life cycle assessments. The most conventional LCA is the *process-based LCA*. In its simplest form, the **process-based LCA** creates an itemized list of inputs and outputs for a single step of a process used in creating a product. The inputs may include materials and energy resources, while the outputs may focus on emissions and waste released into the environment (Green Design Institute, 2012). This simple LCA is easy to visualize for a product like a plastic soda bottle as shown schematically in figure 13.1. The LCA considers the bottle—including the clear plastic used for the bottle, the solid plastic used for the lid, and the paper used for the label—and its contents as inputs and the empty bottle as the output. This LCA can also be expanded into a more detailed process-based LCA for the soda bottle as shown in figure 13.2. In the process-based LCA, the petroleum and machinery used in forming the plastic, the ingredients (including natural resources) used to create the soda, and the transport of this soda bottle for retail distribution are all considered inputs, while the environmental impacts of its slow decomposition are outputs. As you can imagine, the process-based LCA becomes exponentially more complicated as the product itself becomes more complex, as seen with the electric car.

A second type of LCA is the **economic input–output LCA**. This type of LCA typically tracks monetary transactions that occur between different industry sectors when goods and services are consumed by other industries (Green Design Institute, 2012). An economic input–output LCA is

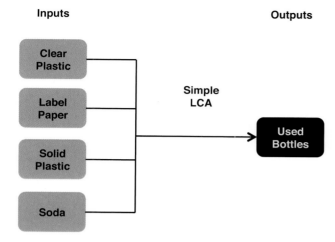

Inputs

Outputs

Simple LCA

- Clear Plastic
- Label Paper
- Solid Plastic
- Soda

→ Used Bottles

FIGURE 13.1 Simple life cycle assessment for a bottled soda where only basic inputs and outputs are considered (source: Green Design Institute, 2012).

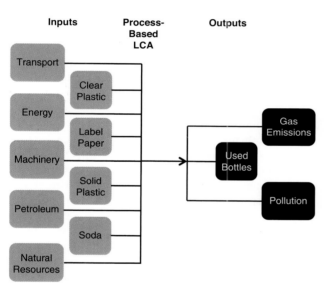

Inputs

Process-Based LCA

Outputs

- Transport
- Energy
- Machinery
- Petroleum
- Natural Resources

- Clear Plastic
- Label Paper
- Solid Plastic
- Soda

→ Used Bottles

- Gas Emissions
- Pollution

FIGURE 13.2 Process-based life cycle assessment for a bottled soda. This more complex life cycle analysis adds energy, transport, machinery, and natural resources as inputs and environmental effects as outputs to give a clearer picture of the true environmental impacts of a bottle of soda (source: Green Design Institute, 2012).

commonly used when studying the demand of a product and the structure of the economy because it is designed to account for the economic impact of each variable involved in producing a product. Because the economic input–output LCA focuses more on the money and cost flow of a product, it is less often used when considering the environmental impact of a product such as a biofuel, but could be important in the future when analyzing biofuels to determine if they are cost-competitive with fossil fuels.

The third and final type of LCA is a combination of the process-based LCA and the economic input–output LCA

known as a **hybrid LCA**. Hybrid LCAs can be valuable because they combine the cost flow information important to the bottom line of many businesses with the environmental impacts important for public perception and ultimately the long-term viability of the business.

While all three LCAs have an important role in understanding the value of a product in terms of both monetary value and environmental impact, the focus of the remainder of this chapter will be on the process-based LCA due to its unique importance in comparing the environmental impacts of different sources of biofuel.

The Basis of a Life Cycle Assessment

The process-based LCA is a simple, consistent, and comprehensive accounting of all of the materials, energy, and waste that flow into and out of a product. This accounting can be done for each step involved in producing this product or all of these steps can be combined to form one comprehensive LCA that describes the impact of the entire product's life cycle. Figure 13.3 schematically shows how materials, energy, waste, and pollution in each step of the production process are considered, as well as the potential for recycling, remanufacturing, and reuse of the product or its processing components (Kendall, 2011). Because this type of LCA risks becoming uncontrollably complicated, it is important we understand the key elements of an LCA in order to produce an assessment that is both practically informative and a reflection of reality. The three key elements to consider in producing an LCA include the following: (1) the goal and scope; (2) the life cycle inventory; and (3) the impact assessment (Curran, 2006).

The first element of an LCA is defining the goal and scope of the assessment. Clearly, these assessments can be complex; therefore, narrowing down the focus of the assessment can make it much easier to create as well as much more valuable in deciphering the actual impact of a product. By considering the purpose, the audience, the objectives, and the manner in which the results will be interpreted, a goal for the LCA can be defined. For instance, consider a mobile smartphone. The smartphone has a variety of parts that are made and transported from locations all around the world using a variety of different materials, some of which can be recycled and some of which must be disposed of as waste. Trying to make an LCA for the smartphone could be extremely complex. However, if in the end the goal of your LCA is to understand the environmental impact of the use of rare earth metals within the smartphone, the complexity of the LCA is significantly decreased because the focus or scope is much more limited. The scope of an LCA defines the function, functional units, and system boundaries of the overall LCA.

In understanding the scope of an LCA, the **functional unit** is extremely important. A functional unit defines what products are directly comparable to one another (Curran, 2006). For instance, when thinking about fuels, we

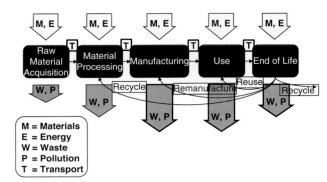

M = Materials
E = Energy
W = Waste
P = Pollution
T = Transport

FIGURE 13.3 Basic flow diagram of a process-based life cycle assessment showing how materials, energy, waste, and pollution for each step of the overall process are considered relative to the potential for recycling, remanufacturing, or reuse of the product or its components (image by Kendall, 2011).

cannot directly compare the volume of ethanol produced with the volume of gasoline produced because these two products have different energy densities. Therefore, the functional unit in this case is the energy value, and both ethanol and gasoline are compared based on their energy content rather than by volume.

The second important component of the scope of an LCA is **system boundaries** (Curran, 2006). Boundaries define the parts of the life cycle included in the study such as rare earth minerals in smartphones. In some cases, processes or phases of a life cycle may be excluded from an analysis based on changing goals for the LCA. This change in system boundaries has been apparent in the biofuel industry. Initial biofuel LCAs omitted carbon dioxide emissions and were only interested in comparing the energy content between the fuels. However, this changed as emissions became a more prominent concern due to biofuels being commonly marketed as environmentally friendly. Today, both direct emissions and indirect land use changes are considered. As different environmental parameters enter our collective awareness and increase in importance, system boundaries of LCAs will change over time.

System boundaries play a role in determining the complexity of the overall LCA. As the boundaries expand, so does the complexity. For instance, let us consider the effects of system boundaries on the complexity of an LCA focused on environmental emissions for corn ethanol as shown in figure 13.4. In the first level of system boundaries (shown within the black dotted line), only the materials, energy, and waste actually used in each step of corn ethanol production such as land preparation, harvest, and transport are considered. This simple scenario gives us a basic idea of the emissions generated during the general life cycle of corn ethanol production, but does not give us the full picture. Now, we can expand out another layer, as shown within the gray dotted line, and emissions from the soil, fields, and coproducts are also included. While this clearly adds to the complexity of the overall LCA, it also gives a more accurate picture of the actual environmental emis-

sions involved in generating corn ethanol. It should be noted that expanding the system boundaries does not always make the overall LCA worse. In the case above, adding emission from the soil and fields adds GHG emission to the process, while adding the coproducts acts as a credit to reduce GHG emissions from the process.

Determining system boundaries is also important for the audience of the LCA. If two products used for the same purpose are compared but have different system boundaries, the results can be misleading as to the true differential of environmental impacts between the two products. This can often be seen when directly comparing LCAs prepared for different types of fuels to one another. A 2006 report compared net energy and net greenhouse gas emitted for published studies on corn ethanol with that of gasoline. The author found that the original published LCAs had very different outcomes, but by making sure all system boundaries between the various studies were consistent, the differences were reduced (Farrell et al., 2006). Therefore, when we read and study LCAs to understand the environmental impacts of biofuels production, it is important that both the functional unit and the system boundaries are carefully calibrated to produce a more practically useful comparison.

Another important element of an LCA is the **life cycle inventory**. The inventory quantifies the inputs and outputs for the production of a product across its entire life cycle (Curran, 2006). Figure 13.5 outlines the components and categories of a life cycle inventory. Inputs that may include both primary and recycled materials as well as energy used are shown on the left as they enter the product system. Outputs that may include the products produced, the waste generated, and the emissions released are shown on the other side of the product system. By considering all of these inputs and outputs, the "cradle-to-grave" inventory can be generated. While this may seem fairly straightforward, determining the inventory can be challenging in certain situations, as when considering the treatment of coproducts in biofuel generation. A coproduct for a biofuel is a product produced in addition to the fuel that can either create more emissions or offset some of the emissions in the life cycle of the primary product. For instance, think back to sugarcane-based ethanol; the ethanol is produced from the sugarcane and the bagasse is used to provide heat for the distillation of the ethanol in the production process. Bagasse is a coproduct of sugarcane-based ethanol that helps offset emissions. In order to understand the role that the coproduct plays in the life cycle of the primary product, the inputs and outputs for each product must be assigned.

Another example where coproducts play a role in the final LCA is in the production of corn-based ethanol. The production of ethanol from cornstarch results in emissions at many levels of its life cycle including in the use of fertilizers on the crops, irrigation practices, harvesting methods, and the actual starch extraction, fermentation, and distilling processes employed. In an LCA, all of these various emissions would be added together to produce the over-

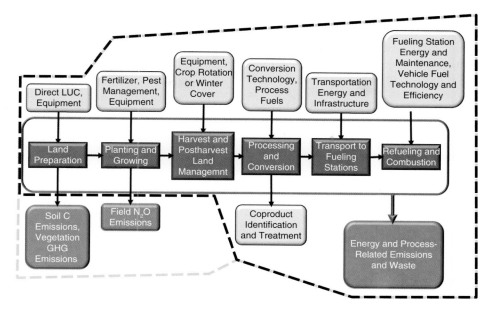

FIGURE 13.4 Flow diagram of the importance of system boundaries. In the first level of the system boundaries (shown within the black dotted line) only the materials, energy, and waste actually used directly in the individual steps of the processing of corn ethanol such as land preparation, harvest, and transport are considered. The second level of system boundaries is shown within the gray dotted line where soil and field emissions are also considered (image by Kendall, 2011).

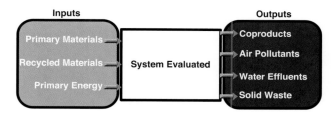

FIGURE 13.5 Flow diagram of the components and categories of a life cycle inventory including inputs such as primary materials, recycled materials, and primary energy as well as outputs including coproducts, air pollutants, water effluents, and solid waste (image by Kendall, 2011).

all emissions for the production of ethanol from cornstarch. This is fairly simple, but there are other products that are also produced from the same corn used to generate the ethanol, such as dry distiller's grain and solubles (DDGS). DDGS is used as an animal feed. This coproduct of ethanol production is actually used to displace soybean meal in cattle feed. Since the production of DDGS prevents the need for the growth and harvest of soybeans, the environmental burdens usually associated with the production of soybeans can be subtracted from the environmental burdens of generating corn ethanol. Thus, the emissions usually produced in the production of soybean meal are credited from the emissions generated during ethanol production, and this lowers the overall environmental impact of corn ethanol generation. The addition of coproducts into an LCA can be quite complex due to a high

number of indirect ways many coproducts can affect the environment.

Once the goals and boundaries of the LCA are established and the inventory assessed, the final element of an LCA is the **impact assessment**. The impact assessment is really what brings all of this information together to make it meaningful to an audience. This assessment tells us what the impact of the product will actually be based on all inputs and outputs, and provides a readout of the overall effect on the environment, energy return, and human health (Curran, 2006). For instance, the impact assessment will look at the impact of inputs such as how the generation of a product affects abiotic and biotic resources like water and minerals, what the impact of this product will be on land use, and the amount of fossil and total primary energy needed to produce the product. This assessment will also look at the impact of outputs, particularly those related to the environment like the potential to contribute to climate change, the ability to deplete stratospheric ozone, and toxicity to both humans and natural ecosystems. In the end, the impact assessment tells the story of a product by bringing into perspective all of the different elements of the LCA related to both environmental and human impact.

Importance of LCAs for Biofuels

With the rising demand for biofuels and increased research and development on various feedstocks, there is a critical need to produce a baseline comparison for the various types of biofuels generated, their EROEI, and their overall impacts on the environment. This comparison is likely to be in the form of an LCA. However, to avoid confusion and misinterpretation of data, it will be critically important that the goals, system boundaries, and inventories of these LCAs are carefully considered to allow for a direct comparison between various products. This will allow for a true understanding of the impact of the production of biofuels and an educated and non-biased opinion about the best and most sustainable method to generate the necessary volumes of biofuels required to displace fossil fuels.

STUDY QUESTIONS

1. Explain why a life cycle assessment is important when comparing different biomass feedstocks and their resulting biofuels.
2. Explain how each of the three elements of an LCA are critical in producing an accurate and comparable assessment. Give examples.

CHAPTER FOURTEEN

Economics and Politics of Biofuels

In recent years, rising costs for oil products, fears of reaching peak oil supply, and environmental concerns about the effects of emissions on global climate change have spurred a huge interest in developing alternative and renewable technologies that alleviate these burdens. Many of these technologies such as wind, solar, and hydropower have proven to be potentially effective methods in reducing fossil fuel usage in the generation of power, particularly the usage of coal and natural gas. However, most of these technologies do not directly affect the use of the third fossil fuel, petroleum. Petroleum is the major feedstock for liquid transportation fuels, with the transportation sector accounting for nearly 30% of all energy consumption within the United States (EIA, 2014a). As stressed in previous chapters, the development of alternative fuels to be utilized in the transportation sector has been slow largely due to the specialized need for energy-dense and portable fuels within this sector. Although still likely considered in its infancy relative to modern large-scale petroleum development, biofuels, particularly liquid biofuels, represent an alternative and renewable fuel source that could prove to be valuable as a liquid transportation fuel substitute. However, the success of these fuels will be highly influenced by the economics and policies surrounding the energy market.

As you have learned throughout this book, biofuels are valuable for a variety of reasons including their ability to replace fossil fuels, lower environmental impacts compared to the burning of fossil fuels, their widespread production in many countries around the world including both developed and developing nations, and their ability to potentially stabilize national securities and economies. With the establishment of the Renewable Fuel Standard mandate, the role of biofuels is also changing in the United States. The production of biofuels is increasing in hopes of lowering the United States' reliance on foreign oil supplies and reducing emission levels, but these increases are also placing added stress to natural resources as well as adding controversy over the balance between food and fuel. In order for liquid biofuels to truly impact the use of petroleum in the United States, it must be developed as a sustainable resource that can be stably maintained in terms of fuel production, cost, and the environment for many generations to come. As with any form of sustainable development, the development of biofuels will largely depend on three criteria: economic, sociopolitical, and environmental impacts (Demirbas, 2009). Previous chapters within this book have discussed the environmental implications of biofuels. In this chapter, we will briefly discuss the economic and political impacts that influence the production of biofuels and their success in the future.

Economics of Energy

In order to grasp the importance of economics relative to biofuels, we must first have a basic understanding of how an economy functions. An economy can generally be defined as the flow of resources between individuals and businesses within a designated region. Ideally, the production and consumption of goods and services derived from these resources such as land, labor, and capital will help control economic growth in the region for future generations. Although this definition may seem complicated, in the most basic sense, the economy is a simple flow of financial and physical resources relative to households, businesses, and markets. Figure 14.1 shows a circular flow diagram used by economists to indicate how households, businesses, and markets are connected. We can begin analyzing this connection by looking at a product. The product's market is involved in two separate flows in this diagram: the financial flow and the physical flow. First, we will consider the financial flow of a product's market. To begin

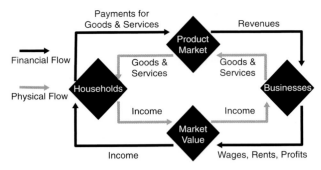

FIGURE 14.1 Basic economic circular flow diagram showing how households, businesses, and markets are connected. As shown, the flow of money (financial flow) generally goes in the opposite direction of the flow of products and services (physical flow).

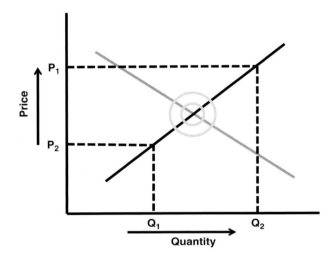

Q_1, P_1 – Quantity demanded is low because price is high
Q_2, P_1 – Quantity supplied is high because price is high

Q_2, P_2 – Quantity demanded is high because price is low
Q_1, P_2 – Quantity supplied is low because price is low

Demand Curve

Higher the price, lower the quantity demanded

Supply Curve

Higher the price, higher the quantity supplied

Equilibrium

When supply and demand are equal

FIGURE 14.2 Basic example of the economic concept of supply and demand. The supply curve shown in solid black increases as the price increases. The demand curve shown in dark gray decreases as the quantity produced increases. A quantity supplied may be low (Q_1) because the price obtained by the producer for the product is low (P_2) or the quantity supplied could be high (Q_2) if the producer can receive a high price (P_1). The meeting point between the two curves, shown with light gray circles, represents the theoretical optimal balance of quantity and price for a product. Sometimes externalities can impact the price of a product and this may shift the supply curve upward, leading to the socially acceptable price for a product actually being higher than the theoretically balance price.

with, the product will be purchased and a revenue will be created. The revenue will flow to its business that then uses this revenue to pay for business expenses including wages, rents, and profits. Some of these expenses like wages will become income for households. This income is then used to pay for goods and services like the original product, thus completing the financial flow cycle.

The second flow shown in the diagram is the physical flow. The physical flow for a product runs in the opposite direction of the financial flow. In this flow, individuals or households will give inputs such as resources or land use rights to a business, allowing them to access the goods and services they need to produce a product. Once generated, the household can use the product or services to complete the physical flow cycle.

These financial and physical flows are critical to the economy because they determine both the availability of products and household incomes that can increase or decrease overall standard of living. To better understand this concept, let us consider a hypothetical example involving shale-based oil. If shale-based oil suddenly became significantly more expensive than traditional imported oil in the United States, then households would generally choose to purchase the cheaper products from the traditional oil sources. This would lower the revenues for the shale oil businesses, preventing them from providing the same level of wages to their workers. Ultimately, this would lower the financial means important in determining the standard of living for these households. On a larger scale, this negative financial impact could also shrink the overall economy for an entire region, particularly in this case a region where shale oil is a dominant product.

Now that you have a basic understanding for how households, businesses, and markets are connected to build the economy, let us begin discussing the factors that influence the economy. There are three key concepts that affect the flow of the economic cycle in the energy market. These concepts include supply and demand, elastic and inelastic demand, and the overall economy.

Supply and demand is a basic economic concept that focuses on the amount of a product that a consumer will buy relative to the amount of a product that a producer will provide. This is usually best understood by considering a supply and demand curve as shown in figure 14.2. A simple demand curve as shown in gray assumes that as the price of a product goes down, the quantity purchased of that product will increase. This illustrates the idea that additional benefits of consumption diminish as you consume more, so consumers will only demand high quantities if the price is low. The supply curve as shown in black shows that quantity supplied goes up as the price of the product goes up. Producing large quantities of goods becomes more expensive, so producers are only willing to supply large quantities if they can charge higher prices. Ideally, the consumer wants the lowest price and the producer wants the highest price; therefore, where these two curves cross is usually indicative of where supply and demand are in equilibrium and represents the optimal price and production level for both the consumer and the producer. However, this equilibrium can move and shift based on changes in factors related to the product such as taxes, price of resources, and competition.

The supply and demand curves can also change based on externalities. An externality represents a variable related to the consumption or production of a product where an activity of one economic agent affects the outcomes of another economic agent in ways that are not reflected in market prices. We can consider the impact of an externality by looking at prescription drugs. A significant amount of research, testing, and discovery must go into the development of a new drug. Because of the high cost of innovation, companies may be reluctant to put large amounts of effort and money into producing new drugs and opt for a level of research and development that only takes into account their own costs and perceived benefits. What this company may not realize is the benefits of technology spillover. Once one company has invested in research and development for a new drug, the new technology may bring down the costs to other producers. In terms of the graph in figure 14.2, an externality can add to the price of producing a product, shifting the socially acceptable supply curve upward. Therefore, the intersection of the company's own supply curve with the demand curve as shown indicates the quantity that the company would provide, while the intersection of the social supply curve with the demand curve would indicate the optimal level.

We can apply these same supply and demand curves to the energy market, particularly by looking at the supply and demand of oil. Figure 14.3 compares the supply (production), demand (consumption), and price for petroleum over the past decade. Between 2001 and 2005, the demand for oil increased, leading to oil companies slightly decreasing the production (supply) of oil and thus resulting in an increase in price. Beginning in 2006, the increase in price likely led to the lowering of the consumption of oil, leading oil companies to increase their production levels again. Unfortunately for the consumer, this only led to a short-term decrease in price (2008–2009) before the price began increasing again likely due to another increase in the overall global demand for oil.

Just as supply and demand are important in the oil market, they are also important economic concepts to consider in the development of biofuels. We can study an example of supply and demand in biofuels by examining corn-based ethanol. Although ethanol was one of the first fuels for motor vehicles, it has not been used as a standard fuel source recently due to the availability of cheap petroleum-based oil. Instead, supply and demand for ethanol have been based largely on the use of ethanol as an additive to petroleum-based fuel for many years. However, the role of ethanol in the fuel market is changing. The demand for ethanol has increased, particularly since the Renewable Fuel Standard mandates were implemented. Because most of the ethanol produced in the United States is currently derived from corn, the rise in the demand for ethanol in turn resulted in a rise in demand for corn. Normally, the supply and demand curves for the corn market would naturally reach an economic equilibrium. However, the rising demand for corn due to the ethanol industry combined with government subsidies for corn ethanol have inflated the demand beyond a socially acceptable level, leading to higher food prices and more ethanol production. Ultimately, the competition between food and fuel has created a corn market that is not at a social equilibrium.

Externalities are another important concept in economics. However, based on their very definition, they do not directly impact market prices. An externality is a situation where the cost or benefit incurred by goods and services does not take into account the total social costs or benefits. The cost of oil (or the products of oil) is an excellent example of how externalities should, but do not, influence final costs. When considering the cost of a gallon of gasoline, the social and economic costs of oil are largely ignored. These social and economic costs come in many forms such as security and environmental damage. In terms of security, the United States spends billions of dollars on its military budget and a percentage goes to protecting oil tankers and pipelines to prevent the disruption of the flow of oil. At the same time, there is a significant social cost for the soldiers that are trained and sometimes lose their lives protecting the energy supply (Nicolson, 2014).

Another externality in the cost of oil is environmental damage. Oil spills can have an impact on both environmental and human health, and decrease the livelihood for households in the region where an oil spill occurs. Oil and the combustion of oil products also result in significant levels of pollution including nitric oxides, volatile organic compounds, particulate materials, and carbon dioxide. There is no direct cost added to the price of oil to reflect the cost associated with the impacts of these pollutants, including the high costs that may be associated with the future of global warming and climate change. The economic and social costs of security and the environmental costs of pollutants are externalities that would likely significantly increase the cost of a gallon of gasoline if taken into account.

Oil is just one example, but all fossil fuels and their products come with externalities not considered in their price that help maintain their relatively low costs. Importantly, renewable energy resources such as biofuels can alleviate some of these additional social, economic, and environmental costs. For instance, wind and solar resources that can be used in many regions around the world have a low theft risk, which lowers their security cost (Nicolson, 2014). Also, a biofuel feedstock should sequester at least as much carbon dioxide as is released through the combustion of its products. This sequestration results in lower environmental costs associated with increasing atmospheric greenhouse

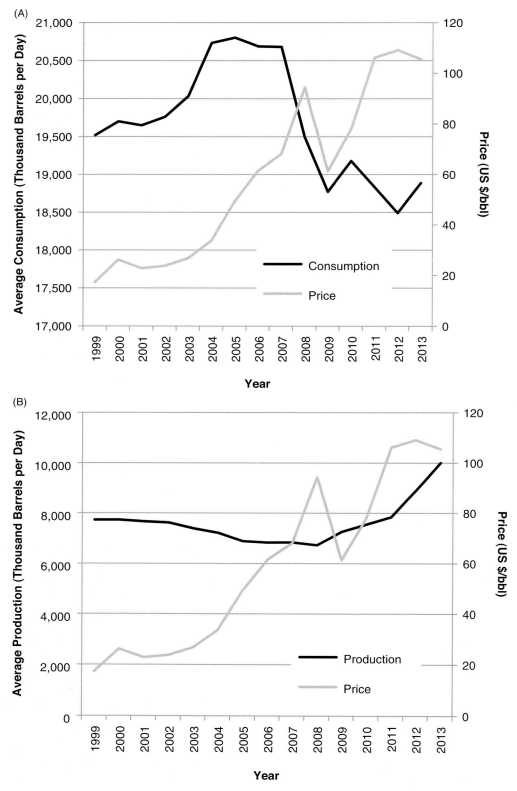

FIGURE 14.3 Comparison of the supply (production), demand (consumption), and price of petroleum over the past decade. (A) US crude oil consumption and price, 1999–2013. (B) US crude oil production and price, 1999–2013. As the consumption or demand for oil increases, oil companies will keep steady or slightly decrease the production or supply of oil. The increased demand combined with a steady quantity of oil led to an increase in the price of oil. When the demand for oil decreased, oil companies increased production and lowered the price, but this lower price was short-lived likely due to the overall global increase in the demand for oil (data from British Petroleum, 2014).

DIRECT REGULATION AND PRICES

Travis L. Johnson

Beyond using alternative energy like biofuels, society can also decrease global warming by using less carbon-intensive fossil fuel sources. But how do we encourage industries and consumers to choose these options that reduce the externality (cost to others) of pollution? From an economist's perspective, it comes down to direct regulation or prices (tax). Direct regulation is when a government explicitly limits or supports an activity in order to affect the externality. An example of this is the multitude of government policies implemented to reduce gasoline consumption in the United States. To name a few, there are fuel efficiency mandates like the Corporate Average Fuel Economy and Low Carbon Fuel standards; subsidies for alternative fuel cars like electric and hydrogen vehicles; and subsidies for public transportation, carpools, and housing developments near workplaces to decrease commuting. The reason why there are so many regulations surrounding gasoline is because there are so many ways to reduce its use. Because of this, it is not feasible or efficient to implement regulations for all of the possible conservation efforts. Prices, on the other hand, come in the form of taxes or fees and affect behavior through economics. Throughout history, industry has been very good at finding alternatives when an input in production causes the cost to increase, and so this same effect can be applied if there is a tax on pollution. If gasoline is more expensive because the pollution caused by its combustion is taxed, consumers will have an incentive to find alternatives or use less gasoline. This effect would ripple through the economy and cover all gasoline conservation efforts—even the ones not covered today. Prices are broad, efficient, and encourage every opportunity to conserve, so from an economist's perspective, these are the most efficient.

SOURCES: Jacobsen (2014); eConPort (2015).

gas concentrations and their long-term impact on climate change. In the future, the overall economics of the energy market will need to consider the cost incurred by society from all of these externalities. When considered, the additional costs associated with fossil fuels could raise their prices enough to make many renewable resources more competitive on the open energy market.

The second key concept to consider in energy economics is the elastic and inelastic demand for products. Elasticity of a product is based on how sensitive demand is for a product to changes in the price of that product. An elastic product is usually a nonessential product where a small change in price results in a large change in demand. For instance, candy bars have elastic demand; a 10% increase in the price of a candy bar will likely cause the quantity demanded of that candy bar to fall more than 10% because many consumers will choose to purchase less candy if it becomes more expensive. An inelastic product is a product that is considered essential and a change in price usually does not significantly affect the change in demand such as with water or gasoline. In the United States, transportation is generally associated with inelastic demand. If the price of a gallon of gasoline goes up 10%, the quantity of gas consumed may fall but by less than 10%. Commuters still need to purchase the same amount of gasoline to get to work regardless of the price. And for small price changes, very few people will switch to an alternative type of transportation if it is even available. Three factors that affect the elasticity of a product include the availability of substitutes, the amount of income available to the consumer, and the amount of time given to the consumer to adjust to the change in price. All three of these are directly related to what the consumer will endure to have that product before making a change, if a change is possible.

In its current state, oil is an inelastic product because it is essential to our daily lives and it currently has no substitute. When you buy a gallon of gasoline, 66% of the purchase price goes directly to the price of crude oil, compared to 14% for refining, 8% for distribution, marketing, and retail, and 12% for taxes (EIA, 2014b). Since there is no replacement product for 66% of this cost, we continue to go to the gas pump during the summer, even when the cost of oil skyrockets. However, the large-scale commercial production of biofuels could offer a substitute for oil, decreasing its demand and increasing its elasticity. Since the global supply of petroleum is expected to decrease, the cost of gasoline may not decrease significantly in the future; however, by substituting with a different fuel, the consumer may not be forced to bear the risk of rising costs.

The final concept we will consider as important in energy economics is the economy itself. A national economy is usually measured by gross domestic product (GDP), which is determined by the products produced in that country, the income generated by the citizens, or the total value of sales within that country. The global GDP is about $87.25 trillion with the United States having a GDP of $16.7 trillion (CIA, 2014). The GDP of a country is usually indicative of the standard of living within that country. Obviously, the standard of living within the United States is high as we are used to having many products readily available to us and having the income to afford these products. Sometimes the production and consumption of a product will greatly affect GDP. Oil is an example of one of these products. Figure 14.4 compares how the supply of crude oil, the price of crude oil, and the global GDP have all risen steadily since 1995. The availability of energy resources is undoubtedly linked to an individual country's economy as well as the overall global economy. (Refer to figure 1.3 to be reminded of how energy use and GDP correlate for countries around the world.)

Biofuels have the potential to offer a low-cost replacement fuel source that can reduce the economic impact of petroleum use on both the consumer and the economy. These renewable fuels could add value to a suffering farm industry by adding value to feedstocks, increase jobs in rural areas, lower reliance on foreign energy sources, and reduce overall emissions. However, the success of biofuels will ultimately depend on the final cost of the fuel in comparison to the currently cheaper fossil fuel sources. The overall cost of biofuels will depend on capital costs, plant capacities, and process technologies, but the most influential and largest component of the overall cost will undoubtedly be the feedstock. Currently, the feedstock used in biofuels production is predicted to represent 75–80% of the total operating cost of a biofuels production facility (Demirbas, 2009). Thus, finding a biofuel feedstock that can be generated and harvested at low cost will be important in lowering the bottom line on the price of these fuels. The overall cost of a biofuel will also depend on its ability to utilize energy infrastructure already available. The replacement of processing and distribution infrastructure would cost trillions, quickly raising the cost of some biofuels to an uncompetitive level. Finally, the success of biofuels may ultimately come down to political decisions and pricing relative to petroleum. By considering externality costs relative to petroleum, the competitiveness of biofuels may also be increased.

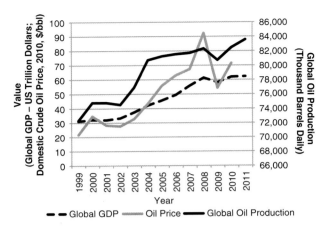

FIGURE 14.4 Comparison of global gross domestic product (GDP), crude oil prices, and global oil production between 1999 and 2011. GDP, oil prices, and oil production have shown a general increasing trend. This trend indicates the availability of oil leads to an increase in GDP and an ability to afford higher oil prices (Data from World Bank Group, 2014; McMahon, 2014; US Department of Commerce, 2014).

Policy and Politics of Biofuels

While the economics of biofuels could prove to be the most important component in the success or failure of these products from the consumers' perspective, the economics of the energy market in the United States does not stand alone but is closely intertwined with past, present, and future political decisions. Already, many policies have impacted the evolution of biofuels. For instance, the tax exemptions, subsidies, and blending requirements implemented for corn ethanol have led to a domination of this alternative fuel on the US energy market despite the recent trends in raising the cost of food and negative environmental impacts. Both corn ethanol and biodiesel markets in the United States have benefitted from previous excise tax credits of $0.45 per gallon for ethanol blended with gasoline and $1.00 per gallon for biodiesel blended with diesel (Pashly et al., 2014).

Over the past 10 years, the single biggest political contribution to biofuels in the United States has been the Renewable Fuel Standard legislation (both RFS1 and RFS2) as well as similar legislation coming from individual states. Originally, RFS1 set a standard requiring that 7.5 billion gallons of renewable fuel be generated in the United States, and we met this goal through the use of corn ethanol. However, RFS2 not only increased the quantity of fuel to 36 billion gallons by 2022 but also put a limit on the amount of the alternative fuel that could come from corn to 13 billion gallons. Thus, 23 billion gallons of alternative fuel mandated to be produced in the United States must come from an alternative feedstock (EPA, 2014).

Unfortunately, the standards set by RFS2 present us with one of the biggest challenges to the future of bioenergy. Even if we used 100% of the corn grown in the United

THE HISTORY OF THE US GOVERNMENT'S ROLE IN ALTERNATIVE ENERGY

Travis L. Johnson

The US government has historically had an enormous impact in accelerating technology through federal funding initiatives. For example, in the 1940s the US government spent around $24 billion over four years on the Manhattan Project to develop atomic bombs, it spent roughly $360 billion over 12 years on its moon landing missions in the 1950s and 1960s, and it spent around $3.4 billion over 14 years on ARPANET, which is the technological foundation of the Internet. It made these investments in order to drive the internet economy and to maintain the lead economically and politically. The US government has also led the use of alternative energy, specifically the US Navy. In the nineteenth century, the Navy transitioned from wind-powered sailing ships to wood-fired and then coal-fired steam engine ships. In the 1910s, it transitioned from coal to oil during World War I because ships were able to go faster and farther with diesel liquid fuel. Then during the mid-1900s, the Navy transitioned to nuclear energy. Large aircraft carriers and submarines need to be out at sea for indefinite periods of time and require a way to power themselves with minimal refueling, so nuclear reactors are now on these ships to provide power. These alternative energy transitions have trickled into civilian systems and helped shape how and what kind of energy we have consumed over the past 100 years.

SOURCE: Zenk (2014).

States to produce bioethanol, we still would not meet the mandate. This means that the mandate specifies not only that only 13 billion gallons can come from corn ethanol, but also that we have to develop an alternative feedstock to even consider meeting the total quantity of the mandate. An emphasis has been placed on meeting this demand with cellulosic ethanol, but progress in creating cost-effective ethanol from cellulosic feedstocks has significantly slowed. In addition, the funding being put into cellulosic ethanol production is limiting the funding available for the research and development of other alternative feedstocks such as algae and jatropha. Therefore, meeting the RFS2 mandate is clearly going to need a significant change in policies and political support to allow for the expansion of alternative fuels from alternative feedstocks.

The policies impacting biofuels are not limited to those related to mandates for fuel quantities and feedstock sources alone but extend to farm policies, trade policies, and carbon policies. Since biofuels are directly produced from feedstocks usually grown using agricultural practices and often depend on available farm resources, farm policies can also greatly impact the production of biofuels. The feedstock itself accounts for nearly three-fourths of the cost of biofuels; therefore, policies that impact farming practices such as taxes and mandates focused on controlling supply can greatly impact the competitiveness of biofuels (Demirbas, 2009). In addition, trade policies can impact the biofuels market. In the United States, most of the ethanol is produced from corn; however, one of the largest ethanol markets is in Brazil where the ethanol is produced from sugarcane. Despite sugarcane being more competitive and having lower environmental impacts, import tariffs on sugarcane ethanol prevent it from being imported into the United States, which protects the corn ethanol market. Reducing this barrier could lead to better competition among these products. Policies impacting the final end use of biofuels can also lead to an increase in the demand for these fuels. For instance, subsidizing the use of flex vehicles drives up the demand for nonpetroleum-based energy sources influencing the overall market of these energy sources (Rajagopal and Zilberman, 2007).

The most important policies to impact the future of biofuels will likely relate to the environment, particularly in regard to carbon emissions. Carbon emissions from human activities and the role that these emissions play in global warming and climate change have become extremely polarized and political topics. Despite mounting scientific evidence for human impacts on global warming and a limited supply of available crude oil in the future, skeptics, assisted by the media, have created a successful campaign to convince many people climate change is not caused by man and that new and alternative methods for extracting oil such as fracking will lead to US energy independence and potentially even result in the United States being an exporter of crude oil. Both of these claims represent an easy solution to the future—a solution that many people "want" to believe, but are likely to be false. Another huge challenge for biofuels both now and in the future will be convincing people biofuels are needed in terms of both meeting energy demand and lowering environmental impacts.

The most obvious and direct policy that could start to convince people of the importance of alternative fuels and lowering fossil fuel consumption is a carbon tax. If a tax were to be placed on carbon, industries with high emissions, such as the fossil fuel industries, would likely have to pay more to produce their product. In deciding on a value for this tax, it would be important to think of the externalities related to the production and use of fossil-fuel-based products. If you consider the costs just associated with a rising sea level, an increase in intense storms such as Hurricane Sandy, and the loss of crops due to changing environmental conditions like droughts and floods, the price tag on carbon emissions is likely to be enormous, possibly reaching $100 billion per year for the next several decades (Chambwera et al., 2014). This type of tax would very likely increase the cost of fossil fuels, a cost that would in turn get passed on to the consumer, and lead to a reduced demand for fossil-fuel-based products and an increase in both the production and interest in the consumption of biofuels.

There is an important role for politics in the development of biofuels as policy and government intervention can balance out market failures. These failures could result from externalities that are not directly considered in the price of the fuel but influence other aspects of the market,

or from simply being an infant industry that needs to be built up to compete on the open market. The need for sound biofuel policy is often overshadowed by alternative political agendas. Ultimately, the implementation of policy will depend on many factors including budget, resource and information availability, costs, and overlying political and economic considerations (Rajagopal and Zilberman, 2007).

The future of renewable liquid fuels will depend on the economic development of biofuels and the influence of a myriad of policies on the overall price. Policies for liquid biofuels will be driven by agriculture, environmental, security, and economic opportunities that reflect the use of energy for the entire country. Unfortunately, it is unlikely that the necessary political and economic interest will be placed on biofuels until the world realizes the expense that will be associated with the future of climate change and when the price of oil reaches prohibitive levels. Therefore, improving public education and understanding of these scientific concepts and the role that biofuels could play in alleviating political and economic burdens in the future are critical for the successful development of these alternative energy sources.

STUDY QUESTIONS

1. Explain the three key concepts of energy economics.
2. How can externalities impact energy economics?
3. Explain the role policies may have on the success of biofuels in the future.

CHAPTER FIFTEEN

Our Energy Future: The Prospects for Developing and Using Sustainable Biofuels

Energy Prospects for the Future

Over the last 100 years we have exploited inexpensive fossil fuels to drive unprecedented growth in the economy, agriculture, and ultimately world population. With the expanding use of fossil fuels also came environmental degradation and more recently climate change. We now face a future that will be very different than what we saw in the last 100 years. Part of this future will be brought about by diminishing fossil fuel resources, and with that decline an ever-increasing cost of fossil energy, and therefore an increasing cost of food. Another part of this future will also be shaped by climate change and the consequences this has on agriculture, weather, and even the migration of people, and we are already seeing some of those changes today. One way to visualize the impact of fossil energy on the world can be seen in figure 15.1, in which world population is overlaid on fossil energy production. When viewed in this way, it seems pretty obvious what fossil energy utilization has done to world population. What is not obvious from this graph is what happens next. Do we find a way to continue to produce and use energy at the amounts we are today, or do we greatly reduce our energy consumption as our fossil reserves decline? And if we reduce our energy consumption, how is that achieved; by increased efficiencies, or reduced consumption, or will we start to slow and even reverse population growth in the decades ahead? Are automobiles and personal transportation a historical anomaly that existed only for a few generations? These are not simple questions and answers are hard to predict, but what is obvious is that we simply cannot follow business as usual in the future; the laws of physics and supply and demand have ruled that out.

The use of fossil fuels, including coal, natural gas, and petroleum, accounts for over 86% of total energy consumption today. While the high percentage of energy coming from fossil sources begins to indicate the importance of fossil fuels to the human population, it is not until we stop to think about how fossil fuels impact every aspect of our daily lives that we can truly grasp the importance of these resources. Imagine a life with no cars, no planes, no air conditioning or heating, no lights or computers, no modern agriculture to stock grocery stores and markets. Over the past 100 years, fossil fuels have been integrated into our society in a way that leaves many human populations completely dependent on these energy resources, and this includes both the consumers and producers of fossil fuels. The price of oil today (2015) is half of what it was just one year ago, but still three times what it was just 12 years ago. These enormous swings in the price of oil have significant impacts on lives around the world. With oil at $50 a barrel many of the oil-producing states like Venezuela, Nigeria, Russia, and even Iran are seeing significant impacts on their economies, leading to social instability. While at $100 a barrel of oil, the economies of the oil-importing countries are negatively impacted, as are food prices. In fact, one could easily argue that $100 a barrel of oil places a significant threat to the bottom 2 billion people on the planet, as food prices rise to levels that they can no longer afford to feed their families.

In 2013 the US Energy Information Administration projected that global energy consumption would increase 56% over the next 30 years (EIA, 2013). Of this increase, more than half was due to the increased consumption of fossil fuels, particularly coal and natural gas. While the increase in energy consumption around the world is allowing for a better standard of living and growing gross domestic product for many countries, it is unlikely to be sustainable. Current estimates predict that based on proved reserves and the current levels of production, global nonrenewable fossil fuel resources are quickly becoming diminished, with coal remaining for 112 years, natural gas for about 64 years,

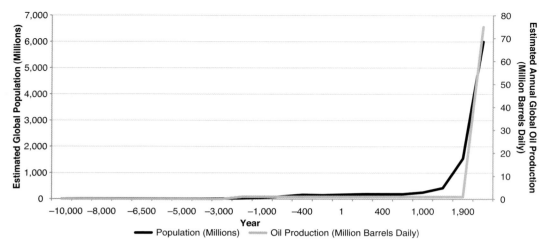

FIGURE 15.1 Diagram of the estimated change in the global population with the estimated global production of oil over the past 12,000 years. This diagram clearly shows how the rapid change in the human population coincides with a rapid increase in the production of crude oil. A question remains as to how long crude oil will last and the impact on the human population when oil is gone (sources: US Census Bureau, 2015; British Petroleum, 2014b).

and petroleum available for about 54 years. If we focus in on the United States alone, the numbers are even more staggering, with 239 years of coal but only 13 years of natural gas and about 11 years of petroleum (British Petroleum, 2014a).

While it may seem that we can simply shift our energy consumption to a more coal-based society in order to extend the period of time we have remaining of available fossil fuels, this is unlikely to be a functional plan in the long term. This increased demand would put added pressure on the reserves of coal, resulting in a more rapid decline of this resource. At the same time, well before we run completely out of these fossil fuels, the world will be faced with economic crisis as the cost of energy soars, creating unavoidable strain not only on the obvious uses of energy like transportation and electricity but also on our most simple of needs—food and water.

As we push forward into the future, it will be critical that we reevaluate our energy consumption and create energy-efficient methods to maintain our lives. At the same time, we must develop alternative sources of energy that are both renewable and sustainable even in light of a growing population and growing energy consumption needs.

Renewing Our Energy Future

Global use of energy is dominated by the use of fossil fuels, with coal and natural gas going mostly to electricity and heating, and petroleum becoming the mainstay of the entire transportation industry. Over the past several decades, many renewable technologies have been and are continuing to be developed to help supplement and replace the use of these fossil fuels. Hydropower, wind energy, and solar power have all seen significant increases in use around the world. Other technologies such as ocean energies like wave and tidal as well as geothermal power are also seeing increased research and development. However, all of these technologies are focused largely on the replacement of power and heat generation, thus the replacement of coal and natural gas. Clearly, with coal being the dirtiest of fossil fuels, these technologies are incredibly important for a sustainable future and their continued development should not be discounted. However, this still leaves the transportation sector, and specifically liquid transportation fuels, as an area in need of an alternative renewable source. The transportation sector makes up nearly 30% of all energy consumption, and energy requirements for fuels are stricter in this sector than in some of the other sectors. Transportation requires fuels that have very high energy densities, are easily transportable, and can be stored for use when demanded. Unfortunately, many of the renewable energy technologies developed thus far lack these important characteristics.

So what are the options for the future of transportation? Can we create a sustainable transportation industry? The answer to this question likely lies with the beginning of the personal transportation industry itself. The first engines were run off renewable natural sources—peanut oil and ethanol! It seems logical that as we stare into the face of a looming fossil energy crisis we consider these original fuel options for the production of renewable transportation fuels. Biofuels are an excellent option for the future as they can provide the highly energy dense, transportable, and storable liquid fuels we need, and they come from agricultural sources, so we understand their production and know-how to scale them to the levels required to truly impact the global liquid fuels markets.

Traditional biofuels, including biodiesel and bioethanol, have been used over the past century in various ways to supplement transportation; however, the increasing of these fuels has put a strain on the source of these fuels—largely soybeans and corn. This strain reaches beyond the price of fuel and has become a significant source of pressure on the food industry, resulting in rising food prices that are placing already impoverished people around the world on the brink of starvation. Our ability to meet the increasing demand for food by a growing population over the past 100 years has depended largely on our innovations and technological advances that have led to huge advances in industrial agriculture. In order to secure our food resources, meet the demand for the future food production, and produce transportation fuels based on biofuels, it will be extremely important that we continue to research and develop novel feedstocks for the production of both foods and fuels.

Three of the most promising of these feedstocks for liquid fuels include lignocellulosic feedstocks for ethanol, new biomass feedstocks like *Jatropha* for biodiesel, and algae as a source for several types of biofuels. The beauty of each of these feedstocks is that none of them place a strain on existing agriculture and the generation of food. Many lignocellulosic sources including corn stover are already available in abundance in the United States and at the same time other sources such as miscanthus and switchgrass are being developed and studied as dedicated energy crops due to their ability to grow on marginal lands as well as their incredible carbon sequestration capabilities. Dedicated oilseed crops like *Jatropha* or *Camelina*, which are able to grow on marginal land, have great potential for development in specific regions around the Earth where their oils can be turned into biodiesel to help with local farming or transportation.

Over the past decade, one feedstock has grown in interest due to its incredible productivity and versatility in the production of biofuels, photosynthetic algae. These algae have been shown to produce almost every type of biofuel including bioethanol, biodiesel, biogas, and biohydrogen. This versatility is likely to set a stage for the production of a combination of different biofuels that could likely lower the cost of individual fuels as well as increase the energy utilization efficiency of the processing while generating these fuels. Another benefit of algae is their ability to grow to very high densities on lands unsuitable for traditional agriculture and in many cases using non-potable water sources. These characteristics give algae a huge potential for the future of biofuels.

We also have to learn to reclaim and recycle energy in every possible way. The use of cooking oil to fry foods by many people around the world results in an unhealthy diet, but it has also opened the door to the production of massive amounts of waste oil and grease. Oil and grease are composed of the same hydrocarbons needed in the production of biodiesels. The development of collection, extraction, and purification methods in the future may very well transform this "waste" oil into a next-generation biofuel. Biogas generation from agricultural and municipal waste must also be integrated into our energy future, as this both provides a source of renewable gas and eliminates a significant source of greenhouse gas emissions from rotting food and fiber put into landfills. Research to transform waste plastics into energy also needs to be advanced, as plastic is simply another form of hydrocarbon and capturing the energy of these waste products can again benefit the environment on multiple levels.

In all likelihood, the future of liquid transportation fuels will depend on a combinatorial approach utilizing many different feedstocks to provide a variety of different biofuels. Because of this, thermochemical technologies to convert organic carbon into liquid fuels are also being considered, and these technologies have the advantage of being less selective when it comes to feedstock use. Pyrolysis, gasification, and Fischer–Tropsch synthesis all represent technologies that could use almost any organic feedstocks to generate liquid fuels. While these technologies may prove valuable in the future, more research is needed to address the poor energy return on energy investment that they produce today.

As research and development continue for the use of various biomass feedstocks for the production of biofuels, it is important that continued efforts are made to improve both the feedstocks themselves and the processes used in the extraction and production of fuel from these sources. Biotechnological applications will be an important factor in these improvements. Just as the Green Revolution saw vast improvements in agricultural crop yields, the current biotechnology revolution could result in these same improvements in biofuels-related crops. Specific emphasis is likely to be placed on the production of higher levels of fuel molecules and the ability for a crop to produce a specific fuel molecule such as a diesel or a jet fuel hydrocarbon. In a time where natural resources like water and fertilizers are becoming scarcer and an impediment to the environment, generating biofuel crops that are engineered to have lower fertilizer, irrigation, and pesticide requirements will be critical.

In addition, biotechnology may offer some advancement in the process of extracting and producing fuel from these feedstocks. While the cost associated with the growth of the feedstock can be high, the harvest, extraction, and purification of fuel can add to the overall cost of a biofuel in terms of both economics and environmental impact. In order to create a cheap and environmentally friendly biofuel, biotechnological advances will be needed to improve harvestability and to increase the versatility of microbial assemblages and enzymes used in the production of fuel molecules themselves, such as the cellulases used in the production of cellulosic ethanols.

Our energy future remains unclear, it almost certainly requires a dramatic change in the way we produce, distribute, use, and conserve energy. Although this future may be uncertain, humans are the most adaptable species to ever roam this planet, and the rate of scientific breakthroughs and biotechnology innovation today allows us to imagine a path to a sustainable energy future. Although the future prospects of renewable energy are good, their successful implementation will depend largely on the continued advancements in science and technology and the support of policies to promote this development and the large-scale deployment of these fuels. Clearly, the diminishing supplies, and the associated price swings, of fossil fuels is one reason to promote biofuels, but in the end the supply of fossil fuels could be overshadowed by the environmental impacts of the combustion of fossil fuels. In a period where climate change is leading to harsh and life-threatening changes in weather patterns including crippling droughts and powerful storms, the increasing consumption of fossil fuels and thus the increasing emissions of carbon dioxide into the atmosphere will undoubtedly exacerbate these environmental impacts. Biofuels represent an opportunity not only to produce a fuel source to replace 30% of our energy utilization, but also potentially to create a sustainable planet for the future. Our energy future, and thus our future, is truly in our hands.

VOCABULARY

Definition of vocabulary words relative to the information found in this book

ACETOGENESIS Biological reaction converting volatile fatty acids into acetic acid, carbon dioxide, and hydrogen.

ACIDOGENESIS Biological reaction converting monomeric biomolecules into volatile fatty acids.

ACTIVE SOLAR ENERGY Solar energy obtained from mechanical structures built to actively absorb, concentrate, and collect the sunlight.

AEROBIC DIGESTION Metabolic process of breaking down organic materials in the presence of oxygen.

AGRICULTURE Distribution of natural resources for the cultivation of living organisms, particularly animals and plants that culminate in the production of food and fiber, products important in sustaining life.

ANAEROBIC DIGESTION Metabolic process of breaking down organic materials in the absence of oxygen usually by microorganisms.

ANNOTATION Method of assigning a function to a sequenced region of DNA.

ANTHRACITE Most energy-dense and hardest form of coal containing greater than 90% carbon, final product of coalification occurring after the formation of bituminous coal.

ANTHROPOGENIC Man-made.

AQUATIC BIOMASS All organisms and their by-products that utilize photosynthesis within an aquatic environment such as a lake, river, pond, or ocean.

BAND SATURATION EFFECT Effect that occurs when greenhouse gases have a varying degree of potential radiative force due to their concentration in the atmosphere.

BIO-OIL Liquid fuel product of biomass pyrolysis.

BIOCHAR Any type of biomass feedstock that can be burned to produce energy usually heat; often include wood and dung.

BIOCHEMICAL CONVERSION Use of microbial organisms to convert plant feedstocks into various forms of biofuels.

BIODIESEL Diesel fuel derived from the fatty acids found in plant- and animal-based lipids.

BIOENERGY The use of the photosynthetic process to transfer electromagnetic energy into chemical energy stored in the form of biomass.

BIOETHANOL Clear, colorless volatile liquid that like all alcohols contains a hydroxyl group, generally obtained through the fermentation of sugars from plant feedstocks.

BIOFUELS Renewable sources of energy that derive from biomass.

BIOGAS Gas produced under anaerobic conditions (oxygen free) from the breakdown of biological materials.

BIOHYDROGEN Gaseous hydrogen produced biologically generally from microorganisms.

BIOMASS Biological material that is derived from either a living or a recently living organism or a metabolic by-product of a living organism.

BIOPROSPECTING Collection of organisms directly from the environment to find novel species and traits that may have better characteristics for growth, survival, and metabolic production.

BIOREMEDIATION The use of organisms to remove pollutants from a contaminated site.

BIOTECHNOLOGY The use of biological systems to develop products.

BITUMINOUS Intermediate type of coal containing between 50% and 90% carbon; third product of coalification forming between subbituminous and anthracite coal; most abundant form of coal.

BREED REACTION Nuclear reaction where the products of a fission reaction create more fissionable material that can be used for downstream reactions.

CARBON CYCLE Biogeochemical cycle based on the movement of carbon between the biosphere, atmosphere, ocean, and geological features.

CARBON NEUTRAL When the release of carbon such as in carbon dioxide from the combustion of a product is equivalent to the amount of carbon required to produce that product.

CATAGENESIS Stage of organic diagenesis where the process of cracking or breaking down complex geopolymers into simpler units occurs; contains the oil window.

CELLULOSE Main structural component of plant cell walls made of long chains of glucose molecules.

CENTRAL DOGMA OF MOLECULAR BIOLOGY Method of transfer of genetic information from DNA to protein function.

CETANE NUMBER Measure of the quality of a diesel fuel for ignition.

CHARCOAL Combustible carbon material obtained from removing water from organic material.

CHEMICAL ENERGY Energy that occurs due to chemical transformations or chemical reactions.

CHEMOSYNTHESIS Metabolic conversion of carbon dioxide or methane into biomass using inorganic molecules such as chemicals as the source of energy.

CHLOROPLAST Plant organelle that houses all of the machinery needed for photosynthesis.

CLAUS PROCESS Step during the refinement process of crude oil that removes sulfur contaminants causing the refined oil to be sweeter.

CLIMATE The way the atmosphere behaves in the long term referring to timescales from years and decades to even centuries.

CLOUD POINT Temperature where diesel fuel begins to precipitate.

COAL Combustible solid form of fossil energy usually appearing as a black or brown rock.

COLD FLOW Temperature at which the flow of a fuel begins to slow.

COMBUSTION REACTION Exothermic chemical reaction requiring a fuel and an oxidant that when burned transfers the energy into heat and a different chemical product.

CONSUMPTION The use of an energy product.

CRACKING Process of breaking down complex organic molecules into simpler molecules such as short hydrocarbons.

CROP ROTATION Process of planting different crops on the same field during subsequent seasons in order to recover lost or depleted nutrient resources.

CYANOBACTERIA Photosynthetic prokaryotic bacteria thought to be responsible for much of the original oxygenation of the planet.

DARK FERMENTATION Conversion of organic materials into biohydrogen through fermentation by bacteria.

DEAD ZONE Low oxygen areas of the oceans and large lakes where excess nutrients from human-derived runoff have resulted in a change in the water chemistry.

DENSITY Mass per unit volume.

DIAGENESIS Set of abiotic reactions due to changes in heat and pressure that occur to organic material as it sinks deeper into the Earth.

DIRECT PHOTOLYSIS The direct production of biohydrogen from photosynthetic cells.

DISTILLATION Method of separating volatile compounds based on their variations in boiling temperatures.

DOMESTICATION Process of taking a wild organism and teaching it to grow and thrive within a semi-controlled and confined environment.

DRY GRIND Process used to break open the outer kernel of a seed to allow access to the other component usually requiring manual grinding.

EARTH'S ENERGY BUDGET Comparison of the amount of incoming solar radiation with the amount of outgoing radiation energy.

ECONOMIC INPUT–OUTPUT LCA Assessment tracking the monetary transactions that occur between different industry sectors when goods and services are consumed by other industries.

ECONOMIES OF SCALE Microeconomic concept where cost advantages result from businesses due to their size, output, or scale of production.

ECONOMY Methods a country uses to maximize their levels of production and promote growth for future generations through the allocation of resources such as land, labor, and capital in order to increase total output.

ELASTICITY Flexibility of the price of a product before the demand for the product will change.

ELECTRICAL ENERGY Energy from the flow of electrons.

ELECTRICITY Flow of negatively charged electrons from one atom to another usually induced by a magnetic field.

ELECTROMAGNETIC ENERGY (RADIATION) Energy of light.

ELECTROMAGNETIC RADIATION SPECTRUM Range of energy emitted by charged particles as they move through space in a wave-like motion.

ENDOTHERMIC REACTION A chemical reaction that requires energy to complete.

ENERGY Ability to do work on a physical system.

ENERGY CONTENT Measure of the efficiency of a diesel fuel for producing power.

ENERGY RETURN ON ENERGY INVESTMENT (EROEI) Ratio of useable energy acquired from an energy resource to the amount of energy consumed to produce that energy resource.

ENTERIC FERMENTATION Digestion of carbohydrates in animals by microorganisms, often resulting in the release of methane gas.

ENTHALPY Measure of the change of internal energy within a system.

ESSENTIAL FERTILIZERS Nutrients that are required for the productivity of plants.

EUTROPHICATION Enrichment of an aquatic ecosystem with nutrients such as nitrogen and phosphorus.

EVAPOTRANSPIRATION Transport of water into the atmosphere from surfaces including plant leaves.

EXOTHERMIC REACTION A chemical reaction that releases energy during the reaction.

EXTERNALITY Variable related to the consumption or production of a product that is not directly calculated within the price.

FATTY ACID SYNTHASE Enzyme responsible for the biosynthesis of fatty acids.

FERMENTATION Metabolic process converting carbohydrates to alcohols usually by yeast or bacteria.

FERTILIZER Organic or inorganic materials that are supplied to soils to replenish nutrients.

FIRST LAW OF THERMODYNAMICS Physics law stating that energy cannot be created nor destroyed; it can only be transformed from one form to another.

FISCHER ESTERIFICATION Chemical reaction between a molecule containing a carboxylic acid and an alcohol in the presence of an acid, resulting in an esterified molecule.

FISCHER–TROPSCH SYNTHESIS Chemical method for upgrading carbon monoxide into an extended hydrocarbon chain.

FLASHPOINT Measure of the volatility of a diesel fuel toward burning.

FLOCCULATION Process in which suspended solids aggregate together.

FOSSIL FUELS Hydrocarbons formed from the natural degradation of the organic remains from plants and animals buried in the Earth and exposed to high temperatures and pressures for millions of years.

FUNCTIONAL UNIT Defines what products are directly comparable to one another.

FUNGIBLE FUEL Drop-in fuels meaning that the oil they produce, once extracted and purified, can actually be used as a direct replacement for gasoline and petrodiesel without significant modification to current combustion engines.

GASIFICATION Thermochemical transformation of a material into a gaseous mixture.

GENETICALLY MODIFIED ORGANISM (GMO) Any organism whose genetic material has been altered through these types of metabolic engineering technologies with the genetic material from another organism.

GENOMICS The study of genomes of organisms.

GEOTHERMAL ENERGY Energy resulting from the transfer of heat created at the core of the Earth to its surface.

GLOBAL CLIMATE CHANGE Changes in the Earth's climate due to changes in radiative forcing caused by rising levels of greenhouse gases, particularly carbon dioxide.

GREEN REVOLUTION Research and development advances to improve crop yields during the twentieth century.

GREENHOUSE EFFECT Trapping of heat energy within the atmosphere of the Earth, causing an increase in Earth's surface temperature.

GREENHOUSE GASES Atmospheric gases responsible for the greenhouse effect in the Earth's atmosphere.

GROSS DOMESTIC PRODUCT Measure of a country's standard of living determined by the products produced in that country, the income generated by the citizens, or the total value of sales within that country.

HABER–BOSCH PROCESS Industrial reaction of nitrogen gas and hydrogen gas to produce ammonia.

HALOPHILE Organisms that thrive in high salt concentrations.

HEMICELLULOSE Structural component of plants containing 5- and 6-carbon monosaccharide units; second most abundant biopolymer in plants.

HOMOLOGOUS RECOMBINATION Genetic recombination approach used to insert foreign DNA into the chromosome or a vector within a host organism.

HYBRID LCA Assessment combining the money and cost flow information important to the bottom line of many businesses with the environmental impacts that may be important for public perception.

HYDRAULIC FRACTURING (FRACKING) Process in which areas containing natural gas are flooded with high-pressure water containing tiny sand or glass particles to fracture the Earth and allow the gas to escape toward the surface.

HYDROCARBONS Chemical molecules composed of hydrogen and carbon.

HYDROLOGICAL CYCLE The Earth's water cycle describing the movement of water on, above, and within the Earth.

HYDROLYSIS Chemical cleavage of molecules through the addition of a water molecule.

HYPOXIA Condition that occurs when the influx of excessive nutrients to an aquatic system causes a massive algal bloom that results in microbial degradation using up the oxygen in the water column.

IMPACT ASSESSMENT Part of assessment that tells us what the impact of the product will actually be based on categories of inputs and outputs and the overall effect on the environment and human health.

INDIRECT PHOTOLYSIS The indirect production of biohydrogen due to a separation of the photosynthetic machinery from the hydrogen-producing machinery in cells.

IRRIGATION Supplementation of natural rain levels with storage and transportation of water sources from other areas.

KEROGEN Irregular polymer formed when complex biological products like carbohydrates, proteins, and lipids are broken down into their monomeric units.

KINETIC ENERGY Energy of an object in motion.

LIFE CYCLE ASSESSMENT Method to assess the environmental impacts of a product at all stages of its production and consumption.

LIFE CYCLE INVENTORY Quantifies the inputs and outputs for the production of a product across its entire life cycle.

LIGNIN Three-dimensional polymer of phenylpropanoid units that function in giving a plant strength, stiffness, and resistance to invasion by foreign substances.

LIGNITE Type of coal formed at shallow depths and lower temperatures; first product of coalification forming between peat and subbituminous coal.

LIQUEFACTION Process of converting solid substances into a more liquid state.

MACROALGAE Macroscopic, multicellular eukaryotic algae inhabiting aquatic environments that more closely resemble a traditional terrestrial plant.

MECHANICAL ENERGY Combined energy from potential and kinetic energy.

MESOPHILIC An organism that grows best at moderate temperatures ranging from 20 to 40 degrees Celsius.

METABOLIC ENGINEERING Optimizing conditions within cells to maximally produce a metabolic product.

METABOLOMICS Study of metabolites and the chemical processes involved in their production.

METAGENESIS Stage of organic diagenesis where geopolymers are broken into the simplest hydrocarbon methane.

METHANOGENESIS Biological reaction converting acetic acid into methane and carbon dioxide while consuming hydrogen.

MICROALGAE Photosynthetic aquatic microscopic organisms including true algae, diatoms, and cyanobacteria.

NATURAL GAS Combustible gaseous form of fossil energy made primarily of hydrocarbons containing less than five carbons.

NITROGENASE Enzyme responsible for the conversion of gaseous nitrogen into ammonia and other organic forms of nitrogen; also important in the production of biohydrogen.

NONPOLAR A molecule where the center of the positive charge is balanced with the center of the negative charge.

NUCLEAR ENERGY Form of energy that results from mass-to-energy conversion within the nucleus of an atom.

NUCLEAR FISSION Reactions occurring when the atomic nuclei of larger atoms are split, forming smaller atoms and large amounts of energy.

NUCLEAR FUSION Reactions occurring when smaller atomic nuclei combine to form a larger atom, producing large amounts of energy.

OCEAN THERMAL ENERGY CONVERSION (OTEC) Energy generated from the temperature differentials between the warm ocean surface waters and the cold deep ocean waters.

OXIDATIVE STABILITY Measure of the susceptibility of a diesel fuel to oxidative degradation.

PASSIVE SOLAR ENERGY Solar energy that does not require any mechanical equipment but rather takes advantage of the inherent properties built into a structure.

PEAK OIL Point in time when a country, region, or the entire planet is producing the maximum amount of oil possible.

PEAT Partially decayed organic matter; precursor to coal.

PETROLEUM (OIL) Combustible liquid form of fossil energy made primarily of hydrocarbons containing more than five carbons.

PHOTOFERMENTATION Conversion of organic materials into biohydrogen through fermentation by photosynthetic bacteria.

PHOTOLYSIS Chemical process that uses light to catabolize larger biomolecules into smaller biomolecules.

PHOTOSYNTHESIS Process by which plants use the energy from sunlight to fix carbon dioxide from the air into a useable form of organic carbon such as a sugar.

PHOTOVOLTAIC EFFECT Creation of electric current directly from light.

PHYTOPLANKTON Microscopic autotrophs that inhabit aquatic systems such as lakes, ponds, rivers, and oceans.

POLAR A molecule where the center of the positive charge is not balanced with the center of the negative charge.

POLYMERASE CHAIN REACTION (PCR) Process of amplifying DNA for genetic engineering purposes.

POTENTIAL ENERGY Energy obtained by an object due to its position.

POWER Energy divided by time.

PRIMARY ENERGY SOURCES Natural energy source that has not undergone any type of conversion or transformation.

PRIMARY PRODUCTION Biological conversion of atmospheric carbon dioxide into biomass.

PROCESS-BASED LCA Assessment creating an itemized list of inputs and outputs for a single step of a process used in creating a product.

PRODUCTION The transformation of energy into a common useable form.

PROTEOMICS The study of proteins and their functions.

PROVED RESERVE Estimated quantity of a natural resource available on Earth based on current technology.

PYROLYSIS Anaerobic thermochemical decomposition of organic material at high temperatures.

RADIATIVE FORCING Changing the net irradiance of solar energy coming into the atmosphere and long-wave radiation leaving the atmosphere.

RENEWABLE Resource that can be exploited from the Earth only to be replaced within a short period of time.

RENEWABLE ENERGY Energy that is derived from natural resources that can be sustainably maintained.

RESERVE-TO-PRODUCTION RATIO Comparison of the proved reserves with the current level of production resulting in the number of years remaining for a natural resource.

SACCHARIFICATION Process of breaking complex carbohydrates such as starch into simple sugars.

SECOND LAW OF THERMODYNAMICS Physics law stating that as useable energy decreases and unusable energy increases, entropy as a gauge of randomness or chaos within a closed system also increases.

SECONDARY METABOLITES Metabolic products produced by an organism for a secondary function like adaptation to a unique environment or protection from a potential predator.

SEDIMENTATION Particulate matter is suspended in waterways, decreasing the quality of the water.

SEQUENCING Technology used to read the nucleotide bases found in a DNA molecule.

SEQUESTRATION The capture and removal of carbon dioxide from the atmosphere.

STEEPING Biological seeds such as a corn kernel soaked in a mixture of water and dilute acid to break down the tough outer shell, allowing better access to the biomaterials inside.

SUBBITUMINOUS Intermediate type of coal containing 42–52% carbon; second product of coalification forming between lignite and bituminous coal.

SUPPLY AND DEMAND Basic economic concept that focuses on the amount of a product that a consumer will buy relative to the amount of a product that a producer will provide.

SUSTAINABLE Resource can be exploited from the Earth without impacting its availability in the future.

SYNTHESIS GAS (PRODUCT OR SYNGAS) A gas resulting from a chemical process that can then be used as a feedstock for the production of other products like fuels, electricity, and chemicals such as the gaseous mixture produced by gasification containing carbon monoxide and hydrogen.

SYNTHETIC BIOLOGY Area of technology based on the design and production of new biological parts.

SYSTEM BOUNDARIES Define the parts of the life cycle that will be included in the study.

THERMOCHEMICAL PROCESSES Chemical methods that convert carbon-containing materials into an energetic product by taking advantage of energy and heat released during the reaction.

THERMONUCLEAR REACTIONS Reactions involving a transformation of the atomic nuclei of atoms, resulting in the release of large amounts of energy and the formation of a new atom.

THERMOPHILIC (THERMOPHILE) An organism that thrives at relatively high temperatures ranging from 41 to 122 degrees Celsius.

TIDAL ENERGY Energy generated by the gravitational forces pulling water across the Earth between the moon and the sun.

TRANSCRIPTION Process of transferring genetic information from DNA to RNA.

TRANSCRIPTOMICS The study of the expression of RNA (gene expression) within an organism.

TRANSESTERIFICATION Reaction between a triacylglyceride and an alcohol to yield glycerol and three alcohol–ester-bound fatty acids.

TRANSFORMATION Process of inserting foreign DNA into an organism during genetic manipulations.

TRANSGENIC Organisms capable of expressing genes from another foreign organism.

TRANSLATION Process of transferring genetic information from RNA to protein.

TRIACYLGLYCERIDE Chemical molecule combining glycerol with three fatty acids; energy storage unit in plants and animals.

VISCOSITY Measure of the thickness of a fuel that effects flow.

WAVE ENERGY Kinetic energy produced by the friction of wind on water, creating surface waves.

WEATHER The way the atmosphere is behaving in the short term, basically referring to timescales ranging from hours and days to weeks.

WET MILLING Process used to separate the kernel of a seed from the other components consisting of soaking the seeds in water with or without acid.

REFERENCES

Chapter 1

APS Panel on Public Affairs (2014) *Energy Units*. American Physical Society, College Park, MD. Available at www.aps.org.

Bloomberg (2014) *Bloomberg Industry Market Leaders*. Available at http://www.bloomberg.com/visual-data/industries/q/market-leaders.

British Petroleum (2011, 2013, 2014) *BP Statistical Review of World Energy 2011, 2013, 2014*. Available at www.bp.com/statisticalreview.

CIA (Central Intelligence Agency) (2012) *The World Factbook*. Available at www.cia.gov.

CIA (2014) *The World Factbook*. Available at www.cia.gov.

EIA (US Energy Information Administration) (2012) *Annual Energy Review*. Available at www.eia.gov.

EIA (2013) *U.S. Imports by Country of Origin*. Available online at www.eia.gov.

EIA (2014a) *Annual Energy Outlook 2014 with Projections to 2035*, DOE/EIA-0383(2014). Available at www.eia.gov.

EIA (2014b) *Primary Energy Consumption by Source and Sector, 2013: Monthly Energy Review*. Available at www.eia.gov.

EIA (2014c) *Historical Production and Consumption of Crude Oil*. Available at www.eia.gov.

EIA (2015) *Total Energy*. Available at www.eia.gov.

Elert, G. (2012) *Chemical Potential Energy: The Physics Hypertextbook*. Available at http://physics.info/energy-chemical/.

IEA (International Energy Agency) (2014) *Key World Energy Statistics 2014*. Available at www.iea.org.

Murphy, T. (2014) *Our Energy Future: Energy by the Numbers*. University of California, San Diego, Coursera. Available at www.coursera.org/learn/future-of-energy.

Nashawi, I.S., Malallah, A., Al-Bisharah, M. (2010) Forecasting world crude oil production using multicyclic Hubbert model. *Energy Fuels* 24:1788–1800.

Roque, C. (2013) *Rockefeller's Unconventional Approach to Getting Rid of Waste*, Attendly. Available at www.attendly.com.

US Census Bureau (2012) *International Data Base: World Population Summary*. US Department of Commerce. Available at www.census.gov.

US Department of Commerce (2012) *The Energy Industry in the United States*. SelectUSA. US Department of Commerce. Available at selectusa.commerce.gov.

US Department of Energy (2011) *Fuel Properties Comparison*. Alternative Fuels Data Center. Available at www.afdc.energy.gov.

US Department of Energy (2014) *Fuel Comparison Chart*. Alternative Fuels Data Center. Available at www.afdc.energy.gov.

Chapter 2

Barnes, M.A., Barnes, W.C., and Bustin, R.M. (1984) Diagenesis 8: chemistry and evolution of organic matter. *Journal of the Geological Association of Canada* 11:103–114.

British Petroleum (2014) *BP Statistical Review of World Energy 2013*. Available at www.bp.com/statisticalreview.

Bryan, P. (2011) *The How's and Why's of Replacing the Whole Barrel*. US Department of Energy. Available at www.doe.gov.

Carmody, R.N. and Wrangham, R.W. (2009) The energetic significance of cooking. *Journal of Human Evolution* 57:379–391.

Coal (2002) *UXL Encyclopedia of Science*. Encyclopedia.com. Available at www.encyclopedia.com. Accessed 2015.

Demirel, Y. (2012) *Energy: Production, Conversion, Storage, Conservation, and Coupling (Green Energy and Technology)*. Springer-Verlag, London.

Diamond, J. (2002) Evolution, consequences and future of plant and animal domestication. *Nature* 418:700–707.

DOI (Department of Interior) (2013) *Draft Resource Management Plan and Environmental Impact Statement for the Buffalo Field Office Planning Area*. Buffalo Field Office, Bureau of Land Management, US Department of the Interior. Available at www.blm.gov.

EIA (US Energy Information Administration) (2013) *Energy Today*. Available at www.eia.gov.

EIA (2014) *Electric Power Monthly*. Available at www.eia.gov.

EIA (2015) *Energy Explained: History of Gasoline*. Available at www.eia.gov.

EPA (US Environmental Protection Agency) (1995) Inorganic chemical industry. In *Compilation of Air Pollution Emission Factors*. Available at www.epa.gov.

EPA (2015) *Hydraulic Fracturing and Its Potential Impact on Drinking Water Resources*. Available at www.epa.gov.

Gleick, P. H. (2014) *The World's Water: The Biennial Report on Freshwater Resources*, Vol. 8. Island Press, Washington, DC.

Leffler, W. L. (2008) *Petroleum Refining in Nontechnical Language*, 4th ed. PennWell Corporation, Tulsa, OK.

Mayfield, S. (2014) *Our Energy Future: Introduction to Energy*. University of California, San Diego, Coursera. Available at www.coursera.org/learn/future-of-energy.

McMurry, J. (2000) *Organic Chemistry*, 5th ed. Brooks/Cole, Pacific Grove, CA.

Murkowski, L. (2013) *Energy 20/20: A Vision for America's Energy Future*. US Senate.

National Geographic (2015) *The Development of Agriculture*. The Genographic Project. Available at Genographic. nationalgeographic.com.

Nersesian, R. L. (2010) *Energy for the 21st Century: A Comprehensive Guide to Conventional and Alternative Sources*. M. E. Sharpe, New York.

Paterson, L. (2015) Oil wells produce even more water than oil. *Marketplace Business*. Available at www.marketplace.org.

Pees, S. T. (2004) *Oil History: Whale Oil*. Petroleum History Institute. Available at www.petroleumhistory.org

Pierce, M. A. (2012) Coal. In *Encyclopedia of Energy*. Salem Press, Ipswich, MA.

Pomeroy, R. (2014) *Our Energy Future: Petroleum*. University of California, San Diego, Coursera. Available at www.coursera.org/learn/future-of-energy.

Rhodes, R. (2007) *Energy Transitions: A Curious History*. Center for International Security and Cooperation, Stanford University. Available at cisac.fsi.stanford.edu/.

Roque, C. (2013) *Rockefeller's Unconventional Approach to Getting Rid of Waste*, Attendly. Available at www.attendly.com.

Tarbell, I. M. (1904) *The History of the Standard Oil Company*. McClure, Phillips, New York.

UK (University of Kentucky) (2012) *Methods of Mining*. Kentucky Geological Survey, University of Kentucky. Available at www.uky.edu/KGS.

Veil, J. A. and Quinn, J. J. (2008) *Water Issues Associated with Heavy Oil Production*, ANL/EVS/R-08/4. Argonne National Laboratory.

Waskey, A. J. (2011) Coal, clean technology. Pp. 80–86. In Mulvaney, D. and Robins, P. (eds): *Green Energy: An A-to-Z Guide*. Sage Publication, London.

Williams, A. B. (2011) Fossil fuels, natural gas. In Pp. 178–183 and 297–303. Mulvaney, D. and Robins, P. (eds): *Green Energy: An A-to-Z Guide*. Sage Publication, London.

Chapter 3

Ahlenius, H. (2007) *Global Outlook for Ice and Snow*. UNEP/GRID-Arendal.

Archer, D. (2012) *Global Warming: Understanding the Forecast*. John Wiley & Sons, New York.

Blasing, T. J. (2014) *Recent Greenhouse Gas Concentrations*. Carbon Dioxide Information Analysis Center. Oak Ridge National Laboratory. Available at cdiac.ornl.gov.

Boden, T., Marland, G., and Andres, B. (2011) *Global CO$_2$ Emission from Fossil Fuel Burning, Cement Manufacture, and Gas Flaring: 1751–2008*. Carbon Dioxide Information Analysis Center, Oak Ridge National Laboratory, Oak Ridge, TN.

Butler, J. H. and Montzka, S. A. (2014) *The NOAA Annual Greenhouse Gas Index (AGGI)*. Global Monitoring Division, Earth System Research Laboratory, National Oceanic and Atmospheric Administration. Available at http://www.esrl.noaa.gov/.

Canright, S. (2011) *Earth's Energy Budget*. NASA Education, National Aeronautics and Space Administration. Available at www.nasa.gov.

Coley, D. (2008) *Energy and Climate Change: Creating a Sustainable Future*. John Wiley & Sons, Chichester, UK.

Core Writing Team, Pachauri, R. K., and Reisinger, A. (eds) (2007) *Climate Change 2007: Synthesis Report*. Intergovernmental Panel on Climate Change, Geneva. Available at www.ipcc.ch.

CSIRO Marine and Atmospheric Research and the Australian Bureau of Meterology (2014) *Cape Grim Greenhouse Gas Data*. Cape Grim Baseline Air Pollution Station. Available at http://www.csiro.au/greenhouse-gases)

Davis, S. C., Diegel, S. W., and Boundy, R. G. (2012) *Transportation Energy Data Book*, 31st ed., ORNL-6987. Center for Transportation Analysis, Oak Ridge National Laboratory, Oak Ridge, TN.

Easterling, D. and Karl, T. (2012) *Global Warming*. National Climatic Data Center, National Oceanic and Atmospheric Administration. Available at www.ncdc.noaa.gov.

EIA (US Energy Information Administration) (2012) *International Energy Statistics: Total Carbon Dioxide Emissions from the Consumption of Energy*. Available at www.eia.gov.

EPA (US Environmental Protection Agency) (2011) *Atmospheric Concentrations of Greenhouse Gases*. Available at cfpub.epa.gov.

EPA (2014) *Climate Change Indicators in the United States, 2014*, 3rd ed., EPA 430-R-14-004. Available at www.epa.gov/climatechange/indicators.

Forester, P., Ramaswamy, V., and Core Writing Team (2007) Changes in atmospheric constituents and in radiative forcing: direct global warming potentials. In Solomon, S., Qin, D., Manning, M., Chen, Z., Marquis, M., Averyt, K. B., Tignor, M., and Miller, H. L. (eds): *Climate Change 2007: The Physical Science Basis. Contribution of Working Group I to the Fourth Assessment Report of the IP CC*. Available at www.ipcc.ch.

Grenci, L. M. and Nese, J. M. (2008) *A World of Weather: Fundamentals of Meterology (A Text/Laboratory Manual)*. Kendall Hunt Publishing, Dubuque, IA. Available at www.ems.psu.edu.

IPCC (2013) Summary for policymakers. Pp. 1–30. In Stocker, T. F., Qin, D., Plattner, G.-K., Tignor, M., Allen, S. K., Boschung, J., Nauels, A., Xia, Y., Bex, V., and Midgley, P. M. (eds): *Climate Change 2013: The Physical Science Basis. Contribution of Working Group I to the Fifth Assessment Report of the Intergovernmental Panel on Climate Change*. Cambridge University Press, Cambridge, UK.

Keeling, C.D. (1974) *Carbon Dioxide Concentration at Mauna Loa Observatory*, Scripps Institution of Oceanography.

Keeling Curve (2012) *Scripps CO₂ Program*. Scripps Institution of Oceanography, La Jolla, CA. Available at scrippsco2.ucsd.edu.

Meehl, G., Stocker, T., and Core Writing Team (2007) Global climate projections. Pp. 747–845. In Solomon, S., Qin, D., Manning, M., Chen, M., Marquis, M., Averyt, K. B., Tignor, M., and Miller, H. L. (eds): *Climate Change 2007: The Physical Science Basis*. Intergovernmental Panel on Climate Change, Cambridge University Press, Cambridge, UK.

Myhre, G., Shindell, D., Bréon, F.-M., Collins, W., Fuglestvedt, J., Huang, J., Koch, D., Lamarque, J.-F., Lee, D., Mendoza, B. et al. (2013) Anthropogenic and natural radiative forcing. Pp. 659–740. In Stocker, T.F., Qin, D., Plattner, G.-K., Tignor, M., Allen, S.K., Boschung, J., Nauels, A., Xia, Y., Bex, V., and Midgley, P.M. (eds.): *Climate Change 2013: The Physical Science Basis. Contribution of Working Group I to the Fifth Assessment Report of the Intergovernmental Panel on Climate Change*. Cambridge University Press, Cambridge, UK.

NASA (National Aeronautics and Space Administration) (2013) *GISS Surface Temperature Analysis GISTEMP*. NASA, Goddard Institute for Space Studies, New York.

Newman, P. (2012) *Regions of the Electromagnetic Spectrum*. Astrophysics Science Division, National Aeronautics and Space Administration. Available at imagine.gsfc.nasa.gov.

Ramanathan, V. (2014) *Our Energy Future: A New Approach for Slowing Down Climate Change and Sea Level Rise in the Near-Term*. University of California, San Diego. Coursera. Available at www.coursera.org/learn/future-of-energy.

Samenow, J. et al. (2010) *Climate Change Indicators in the United States*. EPA Climate Change Division, US Environmental Protection Agency.

Shoemaker, L. (2010) *Stable and Radiocarbon Isotopes of Carbon Dioxide*. Earth System Research Laboratory, National Oceanic and Atmospheric Administration. Available at www.esrl.noaa.gov.

Somerville, R. (2014) *Our Energy Future: Climate Change and the Impact of Carbon Dioxide*. University of California, San Diego. Coursera. Available at www.coursera.org/learn/future-of-energy.

Strom, R.G. (2007) *Hot House: Global Climate Change and the Human Condition*. Copernicus Books, New York.

Tans, P. (2015) *Carbon Tracker*, Earth System Research Laboratory, National Oceanic and Atmospheric Administration. Available at http://www.esrl.noaa.gov/.

Tans, P. and Keeling, R. (2015) *Global Greenhouse Gas Reference Network*. NOAA/ESRL and Scripps Institution of Oceanography. Available at www.esrl.noaa.gov/.

Tyndall, J. (1872) *Contributions to Molecular Physics in the Domain of Radiant Heat: A Series of Memoirs Published in the "Philosophical Transactions" and "Philosophical Magazine," with Additions*. Longman, Greens, London.

United Nations Secretariat (1999) *The World at Six Billion*. Population Division, United Nations, New York.

Viñas, M.J. (2015) *2015 Arctic Sea Ice Maximum Annual Extent Is Lowest on Record*. National Aeronautics and Space Administration. Available at Climate.nasa.gov.

Wong, T., Marvel, T., Lee, S., and Hopson, V. (2014) *The NASA Earth's Energy Budget Poster*. National Aeronautics and Space Administration. Available at www.nasa.gov.

Chapter 4

Alsema, E.A., de Wild-Scholten, M.J., and Fthenakis, V.M. (2006) Environmental impacts of PV electricity generation: a critical comparison of energy supply options. *21st European Photovoltaic Solar Energy Conference*, Dresden, Germany.

Barbier, B. (2012) *Cosmicopia*. Astrophysics Science Division, National Aeronautics and Space Administration. Available at helios.gsfc.nasa.gov.

BLM (US Bureau of Land Management) (2012) *Wind Energy Guide*. Wind Energy Development Programmatic EIS. Available at windeis.anl.gov.

Bodansky, D. (2005) *Nuclear Energy: Principles, Practices and Prospects*. Springer, New York.

BOEM (Bureau of Ocean Energy Management) (2014) *Ocean Wave Energy*. Available at www.boem.gov.

British Petroleum (2014) *BP Statistical Review of World Energy 2013*. Available at www.bp.com/statisticalreview.

Burman, K. and Walker, A. (2009) *Ocean Energy Technology Overview*, DOE/GO-102009-2823. US Department of Energy.

California Energy Commission (2012) Ocean energy. Chap. 14 in *The Energy Story*. California Energy Commission, Sacramento. Available at www.energyquest.ca.gov.

Canright, S. (2011) *Earth's Energy Budget*. NASA Education, National Aeronautics and Space Administration. Available at www.nasa.gov.

Casalenuovo, K. (2011) Hydroelectric power. Pp. 242–246. In Mulvaney, D. and Robins, P. (eds): *Green Energy: An A-to-Z Guide*. Sage Publications, Thousand Oaks, CA.

Coimbra, C. (2014) *Our Energy Future: Photovoltaic and Photothermal Energy Production*. University of California, San Diego. Coursera. Available at www.coursera.org/learn/future-of-energy.

EIA (US Energy Information Administration) (2015) *Monthly Energy Review; Electric Power Monthly; Annual Energy Review; Short-Term Energy and Summer Fuels Outlook*. Available at www.eia.gov.

EPA (US Environmental Protection Agency) (2014) *Renewable Fuels: Regulations & Standards*. Renewable Fuel Standard. Available at www.epa.gov.

Evans, J. and Perlman, H. (2014) *The Water Cycle*. US Geological Survey. Available at Water.usgs.gov/edu/watercycle.html.

FWS (US Fish and Wildlife Service) (2002) *Migratory Bird Mortality*. Available at www.fws.gov.

Gabbard, R.T. (2011) Solar energy. Pp. 403–409. In Mulvaney, D. and Robins, P. (eds): *Green Energy: An A-to-Z Guide*. Sage Publications, Thousand Oaks, CA.

Global Energy Observatory (2014) *Current List of Hydro PowerPlants*. Available at globalenergyobservatory.org.

Isherwood, W. (2011) Geothermal energy. Pp. 197–202. In Mulvaney, D. and Robins, P. (eds): *Green Energy: An A-to-Z Guide*. Sage Publications, Thousand Oaks, CA.

Kagel, A., Bates, D., and Gawell, K. (2007) *A Guide to Geothermal Energy and the Environment*. Geothermal Energy Association. Available at www.geo-energy.org.

Knier, G. (2011) How do photovoltaics work? *NASA Science News*, National Aeronautics and Space Administration. Available at science.nasa.gov.

Lewis, N.S. and Nocera, D.G. (2006) Powering the planet: chemical challenges in solar energy utilization. *Proceedings of the National Academy of Sciences* 103:15729–15735.

Mudd, G. M. (2011) Uranium. Pp. 429–435. In Mulvaney, D. and Robins, P. (eds): *Green Energy: An A-to-Z Guide*. Sage Publications, Thousand Oaks, CA.

Nersesian, R. L. (2010) *Energy for the 21st Century*. M. E. Sharpe, New York.

Olanrewaju, A. O. (2011) Wind power. Pp. 444–448. In Mulvaney, D. and Robins, P. (eds): *Green Energy: An A-to-Z Guide*. Sage Publications, Thousand Oaks, CA.

Perlman, H. (2012) *The Water Cycle: Water Science for Schools*. US Geological Survey. Available at ga.water.usgs.gov.

Philibert, C., Frankl, P., and IEA Renewable Energy Division (2010) *Technology Roadmap: Concentrating Solar Power*. International Energy Agency. Available at www.iea.org.

Roberts, B. J. (2009) *Photovoltaic Solar Resources of the United States*. National Renewable Energy Laboratory, US Department of Energy.

Roberts, B. J. (2014) *Geothermal Power Generation*. National Renewable Energy Laboratory, US Department of Energy.

Sevior, M., Okuniewicz, I., Meehan, A., Jones, G., George, D., Flitney, A., and Filewood, G. (2010) *Advanced Nuclear Fission Technology*. University of Melbourne, Parkville, VIC.

Touran, N. (2012) *Energy Densities of Various Fuel Sources*. What is Nuclear? Available at Whatisnuclear.com.

Tynan, G. (2014) *Our Energy Future: The Future of Nuclear Energy*. University of California, San Diego. Coursera. Available at www.coursera.org/learn/future-of-energy.

US Census Bureau (2013) *Alaska QuickFacts*. United States Census Bureau. US Department of Commerce.

WNA (World Nuclear Association) (2014, 2015) *Outline History of Nuclear Energy: Supply of Uranium*. Available at www.world-nuclear.org.

3Tier by Vaisala (2015) *Power and Energy*. Available at www.3tier.com.

Chapter 5

AgBioWorld (2011) *The Green Revolution and Dr. Norman Borlaug: Towards the "Evergreen Revolution."* Available at www.agbioworld.org.

Amon, R., Kazama, D. B., Wong, T., and Neenan, R. (2008) *California's Food Processing Industry Energy Efficiency Initiative: Adoption of Best Practices*, CEC 400-2008-006. Staff Report. California Energy Commission.

Burney, J. (2014) *Our Energy Future: Climate Change and Food Security*. University of California, San Diego. Coursera. Available at www.coursera.org/learn/future-of-energy.

Cowen, R. (1999) *Exploiting the Earth: Geology 115*. University of California, Davis. Available at mygeologypage.ucdavis.edu.

Diamond, J. (2002) Evolution, consequences and future of plant and animal domestication. *Nature* 418:700–707.

EIA (US Energy Information Administration) (2014a) *How Much Gasoline Does the United States Consume?* Available at www.eia.gov.

EIA (2014b) *Short-Term Energy Outlook and Winter Fuels Outlook (STEO)*. Available at www.eia.gov.

Elhadj, E. (2008) *Saudi Arabia's Agricultural Project: From Dust to Dust*. Rubin Center Research in International Affairs. Available at www.rubincenter.org.

EPA (US Environmental Protection Agency) (2009) *Inventory of U.S. Greenhouse Gas Emissions and Sinks: 1990-2007*, EPA 430-R-09-004. Available at www.epa.gov.

EPA (2013) *Ag 101 Demographics*. Available at www.epa.gov.

Fiala, N. (2009) *How Meat Contributes to Global Warming*. Scientific American. Available at www.scientificamerican.com.

Funke, T., Han, H., Healy-Fried, M. L., Fischer, M., and Schönbrunn, E. (2006) Molecular basis for the herbicide resistance of Roundup Ready crops. *Proceedings of the National Academy of Sciences* 103:13010–13015.

Hazell, P. B. R. (2002) *Green Revolution: Curse or Blessing?* International Food Policy Research Institute. Available at www.ifpri.org.

Hofstrand, D. (2012) More on feeding nine billion people by 2050. *AgMRC Renewable Energy & Climate Change Newsletter*. Available at www.agmrc.org.

Huang, W. (2008) *Factors Contributing to the Recent Increase in U. S. Fertilizer Prices 2002-08*. Economic Research Service, US Department of Agriculture. Available at www.ers.usda.gov.

Jasinski, S. M. (2014) *Mineral Commodity Summaries*. US Geological Survey, US Department of the Interior. Available at minerals.usgs.gov.

Jenkinson, D. S. (2001) The impact of humans on the nitrogen cycle, with focus on temperate arable agriculture. *Plant and Soil* 228:3–15.

Johnston, A. E. (2003) *Understanding Potassium and Its Use in Agriculture*. European Fertilizer Manufacturers Association. Available at www.pda.org.uk.

Karam, S. (2008) Saudi Arabia scraps wheat growing to save water, *Reuters*. Available at www.reuters.com.

Khan, M. F. K. and Nawaz, M. (1995) Karez irrigation in Pakistan. *GeoJournal*. 37:91–100.

Menzel, P. and D'Aluisio, F. (2007) Hungry planet: what the world eats. *TIME Magazine*. Available at www.time.com.

O'Brien, M. and Walton, M. (2010) *Leaf-Cutter Ants: Farmers, Pharmacists and Energy Experts*. Science Nation, National Science Foundation. Available at www.nsf.gov.

Penner, J. E., Lister, D. H., Griggs, D. J., Dokken, D. J., and McFarland, M. (1999) *Aviation and the Global Atmosphere*. Intergovernmental Panel on Climate Change. Available at www.ipcc.ch.

Perry, J., Banker, D. E., and Green, R. (1999) *Broiler Farms' Organization, Management and Performance*. Economic Research Service, US Department of Agriculture. Available at www.ers.usda.gov.

Piroq, R. and Van Pelt, T. (2002) *How Far Do Your Fruit and Vegetables Travel?* Leopold Center for Sustainable Agriculture. Available at www.leopold.iastate.edu.

Science Encyclopedia (2012) *Crop Rotation: History*. Available at science.jrank.org.

Spielmaker, D. (2006) *Historical Timeline: Farm Machinery and Technology*. Growing a Nation: The Story of American Agriculture. Available at www.agclassroom.org.

Tomlinson, I. (2012) *Just Say N_2O: From Manufactured Fertilser to Biologically-Fixed Nitrogen*. Soil Association. Available at www.soilassociation.org.

UM (University of Michigan) (2011) *U. S. Food System Factsheet*. Center for Sustainable Systems. Available at css.snre.umich.edu.

UM (2014) *Carbon Footprint Factsheet*, CSS09-05. Center for Sustainable Systems. Available at css.snre.umich.edu.

UMT (University of Montana) (2012) *The History of Bt Cotton*. The Maureen and Mike Mansfield Center Ethics and Public Affairs Programs. Available at www.umt.edu.

US Census Bureau (2014) *World Population*. US Census Bureau. Available at www.census.gov.

US Department of Transportation (2009) *US Greenhouse Gas Emissions from Domestic Freight Transportation: 1990–2007*. Freight Management and Operations, Federal Highway Administration. Available at Ops.fhwa.dot.gov.

USDA (US Department of Agriculture) (2012) *Feed Grains Database*. Economic Research Service. Available at www.ers.usda.gov.

USDA (2014a) *Feed Grains Database*. Economic Research Service. Available at www.ers.usda.gov.

USDA (2014b) *U.S. Agricultural Imports and Exports*. Economic Research Service. Available at www.ers.usda.gov.

USGS (US Geological Survey) (2014a) *Mineral Commodity Summaries: Phosphate Rock*. USGS, Reston, VA.

USGS (2014b) *Mineral Commodity Summaries: Potash*. USGS, Reston, VA.

Van Kauwenbergh, S.J. (2010) *World Phosphate Rock Reserves and Resources*. International Fertilizer Development Center. Available at pdf.usaid.gov.

Wood, C.W., Mullins, G.L., and Hajek, B.F. (2012) *Phosphorus in Agriculture*. Soil Quality Institute Technical Pamphlet No. 2. US Department of Agriculture. Available at www.nrcs.usda.gov.

Woods, J., Williams, A., Hughes, J.K., Black, M., and Murphy, R. (2010) Energy and the food system. *Philosophical Transactions of The Royal Society B* 365:2991–3006.

The World Bank (2014) *Poverty Overview*. Available at www.worldbank.org.

WWAP (World Water Assessment Programme) (2012) *Facts and Figures: Fact 24 Irrigated Land*. United Nations Educational, Scientific and Cultural Organization. Available at www.unesco.org.

Zadoks, J.C. and Waibel, H. (2000) From pesticides to genetically modified plants: history, economics and politics. *Netherlands Journal of Agricultural Science* 48:125–149.

Chapter 6

ACGF (American Corn Growers Foundation) (2011) *News*. Available at www.acgf.org.

Aden, A. (2007) *Water Usage for Current and Future Ethanol Production*. Southwest Hydrology. Available at www.swhydro.arizona.edu.

Basso, L.C., Basso, T.O., and Rocha, S.N. (2011) Ethanol production in Brazil: The industrial process and its impact on yeast fermentation. Pp. 85–99. In Dos Santos Bernardes, M.A. (Ed.): *Biofuel Production: Recent Developments and Prospects*. InTech, Rijeka, Croatia. Available at cdn.intechweb.org.

Biello, D. (2008) *That Burger You're Eating Is Mostly Corn*. Scientific American. Available at www.scientificamerican.com.

Borglum, G.B. (1980) *Starch Hydrolysis for Ethanol Production*. Pp. 264–269. American Chemical Society. Argonne National Laboratory. Available at web.anl.gov.

BP (British Petroleum) (2014) *BP Statistical Review of World Energy 2014*. Available at www.bp.com/statisticalreview.

Campbell, N.A. and Reece, J.B. (2002) *Biology*, 6th ed. Pearson Education, San Francisco, CA.

Capehart, T. (2014) *Feed Grains Database*. Economic Research Service, US Department of Agriculture. Available at www.ers.usda.gov.

Carroll, A. and Somerville, C. (2009) Cellulosic biofuels. *Annual Reviews of Plant Biology* 60:165–182.

Choi, C. (2012) Humans used fire 1 million years ago. *Discovery News*. Available at news.discovery.com.

Colman, Z. (2014) Changes to EPA gasoline rules may shake U.S. ethanol industry. *Washington Examiner*, September 8. Available at www.washingtonexaminer.com.

Cooper, G. (2012) *How Much Ethanol Can Come from Corn?* National Corn Growers Association. Available at www.cie.us.

DOE (US Department of Energy) (2012) *Fuel Properties*. Alternative Fuels Data Center. Available at www.afdc.energy.gov.

Dwivedi, P., Alavalapati, J.R.R., and Lal, P. (2009) Cellulosic ethanol production in the United States: conversion technologies, current production status, economics and emerging developments. *Energy for Sustainable Development* 13:174–182.

EIA (US Energy Information Administration) (2014a) *Glossary*. Available at www.eia.gov.

EIA (2014b) *Short-Term Energy Outlook and Winter Fuels Outlook (STEO)*. Available at www.eia.gov.

EIA (2014c) *Monthly Energy Review*. Available at www.eia.gov.

Elbheri, A., Coyle, W., and the Inter-Agency Feedstocks Team (2008) *The Economics of Biomass Feedstocks in the United States: A Review of the Literature*. Biomass Research and Development Initiative. Occasional Paper No. 1. Available at www.biomassboard.gov.

EPA (Environmental Protection Agency) (2009) *EPA Lifecycle Analysis of Greenhouse Gas Emissions from Renewable Fuels*, EPA-420-F-09-024.

EPA (2013) *Demographics*, Ag101. Available at www.epa.gov.

EPA (2014) *Renewable Fuel Standard (RFS)*. Available at www.epa.gov.

Ethanol History (2011) *Ethanol History: From Alcohol to Car Fuel*. Available at www.ethanolhistory.com.

EWG (Environmental Working Group) (2012) *Farm Subsidies: Corn Subsidies*. Available at farm.ewg.org.

Gonzalez, R. (2011) Biomass supply chain and conversion economics of cellulosic ethanol. Thesis, North Carolina State University. Available at repository.lib.ncsu.edu.

IowaCorn (2014) *Production and Use*. Available at www.iowacorn.org.

Jahren, A.H. and Schubert, B.A. (2010) Corn content of French fry oil from national chain vs. small business restaurant. *Proceedings of the National Academy of Sciences* 107:2099–2101.

Jaret, P. (2015) Hot spots: bioprospecting for biofuel's mystery bugs. *Bioenergy Connection* 3.1:51.

Johanns, A.M. (2014) *Iowa Corn and Soybean County Yields*. Ag Decision Maker, Iowa State University Extension and Outreach. Available at www.extension.iastate.edu.

Johnson, T., Johnson, B., Scott-Kerr, C. and Kiviaho, J. (2010) *Bioethanol: Status Report on Bioethanol Production from Wood and Other Lignocellulosic Feedstocks*. Beca. Available at www.beca.com.

Khanna, M. (2008) Cellulosic biofuels: are they economically viable and environmentally sustainable. *Choices* 23:16–21. Available at www.choicesmagazine.org.

Klein-Marcuschamer, D., Oleskowicz-Popiel, P., Simmons, B. A., and Blanch, H. W. (2012) The challenge of enzyme cost in the production of lignocellulosic biofuels. *Biotechnology and Bioengineering* 109:1083–1087.

Koundinya, V. (2009) Corn stover. *AgMRC Renewable Energy Newsletter*. Agricultural Marketing Resource Center. Available at www.agmrc.org.

McCord, G. (2014) *Our Energy Future: The Importance of Energy for the Bottom Billion*. University of California, San Diego, Coursera. Available at www.coursera.org/learn/future-of-energy.

Milnes, R., Deller, L., and Hill, N. (2010) *Ethanol Internal Combustion Engine*. Energy Technology Network, International Energy Agency. Available at www.iea-etsap.org.

MLR Solutions (2009) *Ethanol Fuel History*. Fuel Testers Company. Available at www.fuel-testers.com.

Mosier, N. S. and Ileleji, K. (2006) *How Fuel Ethanol Is Made from Corn*, ID-328. BioEnergy, Purdue University. Available at www.extension.purdue.edu.

Perlack, R. D., Wright, L. L., Turhollow, A. F., Graham, R. L., Stokes, B. J., and Erbach, D. C. (2005) *Biomass as Feedstock for a Bioenergy and Bioproducts Industry: The Technical Feasibility of a Billion-Ton Annual Supply*. US Department of Energy and US Department of Agriculture. Available at www1.eere.energy.gov.

Pollan, M. (2007) *The Ominvore's Dilemma: A Natural History of Four Meals*. Penguin, New York.

Ramanathan, V. (2014) *Our Energy Future: A New Approach for Slowing Down Climate Change and Sea Level Rise in the Near-Term*. University of California, San Diego, Coursera. Available at www.coursera.org/learn/future-of-energy.

RFA (Renewable Fuels Association) (2014) *2013 Monthly U.S. Fuel Ethanol Production/Demand*. Available at ethanolrfa.org.

Schnepf, R. (2010) *Cellulosic Ethanol: Feedstocks, Conversion Technologies, Economics, and Policy Options*, R41460. Congressional Research Service Report. Available at nationalaglawcenter.org.

Schnepf, R. and Yacobucci, B. D. (2012) *Renewable Fuel Standard (RFS): Overview and Issues*, R40155. CRS Report for Congress. Congressional Research Service. Available at www.fas.org.

Skinner, R. and Adler, P. (2009) Carbon sequestration potential of a switchgrass bioenergy crop. *Agronomy Abstracts*. Available at www.ars.usda.gov.

SugarCane.org (2014) *Sugarcane Products: Ethanol*. Available at Sugarcane.org.

Taiz, L. and Zeiger, E. (2010) How the Calvin–Benson cycle was elucidated. Chap. 8 in *Plant Physiology*, 5th ed. Available at 5e.plantphys.net.

UN (United Nations) (2013) *Wood Energy*. Food and Agriculture Organization of the United Nations. Available at www.fao.org.

USDA (US Department of Agriculture) (2014) *Feed Grains Database*. Economic Research Service. Available at www.ers.usda.gov.

Chapter 7

Achten, W. M. J., Verchot, L., Franken, Y. J., Mathijs, E., Singh, V. P., Aerts, R., and Muys, B. (2008) Jatropha bio-diesel production and use. *Biomass and Bioenergy* 32:1063–1084.

Ash, M. (2012) *Soybeans & Oil Crops*. Economic Research Service, US Department of Agriculture. Available at www.ers.usda.gov.

Biodiesel Energy Revolution (2014) *The History of Biodiesel*. Available at www.biodiesel-energy-revolution.com.

Canakci, M. and Gerpen, J. V. (2001) Biodiesel production from oils and fats with high free fatty acids. *Transactions of the American Society of Agricultural Engineers* 44:1429–1436.

Canakci, M. and Sanli, H. (2008) Biodiesel production from various feedstocks and their effects on the fuel properties. *Journal of Industrial Microbiology and Biotechnology* 35:431–441.

Collins, H. P., Boydston, R., Alva, A., Hang, A., Fransen, S., and Wanderschnieder, P. (2012) *Biofuel Variety Trials Factsheet*. US Department of Agriculture and Washington State University. Available at www.pacificbiomass.org.

Demirel, Y. (2012) Energy and Energy Types. Pp. 27–70. In *Energy: Production, Conversion, Storage, Conservation, and Coupling (Green Energy and Technology)*. Springer, London.

EIA (US Energy Information Administration) (2014) *Monthly Biodiesel Production Survey*, EIA-22M. Available at www.eia.gov.

Fishbach, M. A. and Walsh, C. T. (2006) Assembly-line enzymology for polyketide and nonribosomal peptide antibiotics: logic, machinery and mechanisms. *Chemical Reviews* 106:3468–3496.

Gerpen, J. V. (2005) Biodiesel processing and production. *Fuel Processing Technology* 86:1097–1107.

Gerpen, J. V., Shanks, B., Pruszko, R., Clements, D., and Knothe, G. (2004) *Biodiesel Production Technology*, NREL/SR-510-36244. National Renewable Energy Laboratory. Available at www.nrel.gov.

Greene, A. K. (2010) Biodiesel synthesis from animal fats using solid catalysts. *RENDER: The International Magazine of Rendering*. Available at rendermagazine.com.

Groschen, R. (2002) *Overview of the Feasibility of Biodiesel from Waste/Recycled Greases and Animal Fats*. Legislative Commission on Minnesota Resources. Available at www.mda.state.mn.us.

Gross, P. (2007) *Federal and State Ethanol and Biodiesel Requirements*. US Energy Information Administration. Available at www.eia.gov.

Gui, M. M. and Lee, K. T. (2008) Feasibility of edible oil vs. non-edible oil vs. waste edible oil as biodiesel feedstock. *Energy* 33:1646–1653.

Habiby, M. (2011) Fuel demand in U.S. rises on highway diesel use, API says, *Bloomberg*. Available at www.bloomberg.com.

Hay, J. (2012) *Soybean as a Biofuel Feedstock*. University of Nebraska-Lincoln. Available at cropwatch.unl.edu.

Herbek, J. H. and Bitzer, M. J. (1997) *Soybean Production in Kentucky part V: Harvesting, Drying, Storage and Marketing*, AGR-132. University of Kentucky. Available at www.ca.uky.edu.

Jingura, R. M., Musademba, D., and Matengaifa, R. (2010) An evaluation of utility of *Jatropha curcas* L. as a source of multiple energy carriers. *International Journal of Engineering, Science and Technology* 2:115–122.

Kotrba, R. (2008) Defining the alternatives. *Biodiesel Magazine*, February 11.

Kumar, A. and Sharma, S. (2008) An evaluation of multipurpose oilseed crop for industrial uses (*Jatropha curcas* L.): a review. *Industrial Crops and Products* 28:1–10.

Lam, M. K., Lee, K. T., and Mohamed, A. R. (2010) Homogenous, heterogenous and enzymatic catalysis for transesterification of high free fatty acid oil (waste cooking oil) to biodiesel: a review. *Biotechnology Advances* 28:500–518.

Lam, M. K., Tan, K. T., Lee, K. T., and Mohamed, A. R. (2009) Malaysian palm oil: surviving the food versus fuel dispute for a sustainable future. *Renewable and Sustainable Energy Reviews* 13:1456–1464.

Melosi, M. V. (2010) *The Automobile and the Environment in American History*. Automobile in American Life and Society. Available at www.autolife.umd.umich.edu.

Moser, B. R. (2009) Biodiesel production, properties, and feedstocks. *In Vitro Cellular Development and Biology – Plant* 45:229–266.

Motor Trend (2005) Diesel Engines 101: new technology, big torque, and better mileage. *Intellichoice*. Available at www.motortrend.com.

Nowatzki, J., Swenson, A., and Wiesenborn, D. P. (2007) *Small-Scale Biodiesel Production and Use*, AE-1344. North Dakota State University. Available at www.ag.ndsu.edu.

Pacific Biodiesel (2014) *History of Biodiesel Fuel*. Available at www.biodiesel.com.

Pienkos, P. T. (2007) *The Potential for Biofuels from Algae*, NREL /PR-510-42414. Algae Biomass Summit, National Renewable Energy Labs.

Pimentel, D. and Patzek, T. W. (2005) Ethanol production using corn, switchgrass, and wood; biodiesel production using soybean and sunflower. *Natural Resources Research* 14:65–76.

Pomeroy, R. (2014) *Our Energy Future: Biodiesel Chemistry and Analysis*. University of California, San Diego, Coursera. Available at www.coursera.org/learn/future-of-energy.

Pradhan, A., Shrestha, D. S., McAloon, A., Yee, W., Haas, M., Duffield, J. A., and Shapouri, H. (2009) *Energy Life-Cycle Assessment of Soybean Biodiesel*. USDA Agricultural Economic Report Number 845.

Schmidt, B. (2014) *Our Energy Future: Jatropha*. University of California, San Diego, Coursera. Available at www.coursera.org/learn/future-of-energy.

Schnepf, R. and Yacobucci, B. D. (2012) *Renewable Fuel Standard (RFS): Overview and Issues*, R40155. CRS Report for Congress. Congressional Research Service. Available at www.fas.org.

Schuchardt, U., Sercheli, R., and Vargas, R. M. (1998) Transesterification of vegetable oils: a review. *Journal of the Brazilian Chemical Society* 9:199–210.

Schultes, R. E. (1993) The domestication of the rubber tree: economic and sociological implications. *The American Journal of Economics and Sociology* 52:479–485.

Singh, S. P. and Singh, D. (2010) Biodiesel production through the use of different sources and characterization of oils and their esters as the substitute of diesel: a review. *Renewable and Sustainable Energy Reviews* 14:200–216.

Stafford, J. (2014) History of the diesel engine, *DieselEngineMotor.com*. Available at www.dieselenginemotor.com.

Taylor, C. R., Lacewell, R. D., and Seawright, E. (2010) *Economic Cost of Biodiesel and Corn Ethanol per Net BTU of Energy Produced*. Bioenergy Policy Brief, Auburn University. Available at sites.auburn.edu.

US Census Bureau (2012) *Statistical Abstract of the United States, 2012*. Table 1376. Available at www.census.gov.

USDA (US Department of Agriculture) (2007) *Indonesia: Palm Oil Production Prospects Continue to Grow*. Foreign Agricultural Service. Available at www.pecad.fas.usda.gov.

USDA (2012) *Rapeseed and Products: World Supply and Distribution*. Foreign Agricultural Service. Available at www.fas.usda.gov.

USDA (2014a) *Oilseeds: World Markets and Trade. Table 01: Major Oilseeds: World Supply and Distribution (Commodity View)*. Foreign Agricultural Service. Available at www.fas.usda.gov.

USDA (2014b) *Oil Crops Yearbook. Table 5: Soybean Oil: Supply, Disappearance, and Price, U.S., 1980/81–2009/10*. Economics, Statistics and Market Research. Available at usda.mannlib.cornell.edu.

USDA (2014c) *Soybeans: Acreage Planted, Harvested, Yield, Production, Value, and loan Rate, U.S., 1960–2009*. Economics, Statistics and Market Research. Available at usda.mannlib.cornell.edu.

The Week (2012) Americans' ton-a-year eating habit: by the numbers. Available at theweek.com.

Yacobucci, B. D. (2012) *Biofuels Incentives: A Summary of Federal Programs*, R40110. CRS Report for Congress. Congressional Research Service. Available at www.fas.org.

Yoon, J. J. (2011) What's the difference between biodiesel and renewable (green) diesel. *Advanced Biofuels USA*.

Chapter 8

Alsalam, J. and Ragnauth, S. (2011) *DRAFT: Global anthropogenic non-CO2 greenhouse gas emissions: 1990–2030*, EPA 430-D-11-003. US Environmental Protection Agency. Available at www.epa.gov.

Börjesson, P. and Mattiasson, B. (2008) Biogas as a resource-efficient vehicle fuel. *Trends in Biotechnology* 26:7–13.

Braun, R., Weiland, P., and Wellinger, A. (2010) *Biogas from Energy Crop Digestion*. IEA Bioenergy, International Energy Agency. Available at biogasmax.co.uk.

British Petroleum (2014) *BP Statistical Review of World Energy 2013*. Available at www.bp.com/statisticalreview

Carrieri, D., Ananyev, G., Costas, A. M. G., Bryant, D. A., and Dismukes, G. C. (2008) Renewable hydrogen production by cyanobacteria: nickel requirements for optimal hydrogenase activity. *International Journal of Hydrogen Energy* 33:2014–2022.

CNN (2009) Cow methane: a trump card in the fight against global warming? *CNN.com/technology*. Available at edition.cnn.com.

Demirbas, A. (2009) Biorenewable gaseous fuels: biofuels securing the planet's future energy needs. Pp. 231–260, Chap. 5. In *Biofuels: Securing the Planet's Future Energy Needs (Green Energy and Technology)*. Springer, London.

DOE (US Department of Energy) (2010) *Fuel Cell Technologies Program: Energy Efficiency and Renewable Energy*. US Department of Energy. Available at www1.eere.energy.gov.

DOE (2013) *Hydrogen Production and Distribution*. Alternative Fuels Data Center, US Department of Energy. Available at www.afdc.energy.gov.

EIA (US Energy Information Administration) (2014a) *Natural gas*. Available at www.eia.gov/naturalgas/.

EIA (2014b) *Hydrogen Explained*. Available at www.eia.gov.

EPA (US Environmental Protection Agency) (2014) *Municipal Solid Waste*. Available at www.epa.gov.

Hamilton, D. W. (2009) *Anaerobic Digestion of Animal Manures: Understanding the Basic Processes*, BAE-1747. Oklahoma Cooperative Extension Service, Oklahoma State University. Available at pods.dasnr.okstate.edu.

Hein, M. (2014) *Our Energy Future: Biogas*. University of California, San Diego, Coursera. Available at www.coursera.org/learn/future-of-energy.

Hockstad, L. and Weitz, M. (2014) *Inventory of U.S. Greenhouse Gas Emissions and Sinks: 1990–2012*, EPA 430-R-14-003. US Environmental Protection Agency. Available at www.epa.gov.

Hoornweg, D. and Bhada-Tata, P. (2011) Waste generation. Pp. 8–12, Chap. 3. In *What a Waste: A Global Review of Solid Waste Management*. The World Bank. Available at web.worldbank.org.

King, R. (1999) *Natural Gas 1998: Issues and Trends*, DOE/EIA-0560(98). Natural Gas and the Environment, US Energy Information Administration. Available at www.eia.gov.

Kotay, S. M. and Das, D. (2008) Biohydrogen as a renewable energy resource: prospects and potentials. *International Journal of Hydrogen Energy* 33:258–263.

Manish, S. and Banerjee, R. (2008) Comparison of biohydrogen production processes. *International Journal of Hydrogen Energy* 33:279–286.

Mathews, K. (2014) *U.S. Cattle and Beef Industry, 2002–2013*. Economic Research Service, US Department of Agriculture. Available at www.ers.usda.gov.

Melis, A. and Happe, T. (2001) Hydrogen production: green algae as a source of energy. *Plant Physiology* 127:740–748.

Weiland, P. (2010) Biogas production: current state and perspectives. *Applied Microbiology and Biotechnology* 85:849–860.

Chapter 9

Bosch, T., Colijn, F., Ebinghaus, R., Körtzinger, A., Latif, M., Matthiessen, B., Melzner, F., et al. (2010) How climate change alters ocean chemistry. In *World Ocean Review: Living with the Oceans 2010*. Available at worldoceanreview.com.

Brennan, L. and Owende, P. (2010) Biofuels from microalgae: a review of technologies for production, processing, and extractions of biofuels and co-products. *Renewable and Sustainable Energy Reviews* 14:557–577.

Cakmak, T., Angun, P., Demiray, Y. E., Ozkan, A. D., Elibol, Z., and Tekinay, T. (2012) Differential effects of nitrogen and sulfur deprivation on growth and biodiesel feedstock production of *Chlamydomonas reinhardtii*. *Biotechnology and Bioengineering* 109:1947–1957.

Capehart, T. (2014) *Feed Grains Database*. Economic Research Service, US Department of Agriculture. Available at www.ers.usda.gov/data-products.

Capehart, T. and Vasavada, U. (2014) *Bioenergy: Findings*. US Energy Information Administration. Available at www.ers.usda.gov.

Chisti, Y. (2007) Biodiesel from microalgae. *Biotechnology Advances* 25:294–306.

Dismukes, G. C., Carrieri, D., Bennette, N., Ananyev, G. M., and Posewitz, M. C. (2008) Aquatic phototrophs: efficient alternatives to land-based crops for biofuels. *Current Opinion in Biotechnology* 19:235–240.

Ducat, D. C., Way, J. C., and Silver, P. A. (2011) Engineering cyanobacteria to generate high-value products. *Trends in Biotechnology*. 29:95–103.

EIA (US Energy Information Administration) (2014) *Petroleum and Other Liquids*. Available at www.eia.gov/petroleum.

Golden, S. (2014) *Our Energy Future: Where Do We Go From Here? Biological Options Cyanobacteria*. University of California, San Diego, Coursera. Available at www.coursera.org/learn/future-of-energy.

Griffiths, M. J. and Harrison, S. T. L. (2009) Lipid productivity as a key characteristic for choosing algal species for biodiesel production. *Journal of Applied Phycology* 21:493–507.

Hannon, M., Gimpel, J., Tran, M., Rasala, B., and Mayfield, S. (2010) Biofuels from algae: challenges and potential. *Biofuels* 1:763–784.

Hargreaves, J. A. (2003) *Pond Mixing*, 4802. Southern Regional Aquaculture Center. Available at srac.tamu.edu.

Hildebrand, M. (2008) Diatoms, biomineralization processes and genomics. *Chemical Reviews*. Vol. 109:4855–4874.

Hildebrand, M., Davis, A. K., Smith, S. R., Trailer, J. C., and Abbriano, R. (2012) The place of diatoms in the biofuels industry. *Biofuels* 3:221–240.

Holland, H. D. (2006) The oxygenation of the atmosphere and oceans. *Philosophical Transactions of the Royal Society B* 361:903–915.

Hu, Q., Sommerfeld, M., Jarvis, E., Ghirardi, M., Posewitz, M., Seibert, M., and Darzins, A. (2008) Microalgal triacylglycerols as feedstocks for biofuel production: perspectives and advances. *The Plant Journal* 54:621–639.

John, R. P., Anisha, G. S., Nampoothiri, K. M., and Pandey, A. (2011) Micro and macroalgal biomass: a renewable source of bioethanol. *Bioresource Technology* 102:186–193.

Jones, C. S. and Mayfield, S. P. (2012) Algae biofuels: versatility for the future of bioenergy. *Current Opinion in Biotechnology* 23:346–351.

Kong, Q., Li, L., Martinez, B., Chen, P., and Ruan, R. (2010) Culture of microalgae *Chlamydomonas reinhardtii* in wastewater for biomass feedstock production. *Applied Biochemistry and Biotechnology* 160:9–18.

Kotay, S. M. and Das, D. (2008) Biohydrogen as a renewable energy resource: prospects and potentials. *International Journal of Hydrogen Energy* 33:258–263.

Li, Y. and Wan, C. (2011) *Algae for Biofuels: Fact Sheet Agriculture and Natural Resources*, AEX-651-11. The Ohio State University Extension. Available at ohioline.osu.edu.

Manish, S. and Banerjee, R. (2008) Comparison of biohydrogen production processes. *International Journal of Hydrogen Energy* 33:279–286.

Mata, T. M., Martins, A. A., and Caetano, N. S. (2010) Microalgae for biodiesel production and other applications: a review. *Renewable and Sustainable Energy Reviews* 14:217–232.

Mayfield, S. (2014) *Our Energy Future: Introduction to Energy.* University of California, San Diego, Coursera. Available at www.coursera.org/learn/future-of-energy

Naked Juice Company (2012) *Green Machine.* Available at www.nakedjuice.com.

Pienkos, P.T. (2007) *The Potential for Biofuels from Algae*, NREL /PR-510-42414. Algae Biomass Summit, National Renewable Energy Labs.

Quintana, N., der Kooy, F.V., Van de Rhee, M.D., Voshol, G.P., and Verpoorte, R. (2011) Renewable energy from cyanobacteria: energy production optimization by metabolic pathway engineering. *Applied Microbiology and Biotechnology* 91:471–490.

Scott, S.A., Davey, M.P., Dennis, J.S., Horst, I., Howe, C.J., Lea-Smith, D.J., and Smith, A.G. (2010) Biodiesel from algae: challenges and prospects. *Current Opinion in Biotechnology* 21:277–286.

Singh, A., Nigam, P.S., and Murphy, J.D. (2011) Renewable fuels from algae: an answer to debatable land based fuels. *Bioresource Technology* 102:10–16.

Singh, J. and Gu, S. (2010) Commercialization potential of microalgae for biofuels production. *Renewable and Sustainable Energy Reviews* 14:2596–2610.

Sorrels, C.M. (2009) Biosynthesis of scytonemin, a cyanobacterial sunscreen. Thesis, University of California, San Diego.

Wageningen UR (2011) *Differences between Micro- and Macroalgae.* Wageningen University. Available at www.algae.wur.nl.

Zimmer, C. (2013) The mystery of Earth's oxygen, *The New York Times.* Available at www.nytimes.com.

Chapter 10

Alper, H. and Stephanopoulos, G. (2009) Engineering for biofuels: exploiting innate microbial capacity or importing biosynthetic potential? *Nature Reviews Microbiology* 7:715–723.

Blatti, J., Beld, J., Behnke, C., Mendez, M., Mayfield, S., and Burkart, M. (2012) Manipulating fatty acid biosynthesis in microalgae for biofuel through protein-protein interactions. *PLOS ONE.* doi: 10.1371/journal.pone.0042949.

Burkart, M. (2014) *Our Energy Future: Metabolic Engineering of Algae.* University of California, San Diego, Coursera. Available at www.coursera.org/learn/future-of-energy.

Clarke, N.D. (2010) Protein engineering for bioenergy and biomass-based chemicals. *Current Opinion in Structural Biology* 20:537–532.

Crick, F. (1970) Central Dogma of Molecular Biology. *Nature* 227:561–563.

Heinzelman, P., Snow, C.D., Wu, I., Nguyen, C., Villalobos, A., Govindarajan, S., Minshull, J., and Arnold, F.H. (2009) A family of thermostable fungal cellulases created by structure-guided recombination. *Proceedings of the National Academy of Sciences* 106:5610–5615.

Li, Y., Han, D., Hu, G., Dauvillee, D., Sommerfeld, M., Ball, S., and Hu, Q. (2010) *Chlamydomonas* starchless mutant defective in ADP-glucose pyrophosphorylaste hyper-accumulates triacylglycerol. *Metabolic Engineering* 12:387–391.

Liu, X., Sheng, J., and Curtiss III, R. (2011) Fatty acid production in genetically modified cyanobacteria. *Proceedings of the National Academy of Sciences* 108:6899–6904.

McBride, R. (2014) *Our Energy Future: Bio-prospecting, Genetics, and Synthetic Biology of Algae.* University of California, San Diego, Coursera. Available at www.coursera.org/learn/future-of-energy.

Peterhansel, C., Niessen, M., and Kebeish, R.M. (2008) Metabolic engineering towards the enhancement of photosynthesis. *Photochemistry and Photobiology* 84:1317–1323.

Schmidt, B. (2014) *Our Energy Future: Jatropha.* University of California, San Diego, Coursera. Available at www.coursera.org/learn/future-of-energy.

Steen, E.J., Chan, R., Prasad, N., Myers, S., Petzold, C.J., Redding, A., Ouellet, M., and Keasling, J.D. (2008) Metabolic engineering of *Saccharomyces cerevisiae* for the production of n-butanol. *Microbial Cell Factories.* 7:36.

Wade, W. (2002) Unculturable bacteria: the uncharacterized organisms that cause oral infections. *Journal of the Royal Society of Medicine* 95:81–83.

Wagner, A., Donaldson, L., Kim, H., Phillips, L., Flint, H., Steward, D., Torr, K., Koch, G., Schmitt, U., and Ralph, J. (2009) Suppression of 4-coumarate-coA ligase in the coniferous gymnosperm *Pinus radiata. Plant Physiology* 149:370–383.

Wilson, D.B. (2009) Cellulases and biofuels. *Current Opinion in Biotechnology* 20:295–299.

Chapter 11

Biomass Energy Centre (2011) *Pyrolysis.* Biomass Energy Centre. Available at http://www.biomassenergycentre.org.uk.

Damartzis, T. and Zabaniotou, A. (2011) Thermochemical conversion of biomass to second generation biofuels through integrated process design: a review. *Renewable and Sustainable Energy Reviews* 15:366–378.

DOE (US Department of Energy) (2005) *Pyrolysis and Other Thermal Processing. Energy Efficiency and Renewable Energy Biomass Program.* Available at www1.eere.energy.gov.

DOE (2012) *Fischer-Tropsch (FT) Synthesis.* Gasifipedia, National Energy Technology Laboratory. Available at www.netl.doe.gov.

EPA (Environmental Protection Agency) (2007) *Emission Facts: Greenhouse Gas Impacts of Expanded Renewable and Alternative Fuels Use*, EPA-420-F-07-035. Office of Transportation and Air Quality, EPA, Ann Arbor, MI.

Herz, R.K. (2014) *Our Energy Future: Thermochemical Conversion of Biomass to Fuel.* University of California, San Diego, Coursera. Available at www.coursera.org/learn/future-of-energy.

Overend, R.P. (2005) Thermochemical conversion of biomass. In Shpilrain, E.E. (ed.): *Renewable Energy Sources Charged with Energy from the Sun and Originated from Earth-Moon Interaction (Encyclopedia of Life Support Systems, Vol. 1).* Available at www.eolss.net.

Sun, S., Tsubaki, N., and Fujimoto, K. (2000) Promotional effect of noble metal to co-based Fischer–Tropsch catalysts prepared from mixed cobalt salts. *Chemistry Letters* 2:176–177.

USDA (US Department of Agriculture) and DOE (2005) *Biomass as Feedstock for a Bioenergy and Bioproducts Industry: The Technical Feasibility of a Billion-Ton Annual Supply*, Technical Report. USDA and US DOE.

Verma, M., Godbout, S., Brar, S. K., Solomatnikova, O., Lemay, S. P., and Larouche, J. P. (2012) Biofuels production from biomass by thermochemical conversion technologies. *International Journal of Chemical Engineering*. doi:10.1155/2012/542426.

Chapter 12

Ahlgren, S. and Di Lucia, L. (2014) Indirect land use changes of biofuel production: a review of modeling efforts and policy developments in the European Union. *Biotechnology for Biofuels* 7 (35):1–10.

British Petroleum (2014) *BP Statistical Review of World Energy 2014*. Available at www.bp.com/statisticalreview

Cone, T. (1997) The vanishing valley. *San Jose Mercury News West Magazine*, June 29, 9–15.

Cornell Waste Management Institute (2015) *Composting*. Cornell University. Available at cwmi.css.cornell .edu.

EIA (US Energy Information Administration) (2014a) *Annual Energy Review*. Available at www.eia.gov.

EIA (2014b) *Fuel Ethanol Overview: Monthly Energy Review*. Available at www.eia.gov.

EPA (US Environmental Protection Agency) (2012a) *Renewable Fuel Standard (RFS)*. Office of Transportation and Air Quality. Available at www.epa.gov.

EPA (2012b) *Future Climate Change*. Available at epa.gov /climatechange.

Gallagher, M. E., Hockaday, W. C., Masiello, C. A., Snapp, S., McSwiney, C. P., and Baldock, J. A. (2011) Biochemical suitability of crop residues for cellulosic ethanol: disincentives to nitrogen fertilization in corn agriculture. *Environmental Science and Technology* 45:2013–2020.

GAO (US Government Accountability Office) (2009) *Biofuels: Potential Effects and Challenges of Required Increases in Production and Use*, GAO-09-446. Report to Congressional Requesters.

Han, F. X., King, R. L., Lindner, J. S., Yu, T. Y., Durbha, S. S., Younan, N. H., Monts, D. L., Luthe, J. C., and Plodinec, M. J. (2011) Nutrient fertilizer requirements for sustainable biomass supply to meet U.S. bioenergy goal. *Biomass and Bioenergy* 35:253–262.

Jasinski, S., Kramer, D., Ober, J., and Searls, J. (1999) *Fertilizers: Sustaining Global Food Supplies*. US Geologic Service Fact Sheet, FS-155-99.

Khanna, M. (2008) Cellulosic biofuels: are they economically viable and environmentally sustainable. *Choices* 23:16–21.

LUMCON (Louisiana Universities Marine Consortium) (2012) *Hypoxia in the Northern Gulf of Mexico*. Louisiana Universities Marine Consortium. Available at www .gulfhypoxia.net/overview/.

Nickerson, C., Ebel, R., Borchers, A., and Carriazo, F. (2011) *Major Uses of Land in the United States*. EIB-89. Economic Research Service, US Department of Agriculture. Available at www.ers.usda.gov/media/188404/eib89_2 _pdf.

NRC (National Research Council of the National Academies) (2008) *Water Implications of Biofuels Production in the United States*. Available at www.nap.edu/catalog/12039 .html.

Pachauri, R. K., Meyer, L., and Core Writing Team (2014) *Climate Change 2014 Synthesis Report: Summary for Policymakers*. Intergovernmental Panel on Climate Change. Available at www.ipcc.ch.

Paerl, H. W. (2006) *Sources and Cycles of Eutrophication*. World Resources Institute. Available at www.wri.org.

Perlack, R. D., Wright, L. L., Turhollow, A. F., Graham, R. L., Stokes, B. J., and Erbach, D. C. (2005) *Biomass as Feedstock for a Bioenergy and Bioproducts Industry: The Technical Feasibility of a Billion-Ton Annual Supply*. US Department of Energy and US Department of Agriculture. Available at www1.eere .energy.gov/biomass/.

Perlman, H. (2012) *The USGS Water Science School*. US Geological Survey. Available at ga.water.usgs.gov.

Qin, Z., Zhuang, Q., and Chen, M. (2012) Impacts of land use change due to biofuel crops on carbon balance, bioenergy production, and agricultural yield, in the conterminous United States. *Global Change Biology Bioenergy* 4 (3):277–288. doi:10.1111/j.1757-1707.2011.01129x.

Rekacewicz, P. (2009) *The Contribution of Climate Change to Declining Water Availability: Vital Water Graphics 2*. GRID-ARENDAL. Available at http://www.grida.no/graphicslib /detail/the-contribution-of-climate-change-to-declining-water-availability_12c2#.

SARE (Sustainable Agriculture Research and Education) (2010) *Conserving Water, Energy and Money on the Texas High (and Dry) Plains. What Is Sustainable Agriculture?* Available at www.sare.org.

Schnepf, R. and Yacobucci, B. D. (2012) *Renewable Fuel Standard (RFS): Overview and Issues*, R40155. CRS Report for Congress, Congressional Research Service. Available at www.fas.org /sgp/crs/misc/R40155.pdf.

Searchinger, T., Heimlich, R., Houghton, R. A., Dong, F., Elobeid, A., Fabiosa, J., Tokgoz, S., Hayes, D., and Yu, T. H. (2008) Use of U.S. croplands for biofuels increases greenhouse gases through emissions from land-use change. *Science* 319:1238–1240.

Theis, T. and Tomkin, J. (2012) Physical resources: water, pollution, and minerals. Chap. 5 in *Sustainability: A Comprehensive Foundation*. University of Illinois Open Source Textbook. Available at cnx.org

USDA (US Department of Agriculture) (2008) *2007 Census of Agriculture: 2008 Farm and Ranch Irrigation Survey*. National Agricultural Statistics Service. Available at www.agcensus .usda.gov.

USDA (2012) *Fertilizer Use and Price*. Economic Research Service. Available at www.ers.usda.gov/.

USGS (US Geological Survey) (2011) *High Plains Regional Ground Water (HPGW) Study*. National Water-Quality Assessment (NAWQA) Program. Available at co.water.usgs.gov/.

Victoria (2011) *Sedimentation of Waterways: A Guide to the Inland Angling Waters of Victoria*. State Government Victoria. Available at www.dpi.vic.gov.au/.

Wu, M., Mintz, M., Wang, M., and Arora, S. (2009) *Consumptive Water Use in the Production of Ethanol and Petroleum Gasoline*, ANL/ESD/09-1. Argonne National Laboratory. Available at www.transportation.anl.gov.

Chapter 13

Curran, M. A. (2006) Lifecycle assessment: principles and practices. Scientific Applications International Corporation. EPA/600/R-06/060. nepis.epa.gov.

Farrell, A., Plevin, R. J., Turner, B. T. Jones, A. D., O'Hare, M., and Kammen, D. M. (2006) Ethanol can contribute to energy and environmental goals. Science. 311:506-508.

Graff Zivin, J. (2014) *Our Energy Future: Economic and Social Impact of Energy Production and Use Economics of Energy.* University of California, San Diego, Coursera. Available at www.coursera.org/learn/future-of-energy.

Green Design Institute (2012) *Economic Input-Output Life Cycle Assessment: Free, Fast, Easy Life Cycle Assessment.* Green Design Institute, Carnegie Mellon University. Available at www.eiolca.net.

Kendall, A. (2011) Biofuel life cycle assessment: critical reviews and research needs. University of California, Davis. Algae Biofuels Symposium Presentation. San Diego, CA.

Chapter 14

British Petroleum (2014) *BP Statistical Review of World Energy 2013.* Available at www.bp.com/statistical review.

Chambwera, M., Heal, G., Hallegatte, S., Leclerc, L., Markandya, A., McCarl, B. A., and Mechler, R. (2014) Economics of adaptation. Chap. 17 in *Climate Change 2014: Impacts, Adaptation, and Vulnerability. Part A: Global and Sectoral Aspects. Contribution of Working Group II to the Fifth Assessment Report of the Intergovernmental Panel on Climate Change.* Available at Ipcc-wg2.gov.

CIA (Central Intelligence Agency) (2014) *The World Factbook.* Available at www.cia.gov.

Demirbas, A. (2009) Political, economic and environmental impacts of biofuels: a review. *Applied Energy* 86:5108–5117.

eConPort (2015) *Direct Regulation.* Available at www.econport.org.

EIA (US Energy Information Administration) (2014a) *Monthly Energy Review,* DOE/EIA-0035(2014/11). Available at www.eia.gov/mer.

EIA (2014b) *What Do I Pay for in a Gallon of Regular Gasoline.* Available at www.eia.gov.

EPA (US Environmental Protection Agency) (2014) *Renewable Fuel Standard (RFS).* Available at www.epa.gov.

Jacobsen, M. (2014) *Our Energy Future: Regulation of Energy Use and Energy Efficiency.* University of California, San Diego, Coursera. Available at www.coursera.org/learn/future-of-energy.

McMahon, T. (2014) *Historic Crude Oil Prices.* Available at Inflationdata.com.

Nicolson, A. (2014) *The Next Revolution: Discarding Dangerous Fossil Fuel Accounting Practices.* Available at RenewableEnergyWorld.com.

Pashly, S., Silva, J., and Windram, T. (2014) *Bringing Clarity to Fuel Excise Taxes and Credits.* The Tax Adviser. Available at www.aicpa.org.

Rajagopal, D. and Zilberman, D. (2007) *Review of Environmental, Economic and Policy Aspects of Biofuels,* WPS4341. The World Bank Development Research Group.

US Department of Commerce (2014) *U.S. GDP.* Bureau of Economic Analysis. Available at www. Bea.gov.

The World Bank Group (2014) *GDP.* Available at data. worldbank.org.

Zenk, T. (2014) *Our Energy Future: Policy Impacts on Renewable Energy Industries.* University of California, San Diego, Coursera. Available at www.coursera.org/learn/future-of-energy.

Chapter 15

British Petroleum (2014a) *BP Statistical Review of World Energy 2013.* Available at www.bp.com/statisticalreview.

British Petroleum (2014b) *BP Statistical Review of World Energy 2014.* Available at www.bp.com/statisticalreview.

EIA (US Energy Information Administration) (2013) *International Energy Outlook 2013,* IEO2013. Available at www.eia.gov.

US Census Bureau (2015) *Historical Estimates of World Population.* Available at www.census.gov.

INDEX